Beyond 3G – Bringing Networks, Terminals and the Web Together

Beyond 3G – Bringing Networks, Terminals and the Web Together

LTE, WiMAX, IMS, 4G Devices and the Mobile Web 2.0

Martin Sauter
Nortel, Germany

A John Wiley and Sons, Ltd, Publication

This edition first published 2009
© 2009 John Wiley & Sons Ltd

Registered office
John Wiley & Sons Ltd, The Atrium, Southern Gate, Chichester, West Sussex,
PO19 8SQ, United Kingdom

For details of our global editorial offices, for customer services and for information
about how to apply for permission to reuse the copyright material in this book
please see our website at www.wiley.com.

Library of Congress Cataloging-in-Publication Data

Sauter, Martin.
  Beyond 3G : bringing networks, terminals and the Web together / Martin Sauter.
     p.   cm.
  Includes bibliographical references and index.
  ISBN 978-0-470-75188-6 (cloth)
  1. Wireless Internet.   2. Smartphones.   3. Mobile computing.   I. Title.
  TK5103.4885.S38 2009
  621.382—dc22
                                                                2008047071

A catalogue record for this book is available from the British Library.

ISBN 978-0-470-75188-6 (H/B)

Set in 10/12pt Times by Integra Software Services Pvt. Ltd. Pondicherry, India
Printed and bound in Great Britain by CPI Antony Rowe, Chippenham, Wiltshire

Contents

Preface   xi

1 Evolution from 2G over 3G to 4G   **1**
  1.1 First Half of the 1990s – Voice-centric Communication   1
  1.2 Between 1995 and 2000: the Rise of Mobility and the Internet   2
  1.3 Between 2000 and 2005: Dot Com Burst, Web 2.0, Mobile Internet 2
  1.4 From 2005 to today: Global Coverage, VoIP and Mobile Broadband 4
  1.5 The Future – the Need for Beyond 3G Systems   5
  1.6 All Over IP   8
  1.7 Summary   11
  References   11

2 Beyond 3G Network Architectures   **13**
  2.1 Overview   13
  2.2 UMTS, HSPA and HSPA+   14
    2.2.1 Introduction   14
    2.2.2 Network Architecture   14
    2.2.3 Air Interface and Radio Network   23
    2.2.4 HSPA (HSDPA and HSUPA)   31
    2.2.5 HSPA+ and other Improvements: Competition for LTE 36
  2.3 LTE   45
    2.3.1 Introduction   45
    2.3.2 Network Architecture   46
    2.3.3 Air Interface and Radio Network   51
    2.3.4 Basic Procedures   65
    2.3.5 Summary and Comparison with HSPA   68
    2.3.6 LTE-Advanced   69
  2.4 802.16 WiMAX   70
    2.4.1 Introduction   70
    2.4.2 Network Architecture   70

2.4.3	*The 802.16d Air Interface and Radio Network*	76
2.4.4	*The 802.16e Air Interface and Radio Network*	79
2.4.5	*Basic Procedures*	83
2.4.6	*Summary and Comparison with HSPA and LTE*	85
2.4.7	*802.16m: Complying with IMT-Advanced*	86
2.4.8	*802.16j: Mobile Multihop Relay*	87
2.5	802.11 Wi-Fi	88
2.5.1	*Introduction*	88
2.5.2	*Network Architecture*	89
2.5.3	*The Air Interface – from 802.11b to 802.11n*	92
2.5.4	*Air Interface and Resource Management*	97
2.5.5	*Basic Procedures*	101
2.5.6	*Wi-Fi Security*	101
2.5.7	*Quality of Service: 802.11e*	103
2.5.8	*Summary*	104
	References	105
3	**Network Capacity and Usage Scenarios**	**107**
3.1	Usage in Developed Markets and Emerging Economies	107
3.2	How to Control Mobile Usage	108
3.2.1	*Per Minute Charging*	109
3.2.2	*Volume Charging*	109
3.2.3	*Split Charging*	109
3.2.4	*Small-screen Flat Rates*	109
3.2.5	*Strategies to Inform Users When Their Subscribed Data Volume is Used Up*	110
3.2.6	*Mobile Internet Access and Prepaid*	110
3.3	Measuring Mobile Usage from a Financial Point of View	111
3.4	Cell Capacity in Downlink	112
3.5	Current and Future Frequency Bands for Cellular Wireless	117
3.6	Cell Capacity in Uplink	118
3.7	Per-user Throughput in Downlink	120
3.8	Per-user Throughput in the Uplink	125
3.9	Traffic Estimation Per User	127
3.10	Overall Wireless Network Capacity	129
3.11	Network Capacity for Train Routes, Highways and Remote Areas	133
3.12	When will GSM be Switched Off?	135
3.13	Cellular Network VoIP Capacity	136
3.14	Wi-Fi VoIP Capacity	140
3.15	Wi-Fi and Interference	141
3.16	Wi-Fi Capacity in Combination with DSL and Fibre	143
3.17	Backhaul for Wireless Networks	148
3.18	A Hybrid Cellular/Wi-Fi Network for the Future	153
	References	155

4 Voice over Wireless   **157**
4.1 Circuit-switched Mobile Voice Telephony   158
   4.1.1 Circuit Switching   158
   4.1.2 A Voice-optimized Radio Network   159
   4.1.3 The Pros of Circuit Switching   159
4.2 Packet-switched Voice Telephony   159
   4.2.1 Network and Applications are Separate in Packet-switched
       Networks   160
   4.2.2 Wireless Network Architecture for Transporting IP packets   160
   4.2.3 Benefits of Migrating Voice Telephony to IP   162
   4.2.4 Voice Telephony Evolution and Service Integration   162
   4.2.5 Voice Telephony over IP: the End of the Operator Monopoly   163
4.3 SIP Telephony over Fixed and Wireless Networks   164
   4.3.1 SIP Registration   164
   4.3.2 Establishing a SIP Call Between Two SIP Subscribers   167
   4.3.3 Session Description   169
   4.3.4 The Real-time Transfer Protocol   171
   4.3.5 Establishing a SIP Call Between a SIP and a PSTN Subscriber   172
   4.3.6 Proprietary Components of a SIP System   174
   4.3.7 Network Address Translation and SIP   175
4.4 Voice and Related Applications over IMS   176
   4.4.1 IMS Basic Architecture   179
   4.4.2 The P-CSCF   181
   4.4.3 The S-CSCF and Application Servers   182
   4.4.4 The I-CSCF and the HSS   184
   4.4.5 Media Resource Functions   186
   4.4.6 User Identities, Subscription Profiles and Filter Criteria   188
   4.4.7 IMS Registration Process   190
   4.4.8 IMS Session Establishment   194
   4.4.9 Voice Telephony Interworking with Circuit-switched Networks   199
   4.4.10 Push-to-talk, Presence and Instant Messaging   203
   4.4.11 Voice Call Continuity   206
   4.4.12 IMS with Wireless LAN Hotspots and Private Wi-Fi
       Networks   209
   4.4.13 IMS and TISPAN   213
   4.4.14 IMS on the Mobile Device   216
   4.4.15 Challenges for IMS Rollouts   219
   4.4.16 Opportunities for IMS Rollouts   222
4.5 Voice over DSL and Cable with Femtocells   224
   4.5.1 Femtocells from the Network Operator's Point of View   226
   4.5.2 Femtocells from the User's Point of View   227
   4.5.3 Conclusion   228
4.6 Unlicensed Mobile Access and Generic Access Network   228
   4.6.1 Technical Background   229
   4.6.2 Advantages, Disadvantages and Pricing Strategies   231
  References   232

5 Evolution of Mobile Devices and Operating Systems   **235**
  5.1   Introduction   235
        5.1.1   *The ARM Architecture*   237
        5.1.2   *The x86 Architecture for Mobile Devices*   238
        5.1.3   *From Hardware to Software*   238
  5.2   The ARM Architecture for Voice-optimized Devices   238
  5.3   The ARM Architecture for Multimedia Devices   241
  5.4   The x86 Architecture for Multimedia Devices   244
  5.5   Hardware Evolution   247
        5.5.1   *Chipset*   247
        5.5.2   *Process Shrinking*   248
        5.5.3   *Displays and Batteries*   249
        5.5.4   *Other Additional Functionalities*   250
  5.6   Multimode, Multifrequency Terminals   252
  5.7   Wireless Notebook Connectivity   255
  5.8   Impact of Hardware Evolution on Future Data Traffic   255
  5.9   The Impact of Hardware Evolution on Networks and Applications   257
  5.10 Mobile Operating Systems and APIs   258
        5.10.1 *Java and BREW*   258
        5.10.2 *BREW*   259
        5.10.3 *Symbian/S60*   260
        5.10.4 *Windows Mobile*   262
        5.10.5 *Linux: Maemo, Android and Others*   262
        5.10.6 *Fracturization*   265
        5.10.7 *Operating System Tasks*   265
  References   271

6 Mobile Web 2.0, Applications and Owners   **273**
  6.1   Overview   273
  6.2   (Mobile) Web 1.0 – How Everything Started   274
  6.3   Web 2.0 – Empowering the User   275
  6.4   Web 2.0 from the User's Point of View   275
        6.4.1   *Blogs*   276
        6.4.2   *Media Sharing*   277
        6.4.3   *Podcasting*   277
        6.4.4   *Advanced Search*   277
        6.4.5   *User Recommendation*   278
        6.4.6   *Wikis – Collective Writing*   278
        6.4.7   *Social Networking Sites*   279
        6.4.8   *Web Applications*   280
        6.4.9   *Mashups*   280
        6.4.10 *Virtual Worlds*   281
        6.4.11 *Long-tail Economics*   281
  6.5   The Ideas Behind Web 2.0   282
        6.5.1   *The Web as a Platform*   282
        6.5.2   *Harnessing Collective Intelligence*   283

        6.5.3 *Data is the Next Intel Inside*   284
        6.5.4 *End of the Software Release Cycle*   284
        6.5.5 *Lightweight Programming Models*   285
        6.5.6 *Software above the Level of a Single Device*   285
        6.5.7 *Rich User Experience*   285
  6.6   Discovering the Fabrics of Web 2.0   286
        6.6.1 *Aggregation*   286
        6.6.2 *AJAX*   289
        6.6.3 *Tagging and Folksonomy*   290
        6.6.4 *Open Application Programming Interfaces*   293
        6.6.5 *Open Source*   295
  6.7   Mobile Web 2.0 – Evolution and Revolution of Web 2.0   296
        6.7.1 *The Seven Principles of Web 2.0 in the Mobile World*   296
        6.7.2 *Advantages of Connected Mobile Devices*   301
        6.7.3 *Offline Web Applications*   304
        6.7.4 *The Mobile Web, 2D Barcodes and Image Recognition*   308
        6.7.5 *Walled Gardens, Mobile Web 2.0 and the Long Tail*   310
        6.7.6 *Web Page Adaptation for Mobile Devices*   311
  6.8   (Mobile) Web 2.0 and Privacy   317
        6.8.1 *On-page Cookies*   318
        6.8.2 *Inter-site Cookies*   320
        6.8.3 *Flash Shared Objects*   320
        6.8.4 *Site Information Sharing, Social Distribution*   321
        6.8.5 *Session Tracking*   322
  6.9   Mobile Applications   322
        6.9.1 *Web Browsing*   323
        6.9.2 *Audio*   324
        6.9.3 *Media Sharing*   328
        6.9.4 *Video and TV*   330
        6.9.5 *Voice and Video Telephony*   332
        6.9.6 *Widgets*   333
        6.9.7 *Social Media*   335
        6.9.8 *Microblogging*   335
        6.9.9 *Location*   338
        6.9.10 *Shopping*   340
        6.9.11 *Mobile Web Servers*   341
  References   343

7 Conclusion   345

Index   349

Preface

In recent years, cellular voice networks have transformed into powerful packet-switched access networks for both voice communication and Internet access. Current 3.5G networks such as UMTS/HSDPA and CDMA 1xEvDO now deliver bandwidths of several megabits per second to individual users, and mobile access to the Internet from handheld devices and notebooks is no longer perceived as slower than a DSL or cable connection. Bandwidth and capacity demands, however, keep rising because of the increasing number of people using the networks and due to new bandwidth-intensive applications such as video streaming and mobile Internet access from notebooks. Thus, network manufacturers and network operators need to find ways to increase capacity and performance while reducing cost.

In the past, network evolution mainly involved designing access networks with more bandwidth and capacity. As we go beyond 3G network architectures, there is now also an accelerated evolution of core networks and, most importantly, user devices and applications. This evolution follows the trends that are already in full swing in the 'fixed-line' Internet world today. Circuit-switched voice telephony is being replaced by voice over IP technologies and Web 2.0 has empowered consumers to become creators and to share their own information with a worldwide audience. In the future, wireless networks will have a major impact on this trend, as mobile phones are an ideal tool for creating and consuming content. The majority of mobile phones today have advanced camera and video capabilities, and together with fast wireless access technologies, it becomes possible to share information with others instantly.

While all these trends are already occurring, few resources are available that describe them from a technical perspective. This book therefore aims to introduce the technology behind this evolution. Chapter 1 gives an overview of how mobile networks have evolved in the past and what trends are emerging today. Chapter 2 then takes a look at radio access technologies such as LTE, HSPA +, WiMAX and the evolution of the Wi-Fi standard. Despite the many enhancements next-generation radio systems will bring, bandwidth on the air interface is still the limiting factor. Chapter 3 takes a look at the performance of next-generation systems in comparison to today's networks, shows where the limits are and discusses how Wi-Fi can help to ensure future networks can meet the rising demand for bandwidth and integrated home networking. Voice over IP is already widely used in fixed line networks today and 'Beyond 3G' networks have enough capacity and performance to bring about this change in the wireless world as well. Chapter 4 thus focuses on Voice over IP architectures, such as the IP Multimedia Subsystem (IMS) and the Session Initiation Protocol (SIP) and discusses the impacts of these systems on future voice and multimedia communication. Just as important as wireless networks are the mobile devices using them, and Chapter 5 gives an overview of current mobile device architectures and their evolution. Finally, mobile devices are only as useful as the applications running on them. So Chapter 6 discusses how 'mobile Web 2.0' applications will change the way we communicate in the future.

No book is written in isolation and many of the ideas that have gone into this manuscript are the result of countless conversations over the years with people from all across the industry. Specifically, I would like to thank Debby Maxwell, Prashant John, Kevin Wriston, Peter van den Broek and John Edwards for the many insights they have provided to me over the years in their areas of expertise and for their generous help with reviewing the manuscript. A special thank-you goes to Berenike for her love, her passion for life and for inspiring me to always go one step further. And last but not least I would like to thank Mark Hammond, Sarah Tilley, Sarah Hinton and Katharine Unwin of John Wiley and Sons for the invaluable advice they gave me throughout this project.

1

Evolution from 2G over 3G to 4G

In the past 15 years, fixed line and wireless telecommunication as well as the Internet have developed both very quickly and very slowly depending on how one looks at the domain. To set current and future developments into perspective, the first chapter of this book gives a short overview of major events that have shaped these three sectors in the previous one-and-a-half decades. While the majority of the developments described below took place in most high-tech countries, local factors and national regulation delayed or accelerated events. Therefore, the time frame is split up into a number of periods and specific dates are only given for country-specific examples.

1.1 First Half of the 1990s – Voice-centric Communication

Fifteen years ago, in 1993, Internet access was not widespread and most users were either studying or working at universities or in a few select companies in the IT industry. At this time, whole universities were connected to the Internet with a data rate of 9.6 kbit/s. Users had computers at home but dial-up to the university network was not yet widely used. Distributed bulletin board networks such as the Fidonet [1] were in widespread use by the few people who were online then.

It can therefore be said that telecommunication 15 years ago was mainly voice-centric from a mass market point of view. An online telecom news magazine [2] gives a number of interesting figures on pricing around that time, when the telecom monopolies where still in place in most European countries. A 10 min 'long-distance' call in Germany during office hours, for example, cost €3.25.

On the wireless side, first-generation analog networks had been in place for a number of years, but their use was even more expensive and mobile devices were bulky and unaffordable except for business users. In 1992, GSM networks had been launched in a number of European countries, but only few people noticed the launch of these networks.

Beyond 3G – Bringing Networks, Terminals and the Web Together: LTE, WiMAX, IMS, 4G Devices and the Mobile Web 2.0   Martin Sauter   © 2009 John Wiley & Sons, Ltd

1.2 Between 1995 and 2000: the Rise of Mobility and the Internet

Around 1998, telecom monopolies came to an end in many countries in Europe. At the time, many alternative operators were preparing themselves for the end of the monopoly and prices went down significantly in the first weeks and months after the new regulation came into effect. As a result, the cost of the 10 min long-distance call quickly fell to only a fraction of the former price. This trend continues today and the current price is in the range of a few cents. Also, European and even intercontinental phone calls to many countries, like the USA and other industrialized countries, can be made at a similar cost.

At around the same time, another important milestone was reached. About 5 years after the start of GSM mobile networks, tariffs for mobile phone calls and mobile phone prices had reached a level that stimulated mass market adoption. While the use of a mobile phone was perceived as a luxury and mainly for business purposes in the first years of GSM, adoption quickly accelerated at the end of the decade and the mobile phone was quickly transformed from a high-price business device to an indispensable communication tool for most people.

Fixed line modem technology had also evolved somewhat during that time, and modems with speeds of 30–56 kbit/s were slowly being adopted by students and other computer users for Internet access either via the university or via private Internet dial-up service providers. Around this time, text-based communication also started to evolve and Web browsers appeared that could show Web pages with graphical content. Also, e-mail leapt beyond its educational origin. Content on the Internet at the time was mostly published by big news and IT organizations and was very much a top-down distribution model, with the user mainly being a consumer of information. Today, this model is known as Web 1.0.

While voice calls over mobile networks quickly became a success, mobile Internet access was still in its infancy. At the time, GSM networks allowed data rates of 9.6 and 14.4 kbit/s over circuit-switched connections. Few people at the time made use of mobile data, however, mainly due to high costs and missing applications and devices. Nevertheless, the end of the decade saw the first mobile data applications such as Web browsers and mobile e-mail on devices such as Personal Digital Assistants (PDAs), which could communicate with mobile phones via an infrared port.

1.3 Between 2000 and 2005: Dot Com Burst, Web 2.0, Mobile Internet

Developments continued and even accelerated in all three sectors despite the dot com burst in 2001, which sent both the telecoms and the Internet industry into a downward spiral for several years. Despite this downturn, a number of new important developments took place during this period.

One of the major breakthroughs during this period was the rise of Internet access via Digital Subscriber Lines (DSL) and TV cable modems. These quickly replaced dial-up connections as they became affordable and offered speeds of 1 MBit/s and higher. Compared with the 56 kbit/s analog modem connections, the download times for web pages with graphical content and larger files improved significantly. At the end of this period, the majority of people in many countries had access to broadband Internet that allowed them to view more and more complex Web pages. Also, new forms of communication like Blogs and Wikis appeared, which quickly revolutionized the creator–consumer imbalance. Suddenly, users were no longer only consumers of content, but could also be creators for a worldwide audience. This is one of the main properties of what is popularly called Web 2.0 and will be further discussed later on in this book.

In the fixed line telephony world, prices for national and international calls continued to decline. Towards the end of this period, initial attempts were also made to use the Internet for transporting voice calls. Early adopters discovered the use of Internet telephony to make phone calls over the Internet via their DSL lines. Proprietary programs like Skype suddenly allowed users to call any Skype subscriber in the world for free, in many cases with superior voice quality. 'Free' in this regard is a relative term, however, since both parties in the call have to pay for access to the Internet, so telecom operators still benefit from such calls due to the monthly charge for DSL or cable connections. Additionally, many startup companies started to offer analog telephone to Internet Protocol (IP) telephone converters, which used the standardized SIP (Session Initiation Protocol) protocol to transport phone calls over the Internet. Gateways ensured that such subscribers could be reached via an ordinary fixed line telephone number and could call any legacy analog phone in the world. Alternative long-distance carriers also made active use of the Internet to tunnel phone calls between countries and thus offer cheaper rates.

Starting in 2001, the General Packet Radio Service (GPRS) was introduced in public GSM networks for the first time. When the first GPRS-capable mobile phones quickly followed, mobile Internet access became practically feasible for a wider audience. Until then, mobile Internet access had only been possible via circuit-switched data calls. However, the data rate, call establishment times and the necessity of maintaining the channel even during times of inactivity were not suitable for most Internet applications. These problems, along with the small and monochrome displays in mobile phones and mobile software being in its infancy, meant that the first wireless Internet services (WAP 1.0) never became popular. Towards 2005, devices matured, high-resolution color displays made it into the mid-range mobile phone segment and WAP 2.0 mobile Web browsers and easy-to-use mobile e-mail clients in combination with GPRS as a packet-switched transport layer finally allowed mobile Internet access to cross the threshold between niche and mass market. Despite these advances, pricing levels and the struggle between open and closed Internet gardens, which will be discussed in more detail later on, slowed down progress considerably.

At this point it should be noted that throughout this book the terms 'mobile access to the Internet' and 'mobile Internet access' are used rather than 'mobile Internet'. This is done on purpose since the latter term implies that there might be a fracture between a 'fixed line' and a 'mobile' Internet. While it is true that some services are specifically tailored for use on mobile devices and even benefit and make use of the user's mobility, there is a clear trend for the same applications, services and content to be offered and useful on both small mobile devices and bigger nomadic or stationary devices. This will be discussed further in Chapter 6.

Another important milestone for wireless Internet access during this timeframe was 3G networks going online in many countries in 2004 and 2005. While GPRS came close to analog modem speeds, UMTS brought data rates of up to 384 kbit/s in practice, and the experience became similar to DSL. Again, network operator pricing held up mass adoption for several years.

1.4 From 2005 to today: Global Coverage, VoIP and Mobile Broadband

From 2005 to today, the percentage of people in industrialized countries accessing the Internet via broadband DSL or cable connections has continued to rise. Additionally, many network operators have started to roll out ADSL2+, and new modems enable download speeds beyond 15 Mbit/s for users living close to a central exchange. VDSL and fiber to the curb/fiber to the home deployments offer even higher data rates. Another trend that has accelerated since 2005 is Voice over IP (VoIP) via a telephone port in the DSL or cable modem router. This effectively circumvents the traditional analog telephone network and traditional network fixed line telephony operators see a steady decline in their customer base.

At the time of publication, the number of mobile phone users has reached 3 billion. This means that almost every second person on Earth now owns a mobile phone, a trend which only a few people foresaw only five years ago. In 2007, network operators registered 1000 new users per minute [3]. Most of this growth has been driven by the rollout of second-generation GSM/GPRS networks in emerging markets. Due to global competition between network vendors, network components reached a price that made it feasible to operate wireless networks in countries with very low revenue per user per month. Another important factor for this rapid growth was ultra-low-cost GSM mobile phones, which became available for less than $50. In only a few years, mobile networks have changed working patterns and access to information for small entrepreneurs like taxi drivers and tradesmen in emerging markets [4]. GSM networks are now available in most parts of the world. Detailed local and global maps of network deployments can be found in [5].

In industrialized countries, third-generation networks continued to evolve and 2006 saw the first upgrades of UMTS networks to High Speed Data Packet Access (HSDPA). In a first step, this allowed user data speeds between 1 and 3 Mbit/s. With advanced mobile terminals, speeds are likely to increase further. Today, such high data rates are mainly useful in combination with notebooks to give users broadband Internet almost anywhere. In the mid term, it is likely that HSDPA will also be very beneficial for mobile applications once podcasts, music downloads and video streaming on mobile devices become mass market applications.

While 3G networks have been available for some time, take-up was sluggish until around 2006/2007, when mobile network operators finally introduced attractive price plans. Prices fell below €40–€50 for wireless broadband Internet access and monthly transfer volumes of around 5 Gbytes. This is more than enough for everything but file sharing and substantial video streaming. Operators have also started to offer smaller packages in the range of €6–15 a month for occasional Internet access with notebooks. Packages in a similar price range are now also offered for unlimited Web browsing and e-mail on mobile phones. Pricing and availability today still vary in different countries. In 2006, mobile data revenue in the USA alone reached a $15.7 billion, of which 50–60% is non-SMS revenue [6]. In some countries, mobile data revenues now accounts for between 20 and 30% of the total operator revenue, as shown in Figure 1.1.

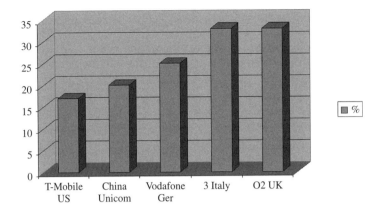

Figure 1.1 Percentage of data revenue of mobile operators in 2007 [6].

While wireless data roaming is still in its infancy, wireless Internet access via prepaid SIM cards is already offered in many countries at similar prices to those for customers with a monthly bill. This is another important step, as it opens the door to anytime and anywhere Internet access for creative people such as students, who favor prepaid SIMs to monthly bills. In addition, it makes life much easier for travelers, who until recently had no access to the Internet while traveling, except for wireless hotspots at airports and hotels. An updated list of such offers is maintained by the Web community on the prepaid wireless Internet access Wiki [7].

1.5 The Future – the Need for Beyond 3G Systems

When looking into the future, the main question for network operators and vendors is when and why Beyond 3G wireless networks will be needed. Looking back only a couple of years, voice telephony was the first application that was mobilized. The Short Message Service (SMS) followed some years later as the first mass market mobile data application. By today's standards comparably simple mobile phones were required for the service and little bandwidth. In a way, the SMS service was a forerunner of other data services like mobile e-mail, mobile Web browsing, mobile blogging, push-to-talk, mobile instant messaging and many others. Such applications became feasible with the introduction of packet-based wireless networks that could carry IP data packets and increasingly powerful mobile devices. Today, the capacity of current 3G and 3.5G networks is still sufficient for the bandwidth requirements of these applications and the number of users. There are a number of trends, however, which are already visible and will increase bandwidth requirements in the future:

- Rising use – due to falling prices, more people will use mobile applications that require network access.
- Multimedia content – while first attempts at mobilizing the Web resulted in mostly text-based Web pages, graphical content is now the norm rather than the exception.

A picture may paint a thousand words, but it also increases the amount of data that has to be transferred for a Web page. Video and music downloads are also becoming more popular, which further increases in bandwidth requirements.

- Mobile social networks – similar to the fixed-line Internet, a different breed of applications is changing the way people are using the Internet. In the past, users mainly consumed content. Blogs, podcasts, picture-sharing sites and video portals are now reshaping the Internet, as users no longer only consume content, but use the network to share their own ideas, pictures and videos with other people. Applications like, for example, Shozu [8] and Lifeblog [9] let users upload pictures, videos and Blog entries from mobile devices to the Web. In particular, picture, podcast and video transfers multiply the amount of data that users transmit and receive.

- Voice over IP – the fixed line world is rapidly moving towards VoIP. It is likely that, five years from now, many of today's fixed line circuit-switched voice networks will have migrated towards IP-based voice transmission. Likewise, on the network access side, many users will use VoIP as their primary fixed line voice service, for example over DSL or TV cable networks. The beginnings can already been observed today, as the circuit-switched voice market is under increasing pressure due to declining subscriber numbers. As a consequence, many operators are no longer investing in this technology. A similar trend can be observed in wireless networks. Here, however, the migration is much slower, especially due to the higher bandwidth requirements for transporting voice calls over a packet-switched bearer. This topic is discussed in more detail in Chapter 1.6.

- Fixed-line Internet replacement – while the number of voice minutes is increasing, revenue is declining in both fixed line and the wireless networks due to falling prices. In many countries, wireless operators are thus trying to keep or increase the average revenue per user by offering Internet access for PCs, notebooks and mobile devices over their UMTS/HSDPA or CDMA networks. Thus, they have started to compete directly with DSL and cable operators. Again, this requires an order of magnitude of additional bandwidth on the air interface.

- Competition from alternative wireless Internet providers – in some countries, alternative operators are already offering wireless broadband Internet access with Wi-Fi or WiMAX/802.16 networks. Such operators directly compete with traditional UMTS and CDMA carriers, who are also active in this market.

- The broadband Internet is not a socket in the wall – this statement combines all previous arguments and was made by Anssi Vanjoki, Executive VP of Nokia's Multimedia division [10], at a press conference. Today, many people already use Wi-Fi access points to create their personal broadband Internet bubble. Thus, broadband Internet is virtually all around them. In the future, people will not only use this bubble with desktop computers and notebooks, but also with smaller devices such as mobile phones with built-in Wi-Fi capabilities. Smaller devices will also change the way we perceive this Internet bubble. No longer is it necessary to sit down at a specific place, for example in front of a computer, in order to communicate (VoIP, e-mail, instant messaging), to get information or to publish information to the Web (pictures, Blog entries, videos, etc.). When the personal broadband bubble is left, mobile devices switch over to a cellular network. As we move into the future, the cellular network will extend into areas not covered today and available bandwidth will have to increase to cope with the rising number of users and their connected applications. Moving between the personal Internet bubble at home and the larger external cellular network will become seamless as devices and services evolve.

A number of wireless technologies are currently under development or in the early rollout phase that are designed to meet these future demands: 3GPP's Long Term Evolution (LTE), HSPA+ and WiMAX. In addition, Wi-Fi is also likely to be an important network technology that is required to meet future capacity demands. All of these technologies will be further discussed in Chapter 2. The question that arises in this context is which of these technologies are 3G and which will be called 4G in the future?

The body responsible for categorizing wireless networks is the International Telecommunication Union (ITU). The ITU categorizes International Mobile Telecommunication (IMT) networks as follows:

- IMT-2000 systems – this is what we know as 3G systems today, for example UMTS and cdma2000. The list of all ITU-2000 systems is given in ITU-R M.1457-6 [11].
- Enhanced IMT-2000 systems – the evolution of IMT-2000 systems, for example HSPA, CDMA 1xEvDo and future evolutions of these systems.
- IMT-Advanced systems – systems in this category are considered to be 4G systems.

At this time, there is still no clear definition of the characteristics of future IMT-Advanced (4G) systems. The ITU-R M.1645 recommendation [12] gives first hints but leaves the door wide open:

> It is predicted that potential new radio interface(s) will need to support data rates of up to approximately 100 Mbit/s for high mobility such as mobile access and up to approximately 1 Gbit/s for low mobility such as nomadic/local wireless access, by around the year 2010 [...] These data rate figures and the relationship to the degree of mobility [...] should be seen as targets for research and investigation of the basic technologies necessary to implement the framework. Future system specifications and designs will be based on the results of the research and investigations.

When comparing current the WiMAX specifications to these potential requirements, it becomes clear that WiMAX does not qualify as a 4G IMT-Advanced standard, since data rates are much lower, even under ideal conditions.

3GPP's successor to its 3G UMTS standard, known as LTE, will also have difficulties fulfilling these requirements. Even with a four-way Multiple Input Multiple Output (MIMO) transmission, data rates in a 20 MHz carrier would not exceed 326 Mbit/s. It should be noted at this point that this number is already a long stretch, since putting four antennas in a small device or on a rooftop will be far from simple in practice.

It is also interesting to compare these new systems with the evolution of current 3G systems. The evolution of UMTS is a good example. With HSDPA and HSUPA, user speeds now exceed the 2 Mbit/s that was initially foreseen for IMT-2000 systems. The evolution of those systems, however, has not yet come to an end. Recent new developments in 3GPP Release 7 and 8 called HSPA+, which include MIMO technology and other enhancements, bring evolved UMTS technology to the same capacity and bandwidth levels as currently specified for LTE on a 5 MHz carrier. HSPA + is also clearly not a 4G IMT-Advanced system, since it enhances a current 3G IMT-2000 radio technology. Thus, HSPA + is categorized as an 'enhanced IMT-2000 system'.

To meet the likely requirements of IMT-Advanced, the WiMAX and LTE standards bodies have started initiatives to further enhance their technologies. On the WiMAX side, the 802.16m task group is working on standardizing an even faster radio interface. On the LTE side, a similar working program has become known as LTE + or Enhanced LTE.

Current research indicates that the transmission speed requirements described in ITU-R M.1645 can only be achieved in a frequency band of 100 MHz or more. This is quite a challenge, both from a technical point of view and also due to a lack of available additional spectrum. Thus, it is somewhat doubtful whether these requirements will remain in place for the final definition of 4G IMT-Advanced.

In practice, several different network technologies will coexist and evolve in the future to meet the rising demands in terms of bandwidth and capacity. It is also likely that a combination of different radio systems, like for example LTE together with Wireless LAN, will be used to satisfy capacity demands.

From a user and service point of view, it does not matter if a network technology is considered 3.5G, 3.9G or 4G. Thus, this book uses the term 'Beyond 3G systems' (B3G), which includes all technologies which will be able to satisfy future capacity demands and which either evolve out of current systems or are a new development.

1.6 All Over IP

While on the radio network side it is difficult to foresee which mix of evolved 3G and 4G technologies will be used in the future, the future of fixed and mobile core networks is much easier to predict. One of the main characteristics of 3G networks is the support for circuit-switched and packet-switched services. The circuit-switched part of the core net-work and circuit-switched services of the radio network were specifically designed to carry voice and video calls. Service control rests with the Mobile Switching Center (MSC), the main component of a circuit-switched network. As subscribers can roam freely in a mobile network, a database is required to keep track of the current location of the subscriber in addition to the subscription information. This database is referred to as the Home Location Register (HLR). To establish a call, a mobile phone always contacts the MSC. The MSC then uses the destination's telephone number to query the HLR for the location of the destination subscriber. The call is then routed to this MSC, which in turn informs the destination subscriber of the incoming call. This process is called signaling. For the speech path, a transparent circuit-switched channel is established between the two parties via the MSCs switching matrix. The signaling required for the call is transferred over an independent signaling network, as the circuit-switched channel only transports the speech signal.

In recent network designs, MSCs are split into an MSC Call Server component that handles the signaling and a media gateway that is responsible for forwarding the voice call as shown in Figure 1.2. Instead of fixed connections, media gateways use packet-switched ATM (Asynchronous Transfer Mode) or IP connections to forward the call. This removes the necessity to transport the voice data via circuit-switched connections in the core network.

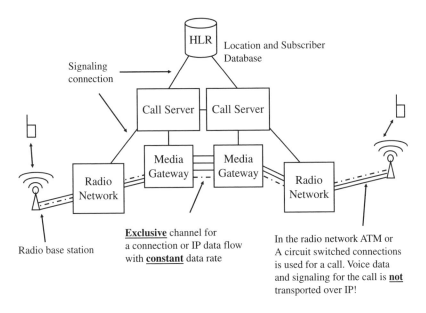

Figure 1.2 Circuit switching with dedicated network components.

While this approach is ideally suited to carry voice and video calls with a constant bandwidth and delay requirements, it performs poorly for a connection to the Internet. Here, all data is transported in data packets. Furthermore, data packets are not only exchanged between two endpoints while a connection is established, but usually between many. An example is a Web browsing session during which a user visits several Web sites, sometimes even simultaneously. While a Web page is transferred, it is desirable to use as much bandwidth as is currently available, rather than be limited to a circuit-switched channel that is designed to carry a digitized narrowband voice or video stream. An Internet connection is often also idle for a substantial duration. During this time, resources are best given to other users. This is also not possible with a circuit-switched connection, because it is an exclusive channel that offers a fixed amount of bandwidth between two parties while it is established.

For these reasons, 3G networks contain a separate core network to forward data packets rather than circuits. This is shown in Figure 1.3. The radio network serves both the circuit-switched and the packet-switched network and the kind of connection established to a user over the air depends on whether a circuit-switched connection or a packet-switched connection is required. Some systems such as UMTS even allow devices to simultaneously use packet and circuit connections so a phone call can be made while being connected to the Internet and transferring data.

Traditional fixed line networks use a similar split for simultaneous voice telephony and Internet access. Since DSL became popular, analog voice service and DSL use the same physical line to the customer's home. A splitter is then used to separate the analog telephone signal from the DSL service as they operate in different frequency bands. In the central exchange office, a similar splitter is used to connect the line of the subscriber to

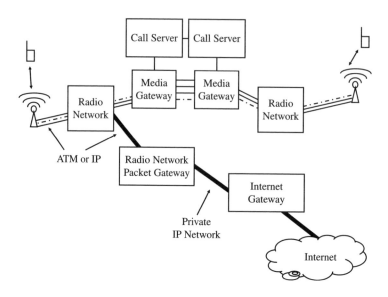

Figure 1.3 Typical circuit-switched and packet-switched dual architecture of 3G networks. The location and subscriber database is not shown.

the local circuit-switched exchange for voice calls and additionally to a DSL Access Multiplexer (DSLAM) for Internet connectivity. Telephone exchanges are then inter-connected via circuit-switched connections, while the DSLAM connects to a packet-switched backbone. In the meantime, however, there is a clear shift to transporting telephone calls over the Internet connection as well. Instead of connecting the analog phone to the splitter, the DSL access device is equipped with a jack for the phone. The DSL access device digitizes the voice signal and sends it as IP packets over the DSL connection. In many cases, an IP-based SIP server and RTP (Real Time Transport Protocol) replace the local circuit-switched telephone exchange. There are several advantages of this approach:

- Only a single type of core network is needed, as the circuit-switched telephone exchanges and the circuit-switched network between them are no longer necessary.
- Using an IP network for voice calls makes it a lot easier for companies other than the local telephone carrier to offer telephony services, as the controlling network element no longer needs to be at the local exchange.
- Voice services can be combined with other services. Since there is more bandwidth available, users can, for example, exchange pictures with each other while being engaged in a voice call or add video at any point during the conversation.

While the trend to VoIP is already fully underway in fixed-line networks, wireless networks have not yet caught up. Here, things are moving more slowly for a number of reasons. The main reason is that 3G mobile networks did not have the necessary bandwidth to support VoIP, which requires a higher data rate than circuit-switched voice calls. The gap has been somewhat reduced by the introduction of 3.5G networks. However, only B3G networks (evolved IMT-2000 and IMT-Advanced) will have enough capacity and an optimized radio network to support VoIP on a large scale.

The challenges are significant, but none of the new B3G network architectures discussed in Chapter 2 have a circuit-switched core network. To be successful, it is essential for B3G wireless network operators to have a fully functioning VoIP solution in place in the future that is able to seamlessly transfer the call to a circuit-switched wireless connection when the user roams out of network coverage. This is discussed in more detail in Chapter 4.

1.7 Summary

This chapter presented how fixed and wireless networks evolved in the past 15 years from circuit-switched voice-centric systems to packet-switched Internet access systems. Due to the additional complexity of wireless systems, enhancements are usually introduced in fixed-line systems first and only some years later in wireless systems as well. To date, fixed-line networks offer data rates to the customer premises of several megabits per second, in some cases already going beyond this. Wireless 3.5G networks are capable of data rates in the order of several megabits per second. In the future, more bandwidth and capacity will be achieved by evolving current wireless network technologies (evolved IMT-2000) and by designing new access networks (IMT-Advanced). This book therefore not only concentrates on 4G systems, but also discusses the evolution of 3G systems. Another important development is the use of packet-switched networks for transporting telephone calls, which is referred to as VoIP. This trend is already fully underway in fixed-line networks and will inevitably also happen in B3G networks, as systems such as WiMAX and LTE have been designed without a circuit-switched core network dedicated to voice calls.

References

1. Background on Fidonet (2008) http://www.fidonet.org.
2. Neuhetzki, T. (December 2005) German long distance tariffs in the 1990s, http://www.teltarif.de/arch/2005/kw52/s19950.html.
3. Sauter, M. (August 2006) 1000 new mobile phone users a minute, http://mobilesociety.typepad.com/mobile_life/2006/08/1000_new_mobile.html.
4. Andersen, T. (19 February 2007) Mobile phone lifeline for world's poor, http://news.bbc.co.uk/1/hi/business/6339671.stm.
5. 2G and 3G coverage maps (2008) http://www.coveragemaps.com.
6. Sharma, C. (September 2007) Global wireless data market, http://www.chetansharma.com/globalmarketupdate1H07.htm.
7. The prepaid wireless Internet access Wiki (2008) http://prepaid-wireless-internet-access.wetpaint.com.
8. Shozu (2008) http://www.shozu.com.
9. Lifeblog (2008) http://r2. nokia.com/nokia/0,71739,00.html.
10. Biography of Anssi Vanjoki Executive VP of Nokia Multimedia (2008) http://www.nokia.com/A4126347.
11. The International Telecommunication Union (2006) Detailed specifications of the radio interfaces of International Mobile Telecommunications-2000 (IMT-2000), ITU-R M.1457-6.
12. The International Telecommunication Union (2003) Framework and overall objectives of the future development of IMT-2000 and systems beyond IMT-2000, ITU-R M.1645.

2

Beyond 3G Network Architectures

2.1 Overview

As discussed in Chapter 1, the general trend in telecommunications is to move all applications to a common transmission protocol, the Internet Protocol. The tremendous advantage of this approach is that applications no longer require a specific network technology but can be used over different kinds of networks. This is important since, depending on the situation, an application might be used best over a cellular network while at other times it is more convenient and cheaper to use a wireless home or office networking technology such as Wi-Fi. The increasing number of multiradio devices supports this trend. Today and even more so in the future, a number of wireless technologies are deployed in parallel. This is necessary as the deployment of a new network requires a considerable amount of time and there are usually only a small number of devices supporting a new network technology at first. It is therefore important that different network technologies are deployed not only in parallel but also at the same location. As well as the introduction of new technologies, existing network technologies continue to evolve to offer improved performance while the new technology is not yet deployed or is just in the process of being rolled out. For these reasons, this chapter looks at a number of different Beyond 3G network technologies with an emphasis on those with the highest market share. In this context, the term 'Beyond 3G networks' is used for cellular networks that offer higher speeds than the original UMTS networks with their maximum data rate of 384 kbit/s per user.

In the cellular world, the Universal Mobile Telecommunication System (UMTS) with its High-speed Packet Access (HSPA) evolution is currently the Beyond 3G system with the broadest deployment. This system, together with its future evolution, HSPA+, is therefore discussed first.

Next, the chapter focuses on the successor technology of HSPA and HSPA+, which is commonly known as Long Term Evolution (LTE). In the standards, LTE is referred to as

Beyond 3G – Bringing Networks, Terminals and the Web Together: LTE, WiMAX, IMS, 4G Devices and the Mobile Web 2.0   Martin Sauter   © 2009 John Wiley & Sons, Ltd

the Evolved Packet System (EPS), which is divided into the Evolved Packet Core (EPC) and the Enhanced-UMTS Terrestrial Radio Access Network (E-UTRAN).

While LTE mainly addresses incumbent wireless operators, there is also great interest from new companies in building wireless networks for Internet access. Many of these companies are attracted by the Worldwide Interoperability for Microwave Access (WiMAX) standard, in particular with the 802.16e air interface. WiMAX is very similar to LTE, but designed from the ground up without the need for backwards-compatibility. Therefore, it is much more suitable for these companies' needs. Since it is expected that both LTE and WiMAX will gain considerable market share, both technologies are discussed to show the similarities and also the differences between the two.

As will be shown throughout this book, 802.11 Wi-Fi networks will play an important role in overall wireless network architectures of the future. Consequently, this chapter also introduces Wi-Fi and the latest enhancements built around the original standard, such as an evolved air interface with speeds of up to 600 Mbit/s, security enhancements for home and enterprise use and quality of service extensions.

To give an initial idea about the performance of each system, some general observations for each system in terms of bandwidth, speed and latency are discussed. Since these parameters are of great importance, and often grossly exaggerated by marketing departments, Chapter 3 will then look at this topic in much more detail.

2.2 UMTS, HSPA and HSPA+

2.2.1 Introduction

Initial drafts of UMTS standards documents appeared in working groups of the Third Generation Partnership Project (3GPP) at the end of 1999, but work on feasibility studies for the system began much earlier. A few UMTS networks were opened to the public in 2003, but it was not until the end of 2004, when adequate UMTS mobile phones became available and networks were rolled out to more than just a few cities, that even early adopters could afford and actually use UMTS. A time frame of five years from a first set of specifications to first deployments is not uncommon due to the complexity involved. This should also be considered when looking at emerging network technologies such as LTE and WiMAX, which are currently in this window between standardization and deployment.

2.2.2 Network Architecture

Figure 2.1 shows an overview of the network architecture of a UMTS network. The upper-left side of the figure shows the radio access part of the network, referred to in the 3GPP standards as the UMTS Terrestrial Radio Access Network (UTRAN).

2.2.2.1 The Base Stations

The UTRAN consists of two components. At the edge of the network, base stations, referred to in the standards as the NodeB, communicate with mobile devices over the air. In cities, a base station usually covers an area with a radius of about 1 km, sometimes

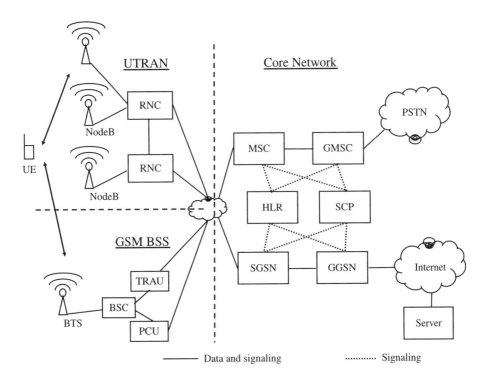

Figure 2.1 Common GSM/UMTS network. (Reproduced from *Communication Systems for the Mobile Information Society*, Martin Sauter, 2006, John Wiley and Sons.)

less, depending on the population density and bandwidth requirements. To increase the amount of data and the number of simultaneous voice calls per base station, the coverage area is usually split into two or three sectors. Each sector has its own directional antenna and transceiver equipment. In the standards, a sector is sometimes also referred to as a cell. A NodeB with three sectors therefore consists of three individual cells. If a user walked around such a base station during an ongoing voice call or while data was exchanged, he would be consecutively served by each of the cells. During that time, the radio network would hand the connection over from one cell to the next once radio conditions deteriorated. From a technical point of view, there is thus little difference between a handover between cells of the same base station and between cells of different base stations. These and other mobility management scenarios will be discussed in more detail in Section 2.2.3. The radio link between mobile devices and the base station is also referred to as the 'air interface' and this term will also be used throughout this book. A device using a UMTS network is referred to in the standard as User Equipment (UE). In this book, however, the somewhat less technical terms 'mobile', 'mobile device' and 'connected mobile device' are used instead.

Today, base stations are connected to the network via one or more 2 Mbit/s links, referred to as E-1 connections in Europe and T-1 connections in the USA (with a slightly lower transmission speed). Each E-1 or T-1 link is carried over a pair of copper cables. An alternative to copper cables is a microwave connection, which can carry several logical E-1 links over a single microwave connection. This is preferred by many operators as they do not have to pay monthly line rental fees to the owner of the copper cable infrastructure. To make full use of the air interface capacity of a multisector base station, several E-1 links are required. The protocol used over these links is ATM (Asynchronous Transfer Mode), a robust transmission technology widely used in many fixed and wireless telecommunication networks around the world today. Figure 2.2 shows a typical base station cabinet located at street level. In practice, base stations are also frequently installed on flat rooftops close to the antennas, as there is often no space at ground level and as this significantly reduces the length and thus the cost for the cabling between the base station cabinet and the antennas.

As technology evolves, using E-1 links over copper cables becomes more difficult since the number of copper cables leading to a base station is limited and, more significantly, the line rental costs per month are high. Network operators have therefore begun using a number of alternative transmission technologies to connect base stations to the network:

- High bandwidth microwave links – a single microwave link can be used to carry several logical E-1 connections. The latest equipment is capable of speeds exceeding 50 Mbit/s [1].
- Fiber links – especially in dense urban areas, many fixed line carriers are currently deploying additional fiber cables for providing very high data rate Internet access to businesses and homes. This infrastructure is also ideal for connecting base stations to the rest of the infrastructure of a wireless network. In practice, however, only a fraction of deployed base stations already have a fiber laid up to the cabinet.
- ADSL/VDSL – a viable alternative to directly using fiber is to connect base stations via a high-speed VDSL link to an optical transmission network. T-Mobile in Germany is one operator that has chosen this solution [2]. In some cases, base stations still require at least one E-1 link for synchronizing the base station with the rest of the network and for carrying voice calls.
- Ethernet – a transmission protocol becoming very popular today in radio access networks is IP over Ethernet. This is reflected in new designs for UMTS/HSPA base stations, which can be equipped with E-1 ATM-based interfaces or alternatively via IP over Ethernet. The Ethernet interface is either based on the standard 100 Mbit/s 100Base-TX twisted pair copper cable interface commonly used with other IT equipment such as PCs and notebooks or via an optical port. In the case of copper cabling, additional equipment is usually required to transport the Ethernet frames over longer distances, as 100Base-TX limits cable length to 100 m.

As the technology used for backhauling data from base stations has a significant impact on the bandwidth and cost of a network, this topic will be discussed in more detail in Section 3.17.

2.2.2.2 The Radio Network Controllers

The second component of the radio access network is the Radio Network Controller (RNC). It is responsible for the following management and control tasks:

- The establishment of a radio connection, also referred to as bearer establishment.
- The selection of bearer properties such as the maximum bandwidth, based on current available radio capacity, type of required bearer (voice or data), quality of service requirements and subscription options of the user.
- Mobility management while a radio bearer is established, that is, handover control between different cells and different base stations of a network.
- Overload control in the network and on the radio interface. In situations when more users want to communicate than there are resources available, the RNC can block new connection establishment requests to prevent other connections from breaking up. Another option is to reduce the bandwidth of established bearers. A new data connection might, for example, be blocked by the network if the load in a cell is already at the limit, while for a new voice call, the bandwidth of an ongoing data connection might be reduced to allow the voice call to be established. In practice, blocking the establishment of a radio bearer for data transmission is very rare, as most network operators monitor the use of their networks and remove bottlenecks, for example by installing additional transceivers in a base station, by increasing backhaul capacity between the base station and the RNC or by installing additional base stations to reduce the coverage area and thus the number of users per base station. Capacity management will be discussed in more detail in Chapter 3.

2.2.2.3 The Mobile Switching Center

Moving further to the right in Figure 2.1, it can be seen that the RNCs of the network are connected to gateway nodes between the radio access network and the core network. In UMTS, there are two independent core network entities. The upper right of the figure shows the MSC, which is the central unit of the circuit-switched core network. It handles voice and video calls and forwards SMS messages via the radio network to subscribers. As discussed in Chapter 1, circuit switching means that a dedicated connection is established for a call between two parties via the MSC that remains in place while the call is ongoing. Large mobile networks usually have several MSCs, each responsible for a different geographical area. All RNCs located in this area are then connected to the MSC. Each MSC in the network is responsible for the management of all users of the network in its region and for the establishment of circuit-switched channels for incoming and outgoing calls. When a mobile device requests the establishment of a voice call, the RNC forwards the request to the MSC. The MSC then checks if the user is allowed to make an outgoing call and instructs the RNC to establish a suitable radio bearer. At the same time, it informs the called party of the call establishment request or, if the called party is located in a different area or different network, establishes a circuit-switched connection to another MSC. If the subscriber is in the same network it might be possible to contact the MSC responsible for the called party directly. In many cases, however, the called party is not in the same network or not a mobile subscriber at all. In this case, a circuit-switched connection is established to a Gateway MSC (GMSC), shown in Figure 2.1 on the top right. Based on the telephone number of the called party, the GMSC then forwards the call to an external fixed or mobile telephone network. In practice, a MSC usually serves mobile subscribers and also acts as a GMSC.

To allow the MSC to manage subscribers and to alert them about incoming calls, mobile devices need to register with the MSC when they are switched on. At the beginning of the registration process, the mobile device sends its International Mobile Subscriber Identity (IMSI), which is stored on a SIM card, to the MSC. If the IMSI is not known to the MSC's Visitor Location Register (VLR) database from a previous registration request, the network's main user database, the Home Location Register, is queried for the user's subscription record and authentication information. The authentication information is used to verify the validity of the request and to establish an encrypted connection for the exchange of signaling messages. The authentication information is also used later on during the establishment of a voice or video call to encrypt the speech path of the connection. Note that the exchange of these messages is not based on the IP protocol but on an out-of-band signaling protocol stack called Signaling System Number 7 (SS7). Out-of-band means that messages are exchanged in dedicated signaling connections, which are not used for transporting circuit-switched voice and video.

2.2.2.4 The SIM card

An important component of UMTS networks, even though it is very small, is the Subscriber Identity Module, the SIM card. It allows the network subscription to be separate from the mobile device. A user can thus buy the SIM card and the mobile device separately. It is therefore possible to use the SIM card with several devices or to use several SIM cards with a single device. This encourages competition between network operators, as users can change from one network to another quickly if prices are no longer competitive. When traveling abroad, it is also possible to buy and use a local prepaid SIM card to avoid prohibitive roaming charges. Separating network subscriptions from mobile devices has the additional benefit that mobile devices can not only be bought from a network operator but also from independent shops, for example electronic stores and mobile phone shops that sell subscriptions for several network operators. This stimulates competitive pricing for mobile devices, which would not happen if a device could only be bought from a single source. A further discussion of this topic can be found in [3].

2.2.2.5 The SMSC

A data service that became very popular long before the rise of current high-speed wireless Internet access technologies is the short message service, used to send text messages between users. As the service dates back to the mid 1990s, it is part of the circuit-switched core network. SMS messages are transported in a store and forward fashion. When a subscriber sends a message, it is sent via the signaling channel, the main purpose of which is to transport messages for call establishment and mobility management purposes, to the Short Message Service Center (SMSC). The SMSC stores the message and queries the Home Location Register database to find the MSC which is currently responsible for the destination subscriber. Afterwards, it forwards the message, again in an SS-7 signaling link, to the mobile switching center. When receiving the text message, the MSC locates the subscriber by sending a paging message. This is necessary, as in most cases the subscriber is not active when a text message arrives and therefore the user's current serving cell is not known to the MSC. On the air interface, the paging message is sent on a broadcast channel that is observed by all devices attached to the network. The mobile device can thus receive the paging message and send an answer to the network despite not having being in active communication with the network. The network then authenticates the subscriber, activates encryption and delivers the text message. In case the subscriber is not reachable, the delivery attempt fails and the SMSC stores the message until the subscriber is reachable again.

2.2.2.6 Service Control Points

Optional, but very important, components in circuit-switched core networks are integrated databases and control logic on Service Control Points (SCPs). An SCP is required, for example, to offer prepaid voice services that allow users to top-up an account with a voucher and then use the credit to make phone calls and send SMS messages. For each call or SMS, the MSC requests permission from the prepaid service logic on an SCP. The SCP then checks and modifies the balance on the user's account and allows or denies the request. Mobile switching centers communicate with SCPs via SS-7 connections. When a prepaid user roams to another country, foreign MSCs also need to communicate with the SCP in the home network of the user. As there are many MSC vendors, the interaction model and protocol between MSCs and SCPs have been specified in the CAMEL (Customized Applications for Mobile Enhanced Logic) standard [4].

For providing the actual service (e.g. prepaid), only signaling connections between SCPs and MSCs are required. Some services, such as prepaid, however, also require an interface to allow a user to check his balance and to top up their account. In practice there are several possibilities. Most operators use some form of scratch card and an automated voice system for this purpose. Therefore, there are usually also voice circuits required between SCP-controlled interactive voice gateways and the MSCs. In addition, most prepaid services also let users top up or check their current balance via short codes (e.g. *100#), which do not require the establishment of a voice call. Instead, such short codes are sent to the SCP via an SS-7 signaling link.

2.2.2.7 Billing

In addition to the billing of prepaid users, which is performed in real time on SCPs, further equipment is required in the core network to collect billing information from the MSCs for subscribers who receive a monthly invoice. This is the task of billing servers, which are not shown in Figure 2.2. In essence, the billing server collects Call Detail Records (CDRs) from the MSCs and SMSCs in the network and assembles a monthly invoice for each user based on the selected tariff. Call detail records contain information such as the identity of the calling party, the identity of the called party, date and duration of the call and the identity of the cell from which the call was originated. Location information is required as calls placed from foreign networks while the user is roaming are charged differently from calls originated in the home network. Some network operators also use location information for zone-based billing, that is, they offer cheaper calls to users while they are at home or in the office. Another popular billing approach is

Figure 2.2   A typical GSM or UMTS base station cabinet.

to offer cheaper rates at certain times. Most operators combine many different options into a single tariff and continuously change their billing options. This requires a flexible rule-based billing service.

2.2.2.8 The Packet-switched Core Network

The core network components discussed so far have been designed for circuit-switched communication. For communicating with services on the Internet, which is based on packet switching, a different approach is required. This is why a packet-switched core network was added to the circuit-switched core network infrastructure. As can be seen in Figure 2.2, the Radio Network Controller connects to both the circuit-switched core network and the packet-switched core network. UMTS devices are even capable of having circuit-switched and packet-switched connections established at the same time. A user can therefore establish a voice call while at the same time using his device as a modem for a PC, or for downloading content such as a podcast to the mobile device without interrupting the connection to the Internet while the voice call is ongoing. Another example of the benefits of being connected to both the packet-switched and circuit-switched networks is that an ongoing instant messaging session is not interrupted during a voice call.

Before a mobile device can exchange data with an external packet-switched network such as the Internet, it has to perform two tasks. First, the mobile device needs to attach to the packet-switched core network and perform an authentication procedure. This is usually done after the device is switched on and once it has registered with the circuit-switched core network. In a second step, the mobile device can then immediately, or at any time later on, request an IP address from the packet-switched side of the network. This process is referred to as establishing a data call or as establishing a PDP (Packet Data Protocol) context. The expression 'establishing a data call' is interesting because it suggests that establishing a connection to the Internet or another external packet network is similar to setting up a voice call. From a signaling point of view, the two actions are indeed similar. The connections that are established as a result of the two requests, however, are very different. While voice calls require a connection with a constant bandwidth and delay that remain in place while the call is ongoing, data calls only require a physical connection while packets are transmitted. During times in which no data is transferred, the channel for the connection is either modified or completely released. Nevertheless, the logical connection of the data call remains in place so the data transfer can be resumed at any time. Furthermore, the IP address remains in place even though no resources are assigned on the air interface. In UMTS, separating the attachment to the packet-switched core network from the establishment of a data connection makes sense when looked at from a historical and practical perspective. The majority of mobile devices today are mostly used for voice communication for which no Internet connection is required. Therefore, the mobile device can perform its main duty without establishing a data call. All other networks that will be discussed in this chapter, however, are only based on a packet-switched core network which is also used for voice calls. As a result, attaching to the network and requesting an IP address is part of the same procedure and the notion of a 'data call' is no longer part of the system design.

2.2.2.9 The Serving GPRS Support Node

The packet-switched UMTS core network was, like the circuit-switched core network, adapted from GSM with only a few modifications. This is the reason why the gateway node to the radio access network is still referred to as the Serving GPRS Support Node (SGSN). GPRS stands for General Packet Radio Service and is the original name of the packet-switched service introduced in GSM networks. Like the MSC, the SGSN is responsible for subscriber and mobility management. To enable users to move between the coverage areas of different RNCs, the SGSN keeps track of the location of users and changes the route of IP packets arriving from the core network when they change their locations. SGSNs are connected to RNCs via ATM links. Since RNCs often control several hundred cells and are physically distant from an SGSN, optical connections are used. In practice, a single RNC is connected to an SGSN via one or more OC-3 optical links with a speed of 155 Mbit/s each or an OC-12 optical connection with a speed of 622 Mbit/s [5]. Above the ATM layer, IP is used as a transport protocol.

The SGSN also has a signaling connection to the Home Location Register of the network that in addition to the data required for the circuit-switched network also contains subscription information for the packet-switched network. This includes, for example, if the user is allowed to use packet-switched services, their quality of service settings such as the maximum transmission speed that is granted by the network and which access points to the Internet they are allowed to use. The record also shows

whether a user has a prepaid contract, in which case the SGSN has to contact a prepaid
SCP before a connection request is granted.

2.2.2.10 The Gateway GPRS Support Node

At the edge of the wireless core network, Gateway GPRS Support Nodes (GGSNs)
connect the wireless network to the Internet. Their prime purpose is to hide the mobility
of the users from routers on the Internet. This is required as IP routers forward packets
based on the destination IP address and a routing table. For each incoming packet, each
router on the Internet consults its routing table and forwards the packet to the next router
via the output port specified in the routing table. Eventually, the packets for wireless
subscribers end up at the GGSN. Here, the routing mechanism is different. As RNCs and
SGSNs can change at any time due to the mobility of the subscriber, a static database
containing the next hop based on the IP address is not suitable. Instead, the GGSN has a
database table which lists the IP address of the SGSN currently responsible for a subscriber with a certain IP address. The IP packet for the subscriber is then enclosed in an IP
packet with the destination address of the SGSN. This principle is called tunneling because
each IP packet to the subscriber is encapsulated in an IP packet to the responsible SGSN.
On the SGSN the original IP packet is restored and once again tunneled to the RNC.

Figure 2.3 shows how tunneling an IP packet in the core network works in practice.
The protocol stack shown uses the Ethernet protocol on layer 2 which is followed by IP
on layer 3. The source and destination address of the IP packet belong to the SGSN and
GGSN of the network. Afterwards, the GPRS Tunneling Protocol (GTP) encapsulates
the original IP packet. Here, the source and destination address represent the subscriber
and a host on the Internet such as a Web server.

In addition to tunneling, the GGSN is also responsible for assigning IP addresses to
subscribers. During the connection establishment, the SGSN verifies the request of the
subscriber with the information from the HLR and then requests the establishment of a

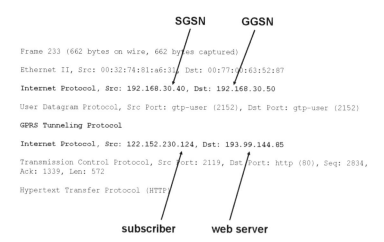

Figure 2.3   Encapsulated IP packet in a packet-switched core network.

tunnel and an assignment of an IP address from the GGSN. The GGSN then assigns an IP address from a pool and establishes the tunnel. In practice, there are two types of IP addresses. Many network operators use private IP addresses that are not valid outside the network. This is similar to the use of private IP addresses in home networks where the DSL or cable router assigns private IP addresses to all devices in the home network. To the outside world, the network is represented with only a single IP address and the DSL or cable router has to translate internal addresses in combination with TCP and UDP port numbers into the external IP address with new TCP and UDP port numbers. This process is referred to as Network Address Translation (NAT). If private IP addresses are used in wireless networks, the same process has to be performed by the GGSN. The advantage for the operator is that fewer public IP addresses are required, of which there is a shortage today. Other operators use public IP addresses for their subscribers by default. For subscribers this has the advantage that they are directly reachable from the Internet, which is required for applications such as hosting a Web server or for remote desktop applications. It should be noted, however, that a public IP address also has disadvantages, especially for mobile devices, as unsolicited packets arriving from the Internet can drain batteries of mobile devices [6]. In practice, a subscriber can have several connection profiles, referred to as Access Point Names (APNs). The operator can therefore choose which subscribers and applications to use private or public IP addresses for.

2.2.2.11 Interworking with GSM

The lower left side of Figure 2.2 shows a GSM radio network, which is also connected to the circuit-switched and packet-switched core network. Depending on the network vendor, the GSM radio network can either be connected to the same MSCs and SGSNs or to separate ones. This is possible since the functionality of these components is very similar for GSM and UMTS. The interfaces to the 2G and 3G radio networks, however, are different. For the connections to the 3G radio network, ATM is used as the lower layer transport protocol while the connection between the SGSN and the 2G radio network is based on the Frame Relay protocol. In a newer version of the standard, an IP-based interface between SGSNs and the 2G radio network has been specified as well. At this point, however, it is not widely used. The MSC is connected to the GSM radio network via circuit-switched links.

2.2.3 Air Interface and Radio Network

2.2.3.1 The CDMA Principle

2G radio systems such as GSM are based on timeslots and channels on different carrier frequencies so a single base station can serve many users simultaneously. While such an approach is well suited for the transmission of circuit-switched voice calls, it only offers limited flexibility for packet-switched data transmission, as the GSM channel bandwidth is only 200 kHz and thus only a limited amount of bandwidth can be assigned to a single user at a time. For 3G systems such as UMTS, it was therefore decided to use a different transmission scheme for the air interface. An alternative transmission technology that overcomes the limitations of a narrow band and was already well understood at the end of the 1990s, when work on UMTS started, was Code Division Multiple Access (CDMA). Instead of a narrow channel bandwidth of 200 kHz as in GSM, the Wideband CDMA (W-CDMA) channel used in UMTS has a bandwidth of 5 MHz. Furthermore, instead of assigning timeslots, all users communicate with the base station at the same time but using a different code. These codes are also referred to as spreading codes, as a single bit is represented on the air interface by a codeword. Each binary unit of the codeword is referred to as a chip to clearly distinguish it from a bit. For a transmission speed of 384 kbit/s, a single bit is encoded in eight chips. The length of the spreading code thus equals 8. The codes used by different users for transmitting simultaneously are mathematically orthogonal to each other. The base station can separate several simultaneous transmissions from mobile devices by applying the reverse algorithm to the one used by the mobile devices, as shown in Figure 2.4, and by knowing the codes used by each mobile device.

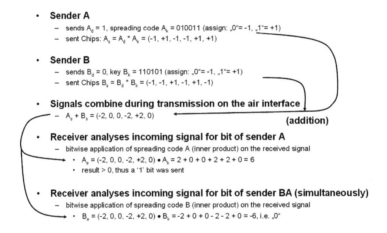

Figure 2.4   Simultaneous data streams of two users to a single base station. (Reproduced from *Communication Systems for the Mobile Information Society*, Martin Sauter, 2006, John Wiley and Sons.)

To accommodate the different bandwidth requirements of different users, the system can use several different spreading code lengths simultaneously. For voice calls, which require a transmission rate of only 12.2 kbit/s, a spreading code length of 128 is used. This means that each bit is represented on the air interface by 128 chips and that up to 128 users can send their data streams simultaneously under ideal conditions. In practice, the number of voice calls per cell is smaller, due to interference from neighboring cells and transmission errors due to nonideal signal conditions. This requires a higher signal-to-noise ratio than could be achieved if there were 128 simultaneous transmissions. Furthermore, some of the codes are reserved for broadcasting system information channels to all devices in the cell, as will be discussed in more detail below. In practice, it is estimated in [7] that each UMTS cell can host up to 60 simultaneous voice calls, excluding subscribers using the cell for packet-switched Internet access.

To increase the data rate of UMTS beyond 384 kbit/s, HSPA and HSPA+ introduce, among other enhancements, the simultaneous use of more than one spreading code per mobile device. This is discussed in more detail below.

2.2.3.2 UMTS Channel Structure

In wired Ethernet networks, which are commonly used in home and business environments today, detecting the network and sending packets is straightforward for devices. As soon as the network cable is plugged in, the network card senses the transmission of packets over the cable and can start transmitting as well, as soon as it detects that no other device is sending a packet. In wireless networks, however, things are more complicated. First of all, there are usually several networks visible at the same time so the device needs to get information about which network belongs to which operator. Also, devices need to detect neighboring cells to be able to quickly react to changing signal levels when the user moves. As keeping the receiver switched on at all times is not very power-efficient, it is also necessary to have a mechanism in place that allows the device to power down the radio in situations where little or no data is transmitted and only wake up periodically to check for new data. Especially for voice calls, it is furthermore important to ensure a certain quality of service in order to prevent the network from breaking down in overload situations. For these reasons, the radio channel is split up into a number of individual channels. Access to these channels is controlled by the network and can be denied during overload situations.

In UMTS networks, a physical channel is represented by a certain spreading code. The following list gives an overview of the most important channels in the downlink direction (network to mobile device) and the uplink direction (mobile device to the network). All of these channels are transmitted simultaneously:

- The Primary Common Control Physical Channel (P-CCPCH) – this channel carries the logical broadcast control channel which is monitored by all mobile devices while they do not have an active connection established to the network. Information distributed via this channel includes the identity of the cell, how the network can be accessed, which spreading codes are used for other channels in the cell, which codes are used by neighboring cells, timers for network access, and so on.
- The Secondary Common Control Physical Channel (S-CCPCH) – this physical channel serves several purposes and thus carries a number of different logical channels. First, it carries paging messages that are used by the network to search for mobile devices for incoming calls, SMS messages or when data packets arrive after a long period of inactivity. The channel is also used to deliver IP packets and control messages to devices which only exchange small quantities of data at a particular time and thus do not need to be put into a more active state.
- Physical Random Access Channel (PRACH) – the random access channel only exists in the uplink direction and is the only channel that a mobile device is allowed to transmit without prior permission of the network. Its purpose is to allow the mobile to request the establishment of a connection to set up a voice call, to establish a data call, to react to a paging message or to send an SMS message. If a data call is already

established and the mobile only sends or receives a small amount of data, the channel can also be used to send IP packets in combination with the S-CCPCH in the downlink direction. However, the data rate in both directions is very low and round trip delay times are high (in the order of 200 ms). The network therefore usually assigns different channels that are dedicated to packet-switched data transfers as soon as it detects renewed network activity from a device.

- Dedicated Physical Data and Control Channels (DPDCH, DPCCH) – once a mobile has contacted the network via the random access channel, the network usually decides to establish a full connection to the mobile. In case of a voice call, the network assigns a dedicated connection. On the air interface, the connection uses a dedicated spreading code that is assigned by the network. In the early days, UMTS networks also used dedicated connections for packet-switched data transmissions. In practice, however, the speed of such dedicated connections was limited to 384 kbit/s. Only a few users could get such a connection, also referred to as a bearer, in a cell simultaneously. Another downside of using a dedicated bearer for packet-switched connections is that the network has to frequently reassign spreading codes to different users. This is necessary, as most packet-switched data transmissions are bursty in nature and there-fore only require a high bandwidth bearer for a limited amount of time. In order to reuse the spreading code for other users, the network needs to change the spreading code length of a connection as soon as it is not fully used any more or even put the connection on a common control channel in order to have the resources available for other users. In practice, this creates a high signaling overhead and a mediocre user experience. It was thus decided that a new concept was needed that offers higher data rates and more flexibility for bursty data transmissions. The outcome of this process is known today as High-speed Packet Access, which many people also refer to as 3.5G. HSPA will be discussed in Section 2.2.4.

2.2.3.3 Radio Resource Control States

To save power and to only assign resources to mobile devices when necessary, a connec-tion to the network can be in one of the following Radio Resource Control (RRC) states:

- Idle state – devices not actively communicating with the network are in this state. Here, they periodically listen to the paging channel for incoming voice or video calls and SMS messages.
- Cell-FACH state – if a mobile device wants to contact the network, it moves to the Cell-Forward Access Channel (Cell-FACH) state. In this state the mobile sends its control messages via the random access channel and the network replies on the forward access channel, which is sent over the S-CCPCH described above. Power requirements also increase as the mobile device now needs to monitor the downlink for incoming control messages.
- Cell-DCH state – once the network decides to establish a voice or data connection, the mobile is instructed to use a dedicated channel and is therefore moved to the Cell Dedicated Channel State (Cell-DCH). If the device is HSPA-capable, the network selects a shared channel instead, as will be discussed in more detail below. From a radio

resource control point of view, however, the mobile is still treated as being in Cell-DCH state. Round trip delay times in Cell-DCH state range from 160 ms with a dedicated bearer to 120 ms for an HSPA connection. In the case of a packet-switched data connection, the network can decide to move a device back to Cell-FACH state if its activity, that is, the amount of data transferred, decreases. Once activity increases again, the network moves the connection back to Cell-DCH state. For even longer periods of inactivity during a data session, the network can even decide to put the mobile device back into idle state to reduce power consumption of the mobile device and to reduce the management processing load for the network. Despite being in Idle state, the mobile device can resume sending IP packets at any time via the random access channel. However, there is a noticeable delay when moving to a more active state. Depending on the radio network vendor, initial round trip delay times between 2 and 4 s can be observed [8].

- Cell-PCH and URA-PCH states – even while in Cell-FACH state, the mobile device requires a considerable amount of power to listen to the forward access channel despite its minimal activity. Because of the noticeable delay when resuming data transmission from the Idle state, two further states have been specified which are between Idle and Cell-FACH state. In Cell-PCH and URA-PCH state, the mobile only needs to periodically listen to the paging channel while the logical connection between the radio network and the device remains in place. If there is renewed activity, the connection can be quickly resumed. The difference between the two states is that in Cell-PCH state the mobile has to report cell changes to the network because the paging message is only sent into one cell, whereas in URA-PCH state the mobile can roam between several cells that belong to the same routing area without reporting a cell change to the network since the paging message is sent into all cells belonging to the routing area. In practice, however, only a few network operators make use of these two additional states.

2.2.3.4 Mobility Management in the Radio Network for Dedicated Connections

An important task in cellular networks is to handle the mobility of the user. In Cell-DCH state, the base station actively monitors the air interface connection to and from a mobile device and adjusts the transmission power of the base station and the transmission power of the mobile device 1500 times per second. In CDMA systems, such very quick power adjustments are necessary, as the transmissions of all mobile devices should be received by the base station with the same power level whether they are very close or very far away. In the direction from the network to a mobile device, as little transmission power as possible should be used as the transmission power of the base station is limited. In practice, data transfers to and from mobile devices close to a base station only require a small amount of transmission power while devices far away or inside buildings require significantly more transmission power. Adjusting the transmission power so frequently also means that, along with the user data, the mobile device has to continuously report to the network how well the user data is received. The network then processes this information and instructs the mobile device, at the same rate, whether it should increase, hold or decrease its current transmission power. For this purpose, a dedicated control channel is always established alongside a dedicated traffic channel. While the content of the dedicated traffic channel is given to higher layers of the protocol stack, which eventually extract a voice packet or an IP packet, the information transported in the control channel remains in the radio software stack and is not visible to higher-layer applications.

When a user moves, they eventually leave the coverage area of a base station. While a DCH is established, the Radio Network Controller takes care that the connection to the mobile device is transferred to a more suitable cell. This decision is based on reception quality of the current cell and the neighboring cells, which is measured by the mobile device and sent to the network. In CDMA-based networks such as UMTS, there are two different kinds of handovers. The first type is referred to as a hard handover. When the network detects that there is a more suitable cell for a mobile device, it prepares the new cell for the subscriber and afterwards instructs the mobile to change to the new cell. The mobile device then interrupts the connection to the current cell and uses the handover parameters sent by the network, which include information on the frequency and spreading codes of the new cell, for establishing a connection. A hard handover is therefore a break-before-make handover, that is, the old connection is cut before the new one is established.

The second type of handover is a soft handover, which is the most commonly used type of handover in CDMA networks. Soft handovers make use of the fact that neighboring cells transmit on the same frequency as the current cell. It is thus possible to perform a make-before-break handover, that is, the mobile device communicates with more than a single cell at a time. A mobile enters the soft handover state when the network sends control information to instruct it to listen to more than a single spreading code in the downlink direction. Each spreading code represents a transmission of a different cell and the mobile combines the signals it receives from the different cells. In the uplink direction, the mobile device continues using only a single spreading code for a dedicated channel. All cells taking part in the soft handover are instructed by the RNC to decode the data sent with this code and forward it to the RNC. During soft handover, the RNC thus receives several copies of the user's uplink data stream. This increases the likelihood that a packet is received correctly and does not have to be retransmitted. The cells which are involved in a soft handover for a dedicated channel are referred to as the active set. Up to six cells can be part of the active set. In practice, however, most networks limit the number of cells in the active set to two or three. Otherwise, the benefit of multiple transmissions that can be combined in both the network and the mobile station is outweighed by the additional capacity required in both the radio network and on the air interface. To improve the capacity of the network, around 30–40% of all dedicated connections in a network are in soft handover state, even if the users are not moving.

Figure 2.5 shows how hard and soft handovers are performed for a case in which not all cells are connected to the same radio network controller. In this example, the uplink and downlink data are collected and distributed by two RNCs. However, only one of the two RNCs, referred to as the Serving-RNC (S-RNC), communicates with the core network. The other RNC, referred to as the Drift-RNC (D-RNC), communicates with the S-RNC.

If the user moves further into the area covered by NodeB 2 and NodeB 3, the mobile will at some point lose contact with NodeB 1. At this point, the involvement of the RNC

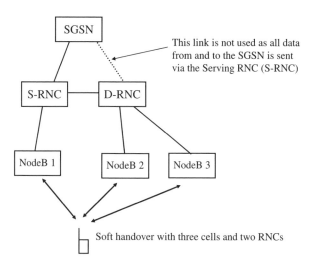

Figure 2.5 Soft handover with several RNCs. (Reproduced from *Communication Systems for the Mobile Information Society*, Martin Sauter, 2006, John Wiley and Sons.)

controlling NodeB 1 is no longer required. The current S-RNC then requests the SGSN to promote the current Drift-RNC as the new Serving RNC.

If the optional interface between the RNCs is not present, a soft handover as shown in Figure 2.5 is not possible. In this case, a hard handover is required when the mobile device moves out of the coverage area of the first NodeB and into the coverage area of the other two NodeB. Furthermore, the change of serving RNC has to be performed at the same time, which further complicates the handover procedure and potentially increases the period during which no data can be exchanged between the mobile device and the network.

When the user moves between the areas controlled by two different SGSNs, it is additionally necessary to change the serving SGSN. To make matters even more complicated, handovers can also take place while both a circuit-switched and a packet-switched connection is active. In such a case, not only the RNCs need to be changed, but also the SGSN and MSC. This requires close coordination between the packet-switched and circuit-switched core network. While in practice typical handover scenarios only involve cells controlled by the same RNC, all other types of handovers have to be implemented as well.

Since 3G/3.5G networks did not have the same coverage area as the already existing 2G GSM networks, and often still lack the same coverage area and indoor penetration today, it is very important to be able to handover active connections to GSM networks. This was and still is especially important for voice calls. Mobile devices, therefore, do not only search for neighboring 3G/3.5G cells but can also be instructed by the network to search for GSM cells. While a dedicated channel is established this is very difficult since in the standard transmission mode the mobile device sends and receives continuously. For cells at the edge of the coverage area, the RNC can activate a compressed transmission mode. The compressed mode opens up predefined transmission gaps that allow the mobile device's receiver to retune to the frequencies of neighboring GSM cells, receive their signal, synchronize to them and return to the UMTS cell to resume communication and to report the result of the neighboring cell signal measurements to the network. Based on these measurements, the network can then instruct the mobile to perform an Inter-Radio Access Technology (Inter-RAT) hard handover to GSM. During an ongoing voice call, such a hard handover is usually not noticed by the user. A data connection handed over to a GSM/GPRS/EDGE network, however, gets noticeably slower.

2.2.3.5 Radio Network Mobility Management in Idle, Cell-FACH and Cell/URA-PCH State

When the user moves and the mobile device is in Idle or Cell/URA-PCH state, that is, no voice call is established and the physical connection of a data call has been removed after a phase of inactivity, the mobile device can decide on its own to move to a cell with a better signal. Cells are grouped into location areas and mobile devices changing to a different cell only need to contact the network to report their new position when selecting a cell in a new location area. This means that paging messages have to be broadcast in all cells belonging to a location area. In practice, a location area contains about 20–30 base stations, providing a satisfactory trade-off between the reduced number of location updates and the increased overhead of sending paging messages into all cells of a location area.

In Cell-FACH state, which is used if the mobile enters a phase of lower activity (i.e. it sends or receives only a few IP packets), the cell changes are also controlled by the mobile and not the network. Here, an efficient transition from one cell to another is deemed to be not important enough to justify the required processing capacity in the network and the higher power consumption in the mobile terminal.

2.2.3.6 Mobility Management in the Packet-switched Core Network

In the core network, the SGSN is responsible for keeping track of the location and reachability of a mobile device. In PMM (Packet Mobility Management) Detached state, the mobile is not registered in the network and consequently does not have an IP address. In PMM Connected state the mobile has a signaling connection to the SGSN (e.g. during a location update to report its new position) and also an IP connection. The SGSN, however, only knows the current RNC responsible for forwarding the packet and not the cell identity. This knowledge is not necessary, as the SGSN just forwards the data packet to the RNC and the RNC is responsible for forwarding it via the current cell to the user. In addition the SGSN does not know if the radio connection to the mobile is in Cell-FACH, Cell-DCH, Cell-PCH or URA-PCH state. This information is not necessary since it is the RNC's task to decide in which state to keep the mobile based on the quality of service requirements of a connection and the current amount of transmitted data. Note that PMM Connected state is also entered for short periods of time during signaling exchanges such as a location update even if no IP connection is established.

If the RNC sets the radio connection to Idle state, it also informs the SGSN that the mobile is no longer directly reachable. The SGSN then modifies its control state to PMM

Idle. For the SGSN this means that, if an IP packet arrives from the Internet later on, the mobile device has to be paged first. The IP packet is then buffered until a response is received and a signaling connection has been set up.

Figure 2.6 shows the different radio and core network states and how they are related to each other while a packet-switched data connection is established. When moving from left to right in the figure, it can be seen that, the deeper the node is in the network, the less information it has about the current mobility state and the radio network connection of the mobile device. Note that the state of a mobile device is also PMM Idle when the mobile has attached to the packet-switched core network after power on but has not yet established a data connection. The radio network mobility management states also apply when the mobile device communicates with the circuit-switched network. During a voice or video call via the MSC, the mobile device is in Cell-DCH state. The mobile is also set into Cell-DCH or Cell-FACH state during signaling message exchanges, for example during a location update with the MSC and SGSN.

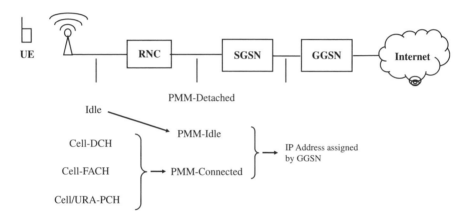

Figure 2.6   Radio network and core network mobility management states for an active packet-switched connection.

2.2.4 HSPA (HSDPA and HSUPA)

Soon after the first deployments of UMTS networks, it became apparent that the use of dedicated channels on the air interface for packet-switched data transmission was too inflexible in a number of ways. As per the specification, the highest data rate that could be achieved was 384 kbit/s with a spreading code length of 8. This limited the number of people who could use such a bearer simultaneously to eight in theory and to two or three in practice, as some spreading codes were required for the broadcast channels of the cell and for voice calls of other subscribers. Because of the bursty nature of many packet-switched applications, the bearer could seldom be fully used by a mobile device and therefore a lot of capacity remained unused. This was countered somewhat by first assigning longer code lengths to the user's connection and then only upgrading the connection when it was detected that the bearer was fully used for some time (e.g. for a file download). Also, short spreading codes were quickly replaced by longer ones once it became apparent to the RNC that the capacity was no longer fully used. Despite these mechanisms, efficiency remained rather low. As a consequence, vendors started to specify a package of enhancements which are now referred to as High-Speed Downlink Packet Access (HSDPA) [9]. The specification of HSDPA started as early as 2001 but it took until late 2006 for the first networks and devices to support HSDPA in practice. Once standardization of HSDPA was a good way on, a number of enhancements were specified for the uplink direction as well. These are referred to as High-Speed Uplink Packet Access (HSUPA). Together, HSDPA and HSUPA are known today as HSPA.

The following paragraphs are structured as follows: first, an overview of the downlink enhancements is given. Afterwards, the performance improvements for the uplink direction are discussed.

2.2.4.1 Shared Channels

To improve the utilization of the air interface, the concept of shared channels has been (re-) introduced. HSDPA introduces a High-Speed Downlink Physical Shared Channel (HS-DPSCH), which several devices can use quasi simultaneously as it is split into timeslots. The network can thus quickly react to the changing bandwidth requirements of a device by changing the number of timeslots assigned to a device. Thus, there is no longer a need to assign spreading codes or to modify the length of a spreading code to change the bandwidth assignment. HS-DPSCH timeslots are assigned to devices via the High-Speed Shared Control Channel (HS-SCCH). This channel has to be monitored by all mobile devices that the RNC has instructed to use a shared channel. The HS-DPSCH uses a fixed spreading code length of 16 and the HS-SCCH uses a spreading code length of 256.

2.2.4.2 Multiple Spreading Codes

To increase transmission speeds, mobile devices can listen to more than a single high-speed downlink shared channel. Category 6 HSDPA devices can listen to up to five high-speed downlink channels simultaneously, and category 8 devices to up to 10. Multiple high-speed downlink channels can also be used to transmit data to several devices simultaneously. For this purpose the network can configure up to four simultaneous shared control channels in the downlink direction and up to four mobile devices can therefore receive data on one or more shared channels simultaneously.

Figure 2.7 shows how user data is transferred through the radio network and which channels an HSDPA device has to decode simultaneously while it is in Cell-DCH state. At the RNC, the user's data packets arrive from the core network in a Dedicated Traffic Channel (DTCH). From there, the RNC repackages the data into a single High-speed Dedicated Shared Channel (DSCH) and forwards it to the NodeB. In addition, the RNC sends control information to the device via a Dedicated Control Channel (DCCH). This channel is not shared between subscribers. It is also possible for the user to establish or receive a voice call during an ongoing HSDPA data transfer. In this case, the RNC additionally uses a dedicated traffic channel to forward the circuit-switched voice channel to the user. At the base station (NodeB), data arriving via the HS-DSCH channel is buffered for a short while and then transmitted over one or several High-speed Physical

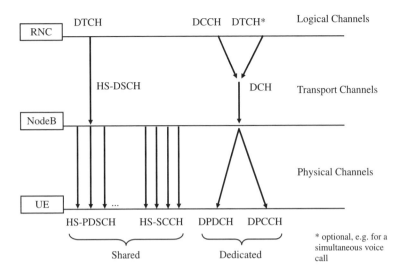

Figure 2.7 Simplified HSDPA channel overview in the downlink direction. (Reproduced from *Communication Systems for the Mobile Information Society*, Martin Sauter, 2006, John Wiley and Sons.)

Downlink Shared Channels (HS-PDSCH). In practice, at least five simultaneous HS-PDSCHs are used and up to 10 simultaneous channels can be used for category 8 devices. In addition to those 5–10 shared channels, the device has to be able to monitor up to four high-speed downlink shared control channels to receive the timeslot assignments for the shared channels. Finally, the mobile has to decode control information arriving on the Dedicated Physical Control Channel (DPCCH) and potentially also the voice data stream arriving on the Dedicated Physical Data Channel (DPDCH).

In the uplink direction, an HSDPA-capable device also has to send a number of different data streams, each with a different spreading code. The first stream is the HSDPA control channel, which is used to acknowledge the correct reception of downlink data packets. In addition, a dedicated channel is required for IP packets transmitted in uplink direction. A further channel is necessary to transmit or to reply to radio resource commands from the RNC (e.g. a cell change command) and for the uplink direction of an ongoing voice call. Finally, the mobile needs to send control information (reception quality of the downlink dedicated channel) to the NodeB.

2.2.4.3 Higher Order Modulation

To further increase transmission speeds, HSDPA introduces a higher order modulation scheme to improve throughput for devices that receive the transmissions of the base station with a good signal quality. Basic UMTS uses Quadrature Phase Shift Keying (QPSK) modulation, which encodes two basic information elements (chips) per transmission step. HSDPA introduces 16QAM (Quadrature Amplitude Modulation) modulation, which encodes four chips per transmission step. This doubles the achievable bandwidth under good reception conditions. Together with using 10 simultaneous downlink shared channels, the theoretical maximum downlink speed is 7.2 Mbit/s. In practice, transmission speeds range between 1.5 and 2.5 Mbit/s for devices that can bundle five shared channels (HSDPA category 6) and 4.5 Mbit/s for devices that can bundle 10 shared channels (HSDPA category 8).

2.2.4.4 Scheduling, Modulation and Coding, HARQ

The higher the transmission speed, the more important it is to detect transmission errors and react to them as quickly as possible. Otherwise, connection-oriented higher-layer protocols such as TCP misinterpret air interface packet errors as congestion and slow down the transmission. To react to transmission errors more quickly, it was decided not to implement the HSDPA scheduler in the RNC, as was done before for dedicated channels, but to assign this task to the base station. HSDPA uses the Hybrid Automated Retransmission Request (HARQ) scheme for this purpose. With HARQ, data frames with a fixed length of 2 ms have to be immediately acknowledged by the mobile device. If a negative acknowledgement is sent, the packet can be repeated within 10 ms. The next data frame is only sent once the previous one has been positively confirmed. The mobile device must be able to handle up to eight simultaneous HARQ processes as the mobile device is allowed up to 5 ms to decode the packet before it has to send the ACK (Acknowledgment) or the NACK (Negative Acknowledgment) to the network. During this time, two further data frames of other HARQ processes can arrive at the mobile station. In case data was not received correctly in one HARQ process, frames received via other HARQ processes have to be stored in the mobile device until all previous frames have been successfully received, so that the data can be forwarded in the correct order to higher protocol layers.

Another reason to implement the scheduler in the base station is to be able to quickly react to changing signal conditions. Based on the feedback of the mobile device, the base station's scheduler decides for each frame which modulation (QPSK or 16QAM) to use and how many error correction bits to insert. This process is referred to as Adaptive Modulation and Coding (AMC). Advanced algorithms in the base station use the knowledge about reception conditions of all mobile devices currently served on the high-speed shared channel. Devices with better signal conditions can be preferred by the scheduler, while devices that are in temporary deep fading situations receive fewer packets. This helps to reduce transmission errors and improves overall throughput of the cell as, on average, frames are transmitted with more efficient modulation and coding schemes. Studies in [10] and [11] have shown that an efficient scheduler can increase overall cell capacity by up to 30% compared with a simple scheduler that assigns time-slots in a round robin fashion.

2.2.4.5 Cell Updates and Handovers

As has been shown above, a mobile device in HSDPA Cell-DCH state receives its packet-switched data via shared channels. Nevertheless, additional dedicated channels are required alongside the shared channels to transport control information in both directions. Furthermore, IP packets in the uplink direction are also transported in a dedicated channel. Finally, a parallel circuit-switched voice call is also transported in a dedicated channel. These dedicated connections can be in soft handover state to improve reception conditions and to better cope with the user's mobility. The shared channels, however, are only sent from one base station of a device's active set. Based on the feedback received from the mobile device, the RNC can at any time promote another base station to forward the IP packets in the downlink. This procedure is referred to as a cell update. Since the procedure is controlled by the network, the interruption of the data traffic in the downlink direction is only very short.

2.2.4.6 HSUPA

In early 3G networks, uplink speeds were limited to 64–128 kbit/s. With the introduction of HSDPA, some vendors also included a radio network software update to allow dedicated uplink radio bearers with speeds of up to 384 kbit/s, if permitted by the conditions on the radio interface to a user. In practice, it can be observed that in many situations the mobile's transmission power is sufficient to actually make use of such an uplink bearer. This is good for sending large files or e-mails with large file attachments, as well as for many Web 2.0 applications (cf. Chapter 6), where user-generated content is sent from a mobile device to a database in the network. As for dedicated downlink transmissions, however, assigning resources in the uplink in this way is not very flexible. The 3GPP standards body thus devised a number of improvements for the uplink direction, which are referred to as High-Speed Uplink Packet Access (HSUPA).

Unlike in the downlink, which was enhanced by using high-speed shared channels, the companies represented in 3GPP decided to keep the dedicated channel approach for uplink transmissions. There were several reasons for this decision. While in the downlink, the base station has an overview of how much data is in the buffers to be sent to all mobile devices, there is no knowledge in the base station about the buffer states of the mobile devices requesting to send data in the uplink. This makes assigning timeslots on a shared channel difficult as mobile devices continuously have to indicate if they have more data. The second reason to re-use the dedicated channel concept was that the soft handover concept is especially valuable for the uplink direction as the transmission power of mobile devices is limited. In the standards, an HSUPA dedicated uplink channel is referred to as an Enhanced-DCH or E-DCH.

A standard uplink dedicated channel is controlled by the RNC and spreading codes can only be changed to reach maximum data transfer rates of 64, 128 or 384 kbit/s. As the RNC is in control of the connection, the bearer parameters are only changed very slowly, for example only every few seconds once the RNC detects that the pipe is too big or too small for the current traffic load or once it detects that signal conditions have significantly changed. With the E-DCH concept, the control of the radio interface channel has been moved from the RNC to the base station, in a similar way to the solution for HSDPA. The base station controls overall scheduling of all E-DCH uplink transmissions by assigning bandwidth grants to all active E-DCH devices. Bandwidth grants are transmitted in the downlink direction via a new shared control channel, the Enhanced Absolute Grant Channel (E-AGCH). Access grants are translated into the maximum amount of transmission power each device is allowed to use. This way, all devices can still transmit their data to the base station at the same time while their maximum transmission speeds can be quickly adjusted as necessary. Staying with the dedicated channel concept and hence allowing all mobile devices to transmit at the same time has the additional benefit of shorter packet delay times compared with an approach where devices have to wait for their timeslot to transmit. Optionally, a second control channel, the Enhanced Relative Grant Channel (E-RGCH) can be used to quickly increase or decrease uplink transmission power step by step. This enables each base station involved in an E-DCH soft handover to decrease the interference caused by a mobile device, if this creates too much interference for communication with other devices.

The HARQ concept first introduced with HSDPA is also used with HSUPA for uplink transmissions. This means that the base station immediately acknowledges each data frame it receives. Faulty frames can thus be retransmitted very quickly. For this purpose, an additional dedicated downlink channel has been created, the Enhanced HARQ information channel (E-HICH). Each HSUPA device gets its own E-HICH while it is in Cell-DCH mode. If an E-DCH connection is in soft handover state, each of the contributing base stations sends its own acknowledgement for a frame to the mobile device. If only one acknowledgement is positive, the transfer is seen as successful and the HARQ process moves on to the next frame.

While it was decided to introduce a new modulation scheme for the downlink to further increase transmission speeds, it was shown with simulations that there would be no such effect for uplink transmissions. Uplink transmissions are usually power limited, which means that a higher order modulation cannot be used as the error rate would become too high. Instead, it was decided to introduce multicode transmissions in a single dedicated channel to allow the mobile to split its data into several code channels which are then transported simultaneously. This concept is similar to the concept of using several (shared) channels in the downlink to increase the data rates. The highest terminal category currently defined can use up to two spreading codes with a length of two and two spreading codes with a length of four. In theory, uplink data rates can reach up to 2 Mbit/s. In practice, data rates are lower in a similar way to those described above for HSDPA.

Figures 2.8 and 2.9 show the channels used in uplink and downlink direction for a terminal that supports both HSDPA and HSUPA for the most complicated case when a voice call is ongoing in parallel. In practice, encoding and decoding so many channels at the same time is very demanding and has only been made possible by the ever increasing processing power of mobile device chipsets, as is further discussed in Chapter 5. Furthermore, it can be observed that early HSPA network implementations fall back to the default dedicated channel approach for the packet-switched connection during a voice call.

2.2.5 HSPA+ and other Improvements: Competition for LTE

The quest to improve UMTS and make it even faster and more power-efficient and to allow more devices to use a cell simultaneously (e.g. for VoIP) has not ended with HSPA. For instance, there are a number of initiatives in Release 7 and Release 8 of the 3GPP standard to further improve the system. Improvements of the air interface are referred to as HSPA+. In addition, the network architecture has also received an optional overhaul to be more efficient with a feature referred to as 'one-tunnel'. The following section gives an overview of these improvements, many of which are likely to be introduced into networks over the next few years.

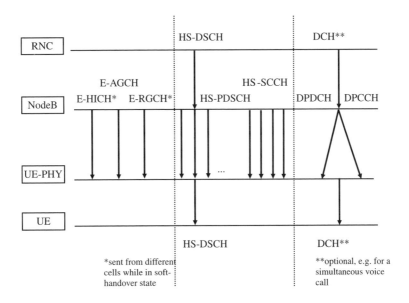

Figure 2.8 Channels for a combined HSDPA and HSUPA transmission in the downlink direction. (Reproduced from *Communication Systems for the Mobile Information Society*, Martin Sauter, 2006, John Wiley and Sons.)

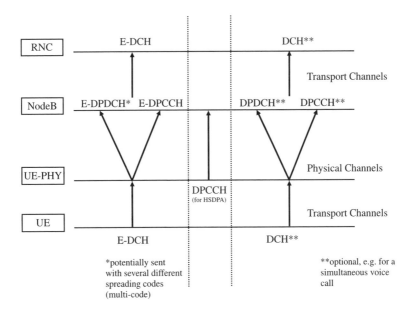

Figure 2.9 Channels for a combined HSDPA and HSUPA transmission in the uplink direction. (Reproduced from *Communication Systems for the Mobile Information Society*, Martin Sauter, 2006, John Wiley and Sons.)

2.2.5.1 Higher Order Modulation

One of the key parameters of a wireless system that is often cited in articles is the maximum transmission speed. So far, HSPA uses QPSK and 16QAM modulation to reach theoretical data rates of up to 14.4 Mbit/s in the downlink direction, which translates to 2–5 Mbit/s in practice under good radio conditions. To increase transmission rates further, 3GPP Release 7 introduces 64QAM modulation in the downlink, which transmits six chips per transmission step compared with four chips with 16QAM. In practice, it is expected that introducing 64QAM will result in a 30% throughput gain for users close to the center of the cell [12]. Most users, however, will not be able to use 64QAM modulation, as their signal-to-interference ratio will be too low. This will be discussed in more detail in Chapter 3. As the overall throughput in the cell increases, these users will nevertheless also benefit from this measure, as the cell will have more time to transmit its data, because data for users closer to the cell center can be sent faster. Furthermore, 64QAM modulation is also beneficial for micro cell deployments in public places such as shopping malls, where users are close to small cells and therefore create little interference.

In the uplink direction, no changes were made to the modulation scheme in 3GPP Release 7. Release 8, however, might include 16QAM modulation despite the earlier conclusion that there would be little benefit. More advanced receivers and a focus on microcell environments in public places may be responsible for changing this opinion.

2.2.5.2 MIMO

Another emerging technology to increase throughput under good signal conditions is Multiple Input Multiple Output, or MIMO for short. In essence, MIMO transmission uses two or more antennas at both the transmitter and the receiver side to transmit two independent data streams simultaneously over the same frequency band. This linearly increases data rates with the number of antennas. Two transmitter antennas and two receiver antennas (2×2) as currently specified for HSPA+ can thus double the data rate of the system under ideal signal conditions. Further technical background on MIMO can be found in Section 2.3. Release 7 of the standards foresees the use of MIMO in combination with 16QAM modulation in the downlink. If the network operator chooses to deploy an extra set of base station antennas, the theoretical data rate of 14.4 Mbit/s is therefore increased to 28.8 Mbit/s.

Depending on the signal conditions, available antennas and device capabilities, the network can choose between a single data stream with 64QAM or two streams with 16QAM. Release 8 of the standards might combine MIMO with 64QAM, which would result in a peak data rate of 43.2 Mbit/s in the downlink. However, it should be once more noted at this point that such data rates can only be achieved by very few users of a cell. Details are given in Chapter 3. As uplink transmissions are usually power limited, MIMO has only been considered for the downlink direction.

2.2.5.3 Continuous Packet Connectivity

Continuous Packet Connectivity (CPC) is a package of features introduced in the 3GPP standards to improve handling of mobile subscribers while they have a packet connection established, that is while they have an IP address assigned. Taken together, they aim at reducing the number of state changes to minimize delay and signaling overhead by introducing enhancements to keep a device on the high-speed channels (in HSPA Cell-DCH state) for as long as possible, even while no data transfer is ongoing. For this, it is necessary to reduce power consumption while mobiles listen to the shared channels and, at the same time, reduce the bandwidth requirements for radio layer signaling to increase the number of mobile devices per cell that can be held in HSPA Cell-DCH state.

CPC does not introduce new revolutionary features. Instead, already existing features are modified to achieve the desired results. To understand how these enhancements work, it is necessary to dig a bit deeper into the standards. 3GPP TR 25.903 [13] gives an overview of the proposed changes and the following descriptions refer to the chapters in the document which have been selected for implementation.

2.2.5.3.1 Feature 1: A new uplink control channel slot format (Section 4.1 of [11])

While a connection is established between the network and a mobile device, several channels are used simultaneously. This is necessary as there is not only user data sent over the connection but also control information to keep the link established, to control transmit power, and so on. Currently, the radio control channel in the uplink direction (the Uplink Dedicated Control Channel, UL DPCCH) is transmitted continuously, even during times of inactivity, in order not to lose synchronization. This way, the terminal can resume uplink transmissions without delay whenever required.

The control channel carries four parameters:

- Transmit Power Control (TPC);
- pilot (used for channel estimation of the receiver);
- TFCI (Transport Format Combination Identifier);
- FBI (Feedback Indicator).

The pilot bits are always the same and allow the receiver to get a channel estimate before decoding user data frames. While no user data frames are received, however, the pilot bits are of little importance. What remains important is the TPC. The idea behind the new slot format is to increase the number of bits to encode the TPC and decrease the number of pilot bits while the uplink channel is idle. This way, additional redundancy is added to the TPC field. As a consequence, the transmission power for the control channel can be lowered without risking corruption of the information contained in the TPC. Once user data transmission resumes, the standard slot format is used again and the transmission power used for the control channel is increased again.

2.2.5.3.2 Feature 2: CQI reporting reduction (Section 4.4 of [11]), uplink discontinuous transmission (Section 4.2 of [11]) in combination with downlink control information transmission enhancements

CQI reporting reduction:to make the best use of the current signal conditions in the downlink, the mobile has to report to the network how well its transmissions are received. The quality of the signal is reported to the network with the Channel Quality Index (CQI) alongside the user data in the uplink. To reduce the transmit power of the terminal while data is being transferred in the uplink but not in the downlink, this feature reduces the number of CQI reports.

UL HS-DPCCH gating (gating = switch off): when no data is being transmitted in either the uplink or downlink, the uplink control channel (UL DPCCH) for HSDPA is switched off. Periodically, it is switched on for a short time to transmit bursts to the network in order to maintain synchronization. This improves battery life for applications such as Web browsing. This solution also lowers battery consumption for VoIP and reduces the noise level in the network (i.e. allowing more simultaneous VoIP users). Figure 2.10 shows the benefits of this approach.

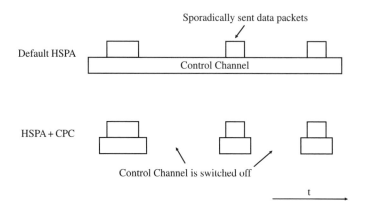

Figure 2.10   Control channel switch-off during times with little activity.

F-DPCH gating: terminals in HSDPA active mode always receive a Dedicated Physical Channel in the downlink, in addition to high-speed shared channels, which carries power control information and Layer 3 radio resource (RRC) messages, for example for handovers, channel modifications and so on. The Fractional-DPCH feature puts the RRC messages on the HSDPA shared channels and the mobile thus only has to decode the power control information from the DPCH. At all other times, that is when the terminal is not in HSDPA active mode, the DPCH is not used by the mobile (thus it is fractional). During these times, power control information is transmitted for other mobiles using the same spreading code. Consequently, several mobiles use the same spreading code for the dedicated physical channel but listen to it at different times. This means that fewer spreading codes are used by the system for this purpose, which in turn leaves more resources for the high-speed downlink channels or allows more users to be kept in HSPA Cell-DCH state simultaneously.

2.2.5.3.3 Feature 3: Discontinuous Reception (DRX) in the Downlink (Based on Section 4.5 of [11])
While a mobile is in HSPA mode, it has to monitor one or more high-speed shared control channels (HS-SCCH) to see when packets are delivered to it on the high-speed shared channels. This monitoring is continuous, that is the receiver can never be switched off. For situations when no data is transmitted, or the average data transfer rate is much lower than that which could be delivered over the high-speed shared channels, the base station can

instruct the mobile to only listen to selected slots of the shared control channel. The slots which the mobile does not have to observe are aligned as much as possible with the uplink control channel gating (switch-off) times. Therefore, there are times when the terminal can power down its receiver to conserve energy. Once more data arrives from the network than can be delivered with the selected DRX cycle, the DRX mode is switched off and the network can once again schedule data in the downlink continuously.

2.2.5.3.4 Feature 4: HS-SCCH-less Operation (Based on Sections 4.7 and 4.8 of [11])

This feature is not intended to improve battery performance but to increase the number of simultaneous real-time VoIP users in the network. VoIP service, for example via the IMS (cf. Chapter 4), requires relatively little bandwidth per user and thus the number of simultaneous users can be high. On the radio link, however, each connection has a certain signaling overhead. Therefore, more users mean more signaling overhead which decreases overall available bandwidth for user data. In the case of HSPA, the main signaling resources are the high-speed shared control channels (HS-SCCH). The more active users there are, the more they proportionally require of the available bandwidth.

HS-SCCH-less operation aims at reducing this overhead. For real-time users who require only limited bandwidth, the network can schedule data on high-speed downlink channels without prior announcements on a shared control channel. This is done as follows: the network instructs the mobile not only to listen to the HS-SCCH but in addition to all packets being transmitted on one of the high-speed downlink shared channels. The terminal then attempts to blindly decode all packets received on that shared channel. To make blind decoding easier, packets which are not announced on a shared control channel can only have one of four transmission formats (number of data bits) and are always modulated using QPSK. These restrictions are not an issue for performance, since HS-SCCH-less operation is only intended for low-bandwidth real-time services.

The checksum of a packet is additionally used to identify for which device the packet is intended. This is done by using the terminal's MAC address as an input parameter for the checksum algorithm in addition to the data bits. If the device can decode a packet correctly and if it can reconstruct the checksum, it is the intended recipient. If the checksum does not match then either the packet is intended for a different terminal or a transmission error has occurred. In both cases the packet is discarded.

In case of a transmission error, the packet is automatically retransmitted since the mobile did not send an acknowledgement (HARQ ACK). Retransmissions are announced on the shared control channel, which requires additional resources but should not happen frequently as most packets should be delivered properly on the first attempt.

2.2.5.4 Enhanced Cell-FACH, Cell/URA PCH States

The CPC features described above aim to reduce power consumption and signaling overhead in HSPA Cell-DCH state. The CPC measures therefore increase the number of mobile devices that can be in Cell-DCH state simultaneously and allow a mobile device to remain in this state for a longer period of time even if there is little or no data being transferred. Eventually, however, there is so little data transferred that it no longer makes sense to keep the mobile in Cell-DCH state, that is it does not justify even the reduced signaling overhead and power consumption. In this case, the network puts the connection into Cell-FACH state as described above or even into Cell-PCH or URA-PCH state to reduce energy consumption even further. The downside of this is that a state change back into Cell-DCH state takes a long time and that little or no data can be transferred during the state change. In Release 7 and 8, the 3GPP standards were thus extended to also use the high-speed downlink shared channels for these states, as described in [14] and [15]. In practice this is done as follows:

- Enhanced Cell-FACH – in the standard Cell-FACH state the mobile device listens to the secondary common control physical channel in the downlink as described above for incoming radio resource control messages from the RNC and for user data (IP packets). With the Enhanced Cell-FACH feature, the network can instruct a mobile device to observe a high-speed downlink control channel or the shared data channel directly for incoming radio resource control messages from the RNC and for user data. The advantage of this approach is that, in the downlink direction, information can be sent much faster. This reduces latency and speeds up the Cell-FACH to Cell-DCH state change procedure. Unlike in Cell-DCH state, no other uplink or downlink control channels are used. In the uplink, the mobile still uses the random access channel to respond to radio resource control messages from the RNC and to send its own IP packets. This limits the use of adaptive modulation and coding since the mobile cannot send frequent measurement reports to the base station to indicate the downlink reception quality. Furthermore, it is also not possible to acknowledge proper receipt of frames. Instead, the RNC informs the base station when it receives measurement information in radio resource messages from the mobile.
- Enhanced Cell/URA-PCH states – in these two states, the mobile device is in a deep sleep state and only observes the paging information channel to be alerted of an incoming paging message which is transmitted on the paging channel. To transfer data, the mobile device is then moved to Cell-FACH or Cell-DCH state. If the mobile device and the network support Enhanced Cell/URA-PCH states, the network can instruct the mobile device not to use the slow paging channel to receive paging information but to use a high-speed downlink shared channel instead. The high-speed downlink channel is then also used for subsequent RRC commands which are required to move the device back into a more active state. Like the measure above, this significantly decreases the wakeup time.

Figure 2.11 shows how this works in practice. While the message exchange to notify the mobile device of incoming data and to move it to another activity state remains the same, using the high-speed downlink shared channels for the purpose speeds up the procedure by several hundred milliseconds.

Which of the described enhancements will make it into networks in the future remains to be seen and will also depend on how quickly LTE and other competing network technologies are rolled out. While CPC and enhanced mobility management states increase the efficiency of the system, they also significantly increase the complexity of the air interface, as both old and new mobile devices have to be supported simultaneously. This rising complexity is especially challenging for the development of devices and networks, as it creates additional interaction scenarios which become more and more

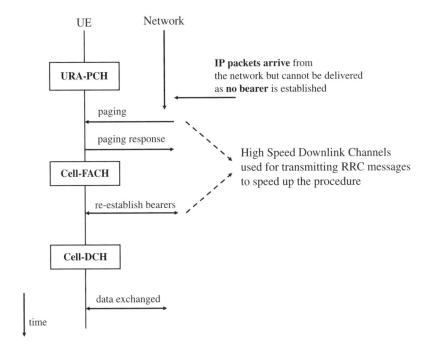

Figure 2.11 Message exchange to move a mobile device from URA-PCH state back to cell-DCH state when IP packets arrive from the network.

difficult to test and debug before. Already today, devices are tested with network equipment of several vendors and different software versions. Adding yet another layer of features will make this even more complex in the future.

2.2.5.5 Radio Network Enhancement: One-tunnel

Figure 2.12 shows the default path of user data between a mobile device and the Internet through the cellular network. In the current architecture, the packet is sent through the GGSN, the SGSN, the RNC and the base station. All user data packets are tunneled through the network as described above, since the user's location can change at any time. The current architecture uses a tunnel between the GGSN and the SGSN and a second tunnel between the SGSN and the RNC. All data packets therefore have to pass through the SGSN, which terminates one tunnel, extracts the packets and puts them into another tunnel. This requires both time and processing power.

Since both the RNC and the GGSN are IP routers, this process is not required in most cases. The one-tunnel approach, now standardized in 3GPP (see [16] and [17]), allows the SGSN to create a direct tunnel between the RNC and the GGSN. It thus removes itself from the transmission chain. Mobility management, however, remains on the SGSN which means, for example that it continues to be responsible for mobility management and tunnel modifications in case the mobile device is moved to an area served by another

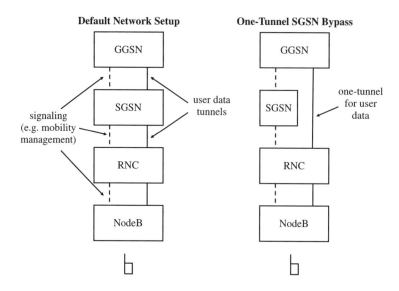

Figure 2.12   Current network architecture vs the one-tunnel enhancement.

RNC. For the user, this approach has the advantage that the packet delay is reduced. From a network point of view, the advantage is that the SGSN requires fewer processing resources per active user, which helps to reduce equipment costs. This is especially important as the amount of data traversing the packet-switched core network is rising significantly.

A scenario where the one-tunnel option is not applicable is international roaming. Here, the SGSN has to be in the loop in order to count the traffic for inter-operator billing purposes. Another case where the one-tunnel option cannot be used is when the SGSN is asked by a prepaid system to monitor the traffic flow. This is only a small limitation, however, since in practice it is also possible to perform prepaid billing via the GGSN.

Proprietary enhancements even aim to terminate the user data tunnel at the NodeB, bypassing the RNC as well. However, this has not found the widespread support of companies in 3GPP and is not likely to be compatible with some HSPA+ extensions, like the enhanced mobile device states.

2.2.5.6 Competition for LTE in 5 MHz

With the enhancements described in this section, it is quite likely that enhanced HSPA networks become a viable alternative to LTE deployments in the short and medium term, as the spectral efficiency in a 5 MHz band of both systems is similar. As will be described in the next section, LTE scores over HSPA when more bandwidth is available, as it is not limited to 5 MHz channels. In practice, it is thus likely that some network operators will choose to improve their HSPA network and only later on move to LTE, while others will prefer to go straight to LTE.

2.3 LTE

2.3.1 Introduction

For several years, there has been an ongoing trend in fixed line networks to migrate all circuit-switched services to a packet-switched IP infrastructure. In practice, it can be observed that fixed line network operators are migrating their telephony services to a packet-switched architecture offering both telephony and Internet access either via DSL or a cable modem. This means, that circuit-switched technology is replaced by VoIP-based solutions, as will be described in more detail in Chapter 4. In wireless networks, this trend has not yet begun. This is mostly due to the fact that current 3G and 3.5G network architectures are still optimized for circuit-switched telephony both in the radio network and in the core network. In addition, today's VoIP telephony implementations significantly increase the amount of data that has to be transferred over the air interface, which means fewer voice calls can be handled simultaneously. Besides these challenges, however, migrating voice telephony to IP offers a number of significant benefits such as cheaper networks and integration with other IP-based applications as discussed in more detail in Chapters 5 and 6.

At the same time, the general trend of ever increasing transmission bandwidths is highlighting the limits of current 3G and 3.5G networks. It was therefore decided in 2005 by the 3GPP standardization body to start work on a next generation wireless network design that is only based on packet-switched data transmission. This research was performed in two study programs. The LTE program focused on the design of a new radio network and air interface architecture. Slightly afterwards, work started on the design of a new core network infrastructure with the Service Architecture Evolution (SAE) program. Later, they were combined into a single work program, the Evolved Packet System (EPS) program. By that time, however, the abbreviation 'LTE' was already dominant in literature and most documents still refer to LTE rather than EPS.

Besides being fully packet-based, the following design goals were set for the new network:

- Reduced time for state changes – in HSPA networks today, the time it takes a mobile device to connect to the network and start communication on a high-speed bearer is relatively long. This has a negative impact on usability, as the user can feel this delay when accessing a service on the Internet after a period of inactivity. It was thus decided that with a new network design it should be possible to move from idle state to being fully connected in less than 100 ms.
- Reduced user plane latency – another downside of current cellular networks is the much higher transmission delay compared with fixed line networks. While one-way delay between a user's computer at the edge of a DSL network to the Internet is around 15 ms today, HSPA networks have a delay of around 50 ms. This is disadvantageous for applications such as telephony and real-time gaming. For LTE, it was decided that air interface delay should be in the order of 5 ms to reach end-to-end delays equaling fixed line networks.
- Scalable bandwidth – HSPA networks are currently limited to a bandwidth of 5 MHz. At some point, higher throughput can only be reasonably achieved by increasing the

bandwidth of the carrier. For certain applications, a carrier of 5 MHz is too large and it was thus decided that the air interface should also be scalable in the other direction.

• Throughput increase – for the new system, a maximum throughput under ideal conditions of 100 Mbit/s should be achieved.

The following sections now describe how these design goals are met in practice.

2.3.2 Network Architecture

2.3.2.1 Enhanced Base Stations

Figure 2.11 shows the main components of an LTE core and radio access network as described in [18]. Compared with UMTS, the radio network is less complex. It was decided that central RNCs should be removed and their functionality has been partly moved to the base stations and partly to the core network gateway. To differentiate UMTS base stations from LTE base stations, they are referred to as Enhanced NodeB, (eNodeB). As there is no central controlling element in the radio network any more, the base stations now perform air interface traffic management autonomously and ensure quality of service. This was already partly the case in UMTS with the introduction of HSPA, as discussed in the previous section. Control over bearers for circuit-switched voice telephony, however, rested with the RNC.

In addition, base stations are now also responsible for performing handovers for active mobiles. For this purpose, the eNodeB can now communicate directly with each other via the X2 interface. The interface is used to prepare a handover and can also be used to forward user data (IP packets) from the current base station to the new base station to minimize the amount of user data lost during the handover. As the X2 interface is optional, base stations can also communicate with each other via the access gateway to prepare a handover. In this case, however, user data is not forwarded during the handover. This means that some of the data already sent from the network to the current base station might be lost, as once a handover decision has been made, it has to be executed as quickly as possible before radio contact is lost. Unlike in UMTS, LTE radio networks only perform hard handovers, that is only one cell communicates with a mobile device at a time.

The interface that connects the eNodeB to the gateway nodes between the radio network and the core network is the S1 interface. It is fully based on the IP protocol and is therefore transport technology agnostic. This is a big difference to UMTS. Here, the interfaces between the NodeB, the RNCs and the SGSN were firmly based on the ATM protocol for the lower protocol layers. Between the RNC and the NodeB, IP was not used at all for packet routing. While allowing for easy time synchronization between the nodes, requiring the use of ATM for data transport on lower protocol layers makes the setup inflexible and complicated. In recent years, the situation has worsened as rising bandwidth demands cannot be satisfied any more with ATM connections over 2 Mbit/s E-1 connections. The UMTS standard was thus enhanced to also use IP as a transport protocol to the base station. LTE, however, is fully based on IP transport in the radio network from day one. Base stations are either equipped with 100 Mbit/s or 1 Gbit/s Ethernet ports, as known from the PC world, or with gigabit Ethernet fiber ports.

2.3.2.2 Core Network to Radio Access Network Interface

As shown in Figure 2.13, the gateway between the radio access network and the core network is split into two logical entities, the Serving Gateway (Serving-GW) and the Mobility Management Entity (MME). Together, they fulfill similar tasks as the SGSN (Serving GPRS Support Node) in UMTS networks. In practice, both logical components can be implemented on the same physical hardware or can be separated for independent scalability.

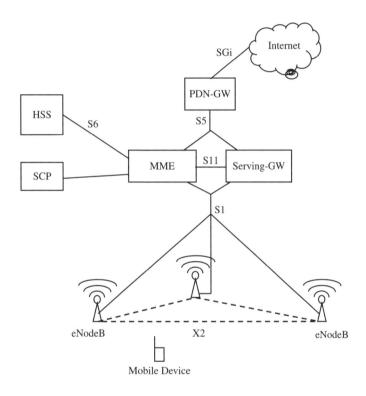

Figure 2.13   Basic LTE network architecture.

The MME is the 'control plane' entity responsible for the following tasks:

- Subscriber mobility and session management signaling. This includes tasks such as authentication, establishment of radio bearers, handover support between different eNodeB and to/from different radio networks (e.g. GSM, UMTS).
- Location tracking for mobile devices in idle mode, that is while no radio bearer is established because they have not exchanged data packets with the network for a prolonged amount of time.
- Selection of a gateway to the Internet when the mobile requests the establishment of a session, that is when it requests an IP address from the network.

The Serving Gateway is responsible for the 'user plane', that is, for forwarding IP packets between mobile devices and the Internet. As already discussed in the section on UMTS, IP tunnels are used in the radio access network and core network to flexibly change the route of IP packets when the user is handed over from one cell to another while moving. The GPRS Tunneling Protocol (GTP) is reused for this purpose and the mechanism is the same as shown for UMTS in Figure 2.3. The difference from UMTS is that the tunnel for a user in the radio network is terminated directly in the eNodeB itself and no longer on an intermediate component such as a radio network controller. This means that the BTS is directly connected via an IP interface to the Serving-Gateway and that different transport network technologies such as Ethernet over fiber or optical cable, DSL, microwave, and so on, can be used. In addition, the S1 interface design is much simpler than similar interfaces of previous radio networks, which relied heavily on services of complex lower layer protocols.

As the S1 interface is used for both user data (to the Serving-GW) and signaling data to the MME, the higher layer protocol architecture is split into two different protocol sets: the S1-C (control) interface is used for exchanging control messaging between a mobile device and the MME. As will be shown below, these messages are exchanged over special 'non-IP' channels over the air interface and then put into IP packets by the NodeB before they are forwarded to the MME. User data, however, is already transferred as IP packets over the air interface and these are forwarded via the S1-U (user) protocol to the Serving-Gateway. The S1-U protocol is an adaptation of the GTP from GPRS and UMTS (cf. Figure 2.3).

If the MME and the Serving Gateway are implemented separately, the S11 interface is used to communicate between the two entities. Communication between the two entities is required, for example for the creation of tunnels, when the user attaches to the network, or for the modification of a tunnel, when a user moves from one cell to another.

Unlike in previous wireless radio networks, where one access network gateway (SGSN) was responsible for a certain number of radio network controllers and each radio network controller in turn for a certain number of base stations, the S1 interface supports a meshed architecture. This means that not only one but several MMEs and Serving-Gateways can communicate with each eNodeB and the number of MMEs and Serving-Gateways can be different. This reduces the number of inter-MME handovers when users are moving and allows the number of MMEs to evolve independently from the number of Serving-Gateways, as the MME's capacity depends on the signaling load and the capacity of the Serving-Gateway depends on the user traffic load. These can evolve differently over time, which makes separation of these entities interesting. A meshed architecture of the S1 interface also adds redundancy to the network. If, for example one MME fails, a second one can take over automatically if it is configured to serve the same cells. The only impact of such an automatic failure recovery is that users served by the failed MME have to register to the network again. How the meshed capabilities of the S1 interface are used in practice depends on the policies of the network operator and on the architecture of the underlying transport network architecture.

2.3.2.3 Gateway to the Internet

As in previous network architectures, a router at the edge of the wireless core network hides the mobility of the users from the Internet. In LTE, this router is referred to as the Packet Data Network (PDN)-Gateway and fulfills the same tasks as the GGSN in UMTS. In addition to hiding the mobility of the users, it also administers an IP address pool and assigns IP addresses to mobiles registering to the network. Depending on the number of users, a network has several PDN-Gateways. The number depends on the capabilities of the hardware, the number of users and the average amount of data traffic per user. As shown in Figure 2.13, the interface between the PDN-GW and the MME/Serving-GWs is referred to as S5. Like the interface between the SGSN and the GGSN in UMTS, it uses the GTP-U (user) protocol to tunnel user data from and to the Serving-Gateways and the GTP-S (signaling) protocol for the initial establishment of a user data tunnel and subsequent tunnel modifications when the user moves between cells that are managed by different Serving-GWs.

2.3.2.4 Interface to the User Database

Another essential interface in LTE core networks is the S6 interface between the MMEs and the database that stores subscription information. In UMTS, this database is referred to as the Home Location Register. In LTE, the HLR is reused and has been renamed the Home Subscriber Server (HSS). Essentially, the HSS is an enhanced HLR and contains subscription information for GSM, GPRS, UMTS, LTE and the IP Multimedia Subsystem (IMS), which is discussed in Chapter 4. Unlike in UMTS, however, the S6 interface does not use the SS-7-based MAP (Mobile Application Part) protocol, but the IP-based Diameter protocol. The HSS is a combined database and it is used simultaneously by GSM, UMTS and LTE networks belonging to the same operator. It therefore continues to support the traditional MAP interface in addition to the S6 interface for LTE and also the interfaces required for the IMS as discussed in Chapter 4.

2.3.2.5 Moving Between Radio Technologies

In practice, most network operators deploying an LTE network already have a GSM and UMTS network in place. As the coverage area of a new LTE network is likely to be very limited at first, it is essential that subscribers can move back and forth between the different access network technologies without losing their connection and assigned IP address. Figure 2.14 shows how this is done in practice when a user roams out of the coverage area of an LTE network and into the coverage area of a UMTS network of the same network operator. When the user moves out of the LTE coverage area, the mobile device reports to the eNodeB that a UMTS (or GSM) cell has been found. This report is forwarded to the MME which contacts the responsible 3G (or 2G) SGSN and requests a handover procedure. The interface used for this purpose is referred to as S3 and is based on the protocol used for inter-SGSN relocation procedures. As a consequence, no software modifications are required on the 3G SGSN to support the procedure. Once the 3G radio network has been prepared for the handover, the MME sends a handover

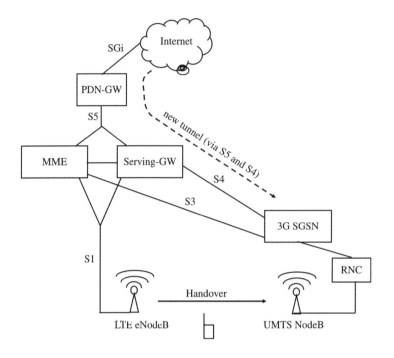

Figure 2.14   LTE and UMTS interworking.

command to the mobile device via the eNodeB. After the handover has been executed, the user data tunnel between the Serving-GW and the eNodeB is re-routed to the SGSN. The MME is then released from the subscriber management, as this task is taken over by the SGSN. The Serving-GW, however, remains in the user data path via the S4 interface and acts as a 3G GGSN from the point of view of the SGSN. From the SGSN's point of view, the S4 interface is therefore considered to be the 3G Gn interface between the SGSN and the GGSN.

2.3.2.6 The Packet Call Becomes History

A big difference of LTE from GSM and UMTS is that mobile devices will always be assigned an IP address as soon as they register to the network. This has not been the case with GSM and UMTS because 2G, 3G and 3.5G devices are still mostly used for voice telephony and so it makes sense to attach to the network without requesting an IP address. In LTE networks, however, a device without an IP address is completely useless. Hence, the LTE network attach procedure already includes the assignment of an IP address. From the LAN/WLAN point of view this is nothing new. From a cellular industry point of view, however, this is revolutionary. The GPRS and UMTS procedure of 'establishing a packet call', a term coined with the old thinking of establishing a circuit-switched connection with a voice call in mind, will therefore become a thing of the past with LTE. Many people in the industry will have to change their picture of the mobile world to accommodate this.

2.3.3 Air Interface and Radio Network

While the general LTE network architecture is mainly a refinement of the 3G network architecture, the air interface and the radio network have been redesigned from scratch. In the 3GPP standards, a good place to start further research beyond what is covered below is TS 36.300 [19].

2.3.3.1 Downlink Data Transmission

For transmission of data over the air interface, it was decided to use a new transmission scheme in LTE which is completely different from the CDMA approach of UMTS. Instead of using only one carrier over the broad frequency band, it was decided to use a transmission scheme referred to as Orthogonal Frequency Division Multiple Access, or OFDMA for short. OFDMA transmits a data stream by using several narrow-band subcarriers simultaneously, for example 512, 1024, or even more, depending on the overall available bandwidth of the channel (e.g. 5, 10, 20 MHz). As many bits are transported in parallel, the transmission speed on each subcarrier can be much lower than the overall resulting data rate. This is important in a practical radio environment in order to minimize the effect of multipath fading created by slightly different arrival times of the signal from different directions. The second reason this approach was selected was because the effect of multipath fading and delay spread becomes independent of the amount of bandwidth used for the channel. This is because the bandwidth of each subcarrier remains the same and only the number of subcarriers is changed. With the previously used CDMA modulation, using a 20 MHz carrier would have been impractical, as the time each bit was transmitted would have been so short that the interference due to the delay spread on different paths of the signal would have become dominant.

Figure 2.15 shows how the input bits are first grouped and assigned for transmission over different frequencies (subcarriers). In the example, 4 bits (representing a 16QAM modulation) are sent per transmission step per subcarrier. A transmission step is also referred to as a symbol. With 64QAM modulation, 6 bits are encoded in a single symbol, raising the data rate further. On the other hand, encoding more bits in a single symbol makes it harder for the receiver to decode the symbol if it was altered by interference. This is the reason why different modulation schemes are used depending on transmission conditions.

In theory, each subcarrier signal could be generated by a separate transmission chain hardware block. The output of these blocks would then have to be summed up and the resulting signal could then be sent over the air. Because of the high number of subcarriers used, this approach is not feasible. Instead, a mathematical approach is taken as follows. As each subcarrier is transmitted on a different frequency, a graph which shows the frequency on the x-axis and the amplitude of each subcarrier on the y-axis can be constructed. Then, a mathematical function called Inverse Fast Fourier Transformation (IFFT) is applied, which transforms the diagram from the frequency domain to the time domain. This diagram has the time on the x-axis and represents the same signal as would have been generated by the separate transmission chains for each subcarrier when summed up. The IFFT thus does exactly the same job as the separate transmission chains for each subcarrier would do, including summing up the individual results.

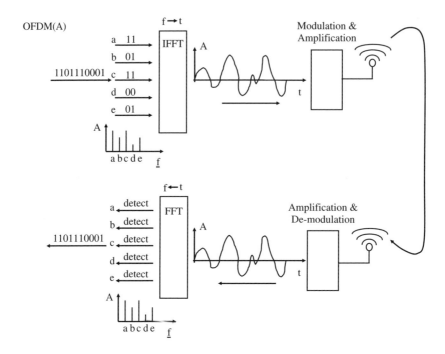

Figure 2.15   Principles of OFDMA for downlink transmissions.

On the receiver side, the signal is first demodulated and amplified. The result is then treated by a fast Fourier transformation function which converts the time signal back into the frequency domain. This reconstructs the frequency/amplitude diagram created at the transmitter. At the center frequency of each subcarrier a detector function is then used to generate the bits originally used to create the subcarrier.

The explanation has so far covered the Orthogonal Frequency Division aspect of OFDMA transmissions. The Multiple Access (MA) part of the abbreviation refers to the fact that the data sent in the downlink is received by several users simultaneously. As will be discussed later, control messages inform mobile devices waiting for data which part of the transmission is addressed to them and which part they can ignore. This is, however, just a logical separation. On the physical layer, this only requires that modulation schemes ranging from QPSK over 16QAM to 64QAM can be quickly changed for different subcarriers in order to accommodate the different reception conditions of subscribers.

2.3.3.2 Uplink Data Transmission

For data transmission in the uplink direction, 3GPP has chosen a slightly different modulation scheme. OFDMA transmission suffers from a high Peak to Average Power Ratio (PAPR), which would have negative consequences for the design of an embedded mobile transmitter; that is, when transmitting data from the mobile terminal to the network, a power amplifier is required to boost the outgoing signal to a level high enough to be picked up by the network. The power amplifier is one of the biggest consumers of energy in a device and should therefore be as power-efficient as possible to increase the battery life of the device. The efficiency of a power amplifier depends on two factors:

- The amplifier must be able to amplify the highest peak value of the wave. Due to silicon constraints, the peak value determines the power consumption of the amplifier.
- The peaks of the wave, however, do not transport any more information than the average power of the signal over time. The transmission speed therefore does not depend on the power output required for the peak values of the wave but rather on the average power level.

As both power consumption and transmission speed are of importance for designers of mobile devices, the power amplifier should consume as little energy as possible. Thus, the lower the difference between the PAPR, the longer is the operating time of a mobile device at a certain transmission speed compared with devices that use a modulation scheme with a higher PAPR.

A modulation scheme similar to basic OFDMA, but with a much better PAPR, is SC-FDMA (Single Carrier-Frequency Division Multiple Access). Due to its better PAPR, it was chosen by 3GPP for transmitting data in the uplink direction. Despite its name, SC-FDMA also transmits data over the air interface in many subcarriers, but adds an additional processing step as shown in Figure 2.16. Instead of putting 2, 4 or

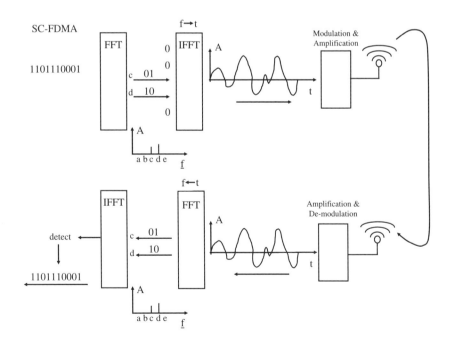

Figure 2.16   SC-FDMA modulation for uplink transmissions.

6 bits together as in the OFDM example to form the signal for one subcarrier, the additional processing block in SC-FDMA spreads the information of each bit over all the subcarriers. This is done as follows: again, a number of bits (e.g. 4 representing a 16QAM modulation) are grouped together. In OFDM, these groups of bits would have been the input for the IDFT. In SC-FDMA, however, these bits are now piped into a Fast Fourier Transformation (FFT) function first. The output of the process is the basis for the creation of the subcarriers for the following IFFT. As not all subcarriers are used by the mobile station; many of them are set to zero in the diagram. These may or may not be used by other mobile stations.

On the receiver side the signal is demodulated, amplified and treated by the fast Fourier transformation function in the same way as in OFDMA. The resulting amplitude diagram, however, is not analyzed straight away to get the original data stream, but fed to the inverse fast Fourier transformation function to remove the effect of the additional signal processing originally done at the transmitter side. The result of the IFFT is again a time domain signal. The time domain signal is now fed to a single detector block which recreates the original bits. Therefore, instead of detecting the bits on many different subcarriers, only a single detector is used on a single carrier.

The differences between OFDM and SC-FDMA can be summarized as follows: OFDM takes groups of input bits (0s and 1s) to assemble the subcarriers which are then processed by the IDFT to get a time signal. SC-FDMA in contrast first runs an FFT over the groups of input bits to spread them over all subcarriers and then uses the result for the IDFT which creates the time signal. This is why SC-FDMA is sometimes also referred to as FFT spread OFDM.

2.3.3.3 Physical Parameters

For LTE, the following physical parameters have been selected:

- Subcarrier spacing, 15 kHz.
- OFDM symbol duration, 66.667 μs;
- Standard cyclic prefix: 4.7 μs. The cyclic prefix is transmitted before each OFDM symbol to prevent inter-symbol interference due to different lengths of several transmission paths. For difficult environments with highly diverse transmission paths a longer cyclic prefix of 16.67 μs has been specified as well. The downside of using a longer cyclic prefix, however, is a reduced user data speed since the symbol duration remains the same.

The selected subcarrier spacing and symbol duration compensate for detrimental effects on the signal such as the Doppler effect (frequency shift) due to the mobility of subscribers. The parameters have been chosen to allow speeds of beyond 350 km/h.

To be flexible with bandwidth allocations in different countries around the world, a number of different channel bandwidths have been defined for LTE. These range from 1.25 MHz on the low end to 20 MHz on the high end. Table 2.1 shows the standardized transmission bandwidths, the number of subcarriers used for each and the FFT size (the number of spectral lines) used at the receiver side to convert the signal from the time to

Table 2.1 Defined bandwidths for LTE.

Bandwidth	Number of subcarriers	FFT size
1.25 MHz	76	128
2.5 MHz	151	256
5 MHz	301	512
10 MHz	601	1024
15 MHz	901	1536
20 MHz	1201	2048

the frequency domain. In practice, it is expected that operators using LTE will deploy networks in the frequency bands already available today for GSM and UMTS but use bandwidths of at least 10 MHz, since there is no speed advantage of using LTE in a 5 MHz band over HSPA+. The smaller bandwidths of 1.25 MHz and 2.5 MHz were specified for operators with little spectrum or for operators wishing to 're-farm' some of their GSM spectrum in the 900 MHz band. In practice, however, it is questionable if this would bring a great benefit since the achievable data rates in such a narrow band are lower than what can be achieved with HSPA today.

In addition to the use of already existing frequency bands, new bands are being made available for B3G wireless technologies. In Europe for example the 2.5 GHz band, also referred to as the IMT extension band, will be opened for LTE and possibly other wireless technologies. As will be shown in Chapter 3, however, there is still sufficient unused bandwidth available in existing bands which might make new bands unattractive for LTE in the near future.

2.3.3.4 From Slots to Frames

Data is mapped to subcarriers and symbols, which are arranged in the time and frequency domain in a resource grid as shown in Figure 2.17. The smallest aggregation unit is referred to as a slot or a resource block and contains 12 subcarriers and seven symbols on each subcarrier in case the default short cyclic prefix is used. The symbol time of 66.67 μs and the 4.7 μs cyclic prefix multiplied by 7 results in a slot length of 0.5 ms. In case the long cyclic prefix has to be used, the number of symbols per slot is reduced to six, again resulting in a slot length of 0.5 ms. The grouping of 12 subcarriers together results in a resource block bandwidth of 180 kHz. As the total carrier bandwidth used in LTE is much larger (e.g. 10 MHz), many resource blocks are transmitted in parallel.

Two slots are then grouped into a subframe, which is also referred to as a Transmit Time Interval (TTI). In case of Time Division Duplex (TDD) operation (uplink and downlink in the same band), a subframe can be used for either uplink or downlink transmission. It is up to the network to decide which subframes are used for which direction. Most networks, however, are likely to use Frequency Division Duplex (FDD), which means that there is a separate band for uplink and downlink transmission. Here, all subframes of the band are dedicated to downlink or to uplink transmissions.

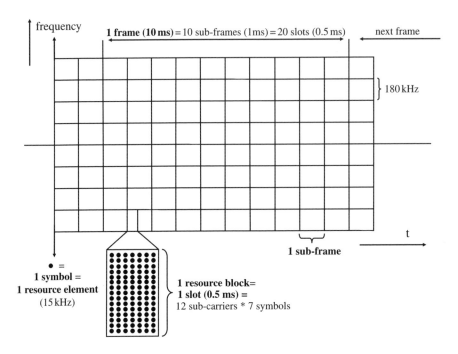

Figure 2.17 The LTE resource grid.

Ten subframes are grouped together to form a single radio frame, which has a duration of 10 ms. Afterwards, the cycle repeats with the next frame. This is important for mobile devices because broadcast information (e.g. uplink bandwidth assignments) is always transmitted at the beginning of a frame.

The smallest amount of resource elements (symbols) that can be allocated to a single mobile device at an instant in time is two resource blocks, which equals one subframe or one transmit time interval. To increase the data rate for the mobile device, the scheduler in the network can concatenate several resource blocks in both the time and the frequency direction. Since there are many resource blocks being transmitted in parallel, it is also possible to schedule several mobile devices simultaneously, each listening to different subcarriers.

2.3.3.5 Reference Symbols, Signals and Channels

Not all resource elements of a resource block are used for transmitting user data. Especially around the center frequency, some resource elements are used for other purposes, as described below.

To enable a mobile device to find the network after power on and when searching for neighboring cells, some resource elements are used for pilot or reference symbols in a predefined way. While data is transferred, pilot symbols are used by the mobile device for downlink channel quality measurements and, since the content of the resource element is known, to estimate how to recreate the original signal that was distorted during transmission.

For the transmission of higher layer data, LTE re-uses the channel concept of UMTS as shown in Figures 2.8 and 2.9. Compared with UMTS, however, all devices use the shared channel on the physical layer. The LTE channel model is therefore much simpler than that of UMTS. LTE also re-uses the concept of logical channels (what is transmitted), transport channels (how is it transmitted) and physical channels (air interface) to separate data transmission over the air interface from the logical representation of data. Figure 2.18 shows the most important channels that are used in LTE and how they are mapped to each other.

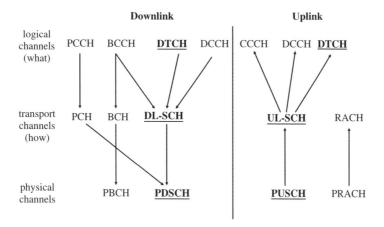

Figure 2.18   LTE uplink and downlink channels.

2.3.3.6 Downlink: Broadcast Channel

While no data is transmitted and the mobile is in idle state, it listens to two logical channels. The logical Broadcast Control Channel (BCCH) is used by the network to transmit system information to mobile devices such as the network and cell, that is, which resource blocks and resource elements to find other channels, how the network can be accessed, and so on. The basic parameters sent on the BCCH are mapped to the BCH transport channel and the Physical Broadcast Channel (PBCH). The PBCH is then mapped to dedicated resource elements in the subchannels of the inner 1.25 MHz of the band. Which resource elements are used for the PBCH is calculated with a mathematical formula which generates a certain pattern and thus distributes the broadcast information between different subcarriers over time [20]. In addition, a number of additional resource elements are used for a Synchronization Channel (SCH), which is not shown in Figure 2.19. As the name implies, these resource elements help mobile devices to synchronize to the cell and to find the resource elements on which the broadcast information can be found. In addition to basic cell parameters, the broadcast channel also carries further information which is necessary, but not essential from the

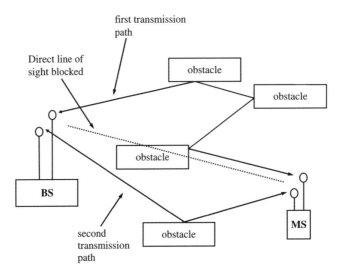

Figure 2.19   Principle of MIMO transmissions. (Reproduced from *Communication Systems for the Mobile Information Society*, Martin Sauter, 2006, John Wiley and Sons.)

start. To save bandwidth and to be flexible in the future, this information is not carried on the PBCH but on the Physical Downlink Shared Channel (PDSCH) instead, which is also used for transferring user data (IP packets). A pointer on the PBCH informs mobiles where to find the broadcast information on the PDSCH.

2.3.3.7 Downlink: Paging Channel

The paging channel is used to contact mobile devices in an idle state when a new IP packet arrives in the core network from the Internet and needs to be delivered to the mobile device. In idle state, which is usually entered after a prolonged period of inactivity, only the tracking area (i.e the identity for a group of cells) where the mobile is located is known. The paging message is then sent into all cells of this group. When the mobile device receives the message, it establishes a connection with the network again, a bearer is set up and the packet is delivered. For services such as instant messaging, push e-mail and VoIP, paging for incoming IP packets is quite common. Such applications, if pro-grammed properly and no network address translation firewalls are used, have a logical connection with a server in the network but are dormant until either the user invokes a new action or the application is contacted by the network-based server, for example because of a new instant message coming in [21]. As can be seen in Figure 2.18, there is no physical channel dedicated for paging messages. Instead, paging messages are sent on the downlink shared channel and a pointer on the logical broadcast channel indicates where and when the paging messages can be found on the shared channel. The broadcast cycle for paging messages a mobile device needs to listen to is usually in the order of 1–2 s. This is a good balance between quick delivery of an incoming packet and power consumption of a mobile device while not being actively used.

2.3.3.8 Downlink and Uplink: Dedicated Traffic and Control Channels and Their Mapping to the Shared Channel

From a logical point of view, user data and RRC control messages are transferred via dedicated traffic channels and the dedicated control channels. Each mobile device has its own dedicated pair of these channels. As can be seen in Figure 2.18, both channels are multiplexed to a single physical downlink or uplink shared channel (PDSCH, PUSCH) which are used for all devices. Higher software layers are therefore independent of the physical implementation of the air interface.

2.3.3.9 Downlink: Physical Layer Control Channels

In addition to the previously mentioned channels, there are a number of additional physical layer control channels which are required to exchange physical layer feedback information. As these channels only carry lower layer control information and are originated by the base station and not the network behind them, they are not shown in Figure 2.18.

To inform mobile devices which resource blocks are assigned to them for transmitting in the uplink direction, the physical downlink shared channel is always accompanied by a Physical Downlink Control Channel (PDCCH). In addition, this channel informs mobile devices about the resource allocation of the PCH and the downlink shared channel.

Since the amount of data carried on the PDCCH varies, the number of OFDM symbols assigned to the physical control channel is broadcast via the Physical Control Format Indicator Channel (PCFICH). Finally, the Physical HARQ Indicator Channel (PHICH) carries acknowledgements for proper reception of uplink data blocks. The HARQ acknowledgment functionality used in LTE is similar to that used in UMTS. For details, see Section 2.2.4.

2.3.3.10 Uplink: Physical Layer Control Channels

In the uplink direction there are two physical layer control channels: The physical uplink control channel (PUCCH) is a per device channel and carries the following information:

- HARQ acknowledgments for data frames received from the network. (cf. Section 2.2.4).
- Scheduling requests from the mobile to inform the network that further uplink transmit opportunities should be scheduled, as there is more data in the output buffer.
- Channel Quality Indications (CQI) to the network, so the base station can determine which modulation and coding to use for data in the downlink direction. CQI information is also important for the scheduler in the base station, as it can decide to temporarily halt data transmission to users in a temporary deep signal fading situation where it is likely that data cannot be received correctly anyway.

The second control channel used in the uplink direction is the Physical Random Access Channel (PRACH). It is used when no bearer is established in the uplink direction to request new uplink transmission opportunities. It is also used when the mobile wants to establish a bearer for the first time, after a long timeout or in response to a paging from the network.

2.3.3.11 Dynamic and Persistent Scheduling Grants

The packet scheduler in the base station decides which resource blocks of the physical downlink and uplink channels are used for which mobile device. This way, the base station controls both uplink and downlink transmissions for each mobile and is therefore able to determine how much bandwidth is available to a mobile device. Input parameters for the scheduler are, for example the current radio conditions as seen from each device, so the data transfer rate can be increased or decreased to mobiles temporarily experiencing exceptionally good or bad radio conditions. Other input parameters are the quality of service parameters for a connection and the maximum bandwidth granted by the operator to a mobile device, based on the user's subscription.

There are two types of capacity grants for uplink data transmissions: dynamic grants are announced once and are valid for one or more transmit time intervals. Afterwards, the network has to issue a new grant for additional transmit opportunities or for the mobile to receiver further data in the downlink direction. Dynamic grants are useful for data that arrives in a bursty fashion, like during Web browsing, for example, and sporadic downloading of content.

Applications such as voice and video calls require a constant bandwidth and as little variation as possible in the time difference between two adjacent packets (jitter). For such applications, the base station can also issue persistent grants which are given once and are then valid for all subsequent transmit time intervals. This way, no signaling resources are required to constantly re-assign air interface resources while a voice or video call is ongoing. This increases the overall efficiency of the cell and increases the amount of bandwidth that is available for user data. The issue arising with persistent grants is how the base station can know when to use this type of assignment. One way to achieve this is to base the decision on the connection's quality of service requirements which are signaled to the network during bearer establishment. This works well for applications which require a constant bandwidth and are based on the IP Multimedia Subsystem. In addition to the Quality of Service (QoS) signaling initiated by the mobile device, the IMS has a connection to the transport network and can influence the bearer as well. This is discussed in more detail in Chapter 4. For Internet-based voice and other applications that are not using the IMS, however, using persistent grants is much more difficult. It is likely that such applications are used over a default bearer which has no guaranteed bandwidth and latency as it is used simultaneously for other applications on the same device such as Web browsing. In practice, it remains to be seen if schedulers will also take a look at the bandwidth usage of a mobile device over time and decide on this basis to use persistent or dynamic grants.

2.3.3.12 MIMO Transmission

So far, this chapter has focused on data transmission via a single spatial stream between a transmitter and receiver. Most wireless systems today operate in this mode and a second transmitter on the same frequency is seen as unwanted interference that degrades the channel. In practice, however, it can be observed that even a single signal is reflected and scattered by objects in the transmission path and that the other end receives several copies of the original signal from different angles at slightly different times. For simple wireless transmission technologies, these copies are also unwanted interference. LTE, however, makes use of scattering and reflection on the transmission path by transmitting several independent data streams via individual antennas. The antennas are spaced at least half a wavelength apart, which in itself creates individual transmissions which behave differently when they meet obstacles in the transmission path. On the receiver side, the different data streams are picked up by independent antenna and receiver chains. Transmitting several independent signals over the same frequency band is also referred to as Multiple Input Multiple Output, and Figure 2.19 shows a simplified graphical representation. In practice, this means that several LTE resource grids, as shown in Figure 2.17, are sent over the same frequency at the same time but via different antennas.

The LTE standard specifies two and four individual transmissions over the same band, which requires two or four antennas at both the transmitter and receiver side respectively. Consequently, such transmissions are referred to as 2×2 MIMO and 4×4 MIMO. In practice, 2×2 MIMO is likely to be used at first, because of size constraints of mobile devices and due to the fact that antennas have to be spaced at least half a wavelength apart. Furthermore, most mobile devices support several frequency bands, each usually requiring its own set of antennas in case MIMO operation is supported in the band. More details on this topic will be discussed in Chapter 3 from a capacity point of view and in Chapter 5 from a mobile hardware point of view. On the network side, 2×2 MIMO transmissions can be achieved with a 'single' cross polar antenna that combines two antennas in a way that each antenna transmits a separate data stream with a different polarization (horizontal and vertical).

While Figure 2.19 depicts the general concept of MIMO transmission, it is inaccurate at the receiver side, as each antenna receives not only a single signal but the combination of all signals as they overlap in space. It is therefore necessary for each receiver chain to calculate a channel propagation that takes all transmissions into account in order to separate the different transmissions from each other. The pilot carriers mentioned above are used for this purpose. The characteristics required for these calculations are the gain, phase and multipath effects for each independent transmission path. A good mathematical introduction is given in [22].

As MIMO channels are separate from each other, 2×2 MIMO can increase the overall data rate by two and 4×4 MIMO by four. This is, however, only possible under ideal signal conditions. MIMO is thus only used for downlink transmissions since the base station transmitter is less power-constrained than the uplink transmitter. In less favorable transmission conditions, the system automatically falls back to single stream transmission and also reduces modulation from 64QAM, to 16QAM or even QPSK. As has been shown in the previous section on HSPA+, there is also a tradeoff between higher order modulation and MIMO use. Under less than ideal signal conditions, MIMO transmission is therefore only used with 16QAM modulation, which fails to double the data rate compared with a single stream transmission using 64QAM.

In the uplink direction, it is difficult for mobile devices to use MIMO due to their limited antenna size and output power. As a result, uplink MIMO is currently not part of the LTE standard. The uplink channel itself, however, is still suitable for uplink MIMO transmissions. To fully use the channel, some companies are thinking about implementing collaborative MIMO in the future, also known as multiuser MIMO [23]. Here, two mobile devices use the same uplink channel for their resource grid. At the base station

; two data streams are separated by the MIMO receiver and treated as two transmissions from independent devices rather than two transmissions from a single device that have to be combined. While this will not result in higher transmission speeds per device, the overall uplink capacity of the cell is significantly increased.

2.3.3.13 LTE Throughput Calculations

Based on the radio layer parameters introduced in this section, the physical layer throughput of an LTE radio cell can be calculated as follows: the transmission time per symbol is 73.167 μs (66.667 μs for the symbol itself + 4.7 μs for the cyclic prefix), the highest modulation order is 64QAM (6 bits per symbol) and there are 1201 subcarriers in a 20 MHz band:

$$\text{Physical speed} = (1/0.000\,073\,167) * 6 * 1201 = 98.487.022 \text{ bit/s} \quad (\text{i.e. about } 100\,\text{Mbit/s})$$

When 2×2 MIMO is used, the physical layer speed doubles to about 200 Mbit/s and in case 4×4 MIMO is used for transmission, the theoretical data transmission speed is 400 Mbit/s, based on a 20 MHz channel.

These values are usually quoted in press releases. However, as already discussed for HSPA, these raw physical layer transmission speeds are not reached in practice for a variety of reasons:

- 64QAM modulation can only be used very close to the base station. For the majority of users served by a cell, 16QAM (4 bits per symbol) or QPSK (2 bits per symbol) is more realistic.
- Error detection and correction bits (coding) are usually added to the data stream as otherwise the bit error rate over the air interface would become too high. Under average signal conditions it is common to see coding rates of 1/3. In practice, the coding overhead is thus in the range of 25–30%.
- Retransmissions – with a very conservative transmission strategy, the coding described above is sufficient to correct most transmission errors. In practice, however, more aggressive transmission strategies are used to make the best use of air interface resources. This usually results in air interface packet retransmission rates in the order of 20%.
- There is a significant overhead from pilot channels and control channels such as the broadcast channel and the dedicated signaling channels per user to acknowledge the correct reception of data packets and to convey signal quality measurement results. Many of those channels are transmitted with a lower order modulation so even devices in very unfavorable signal conditions can receive the information.
- In many cases, less than 20 MHz of bandwidth is available for LTE.
- When using MIMO, the modulation order has to be reduced under less than ideal transmission conditions.
- The overall capacity of the cell has to be shared by all users.
- The interference caused by transmissions of neighboring cells on the same frequency band has a further detrimental effect.

In practice, it is therefore likely that the throughput per cell is only about 30–50% of the theoretical values given above. For a cell with a 10 MHz carrier and 2×2 MIMO, an overall cell capacity on the IP layer of 30 Mbit/s may be achieved. A more detailed capacity analysis can be found in Chapter 3.

2.3.3.14 Radio Resource Control

As in UMTS and HSPA, the LTE network controls access to the air interface resources for both the uplink and the downlink. As there is no longer a central node in the radio network for the administration of resources, the base stations themselves are now responsible for the following tasks:

- Broadcasting of system information.
- Connection management – the mobile devices and the network use control channels such as the random access channel, the paging channel and the dedicated control channels to exchange RRC messages. The first RRC messages exchanged when accessing the network for the first time, or after a long time of inactivity, are connection establishment messages. The eNodeB is then responsible for setting up a logical signaling bearer to the device via the shared uplink and downlink channel or denying the request where the system is overloaded. Connection management also includes the establishment of dedicated bearers, again over the shared physical channel, based on the quality of service parameters of the user's subscription.
- Measurement control – as users move, the radio environment is very dynamic and devices therefore need to report signal strength measurements of the current and neighboring cells to the network.
- Mobility procedures – based on signal measurements of the mobile device, the eNodeB can initiate a handover procedure to another cell or even to another radio network such as UMTS or GSM/GPRS where the LTE coverage area is left.

2.3.3.15 RRC Active State

To minimize the use of resources in the network and to conserve the battery power of mobile devices, there are several connection states. While data is exchanged between the network and a mobile device, the RRC connection is in the active state. This means the network can assign resources to the device on the shared channel at any time and data can be instantly transmitted. The mobile remains in active state even if no data is transferred for some time, for example after the content of a Web page has been fully loaded. This ensures instant package transmission without any further resource control overhead, for example when the user clicks on a link.

While in full active state, the mobile has few opportunities to deactivate its receiver which has a negative impact on the battery capacity. After some time of inactivity, the network can thus decide to activate a Discontinuous Reception Mode (DRX) while the mobile is still in active state. This means that the mobile only has to listen to downlink bandwidth assignments and control commands periodically and can switch off its receiver at all other times. The DRX interval is flexible and can range from milliseconds to seconds.

Even while in DRX mode, mobility is still controlled by the network. This means that the mobile device has to continue sending signal measurement results to the network when a defined high or low signal threshold for the current cell or a neighboring cell is met. The eNodeB can then at any time initiate a handover procedure to another cell if required.

2.3.3.16 RRC Idle State

If no packets have been transmitted for a prolonged amount of time, the eNodeB can put the connection to a user in RRC Idle state. This means that, while the logical connection to the network and the IP address is retained, the radio connection is removed. The MME is informed of this state change as well, as IP packets arriving from the Internet can no longer be delivered to the radio network. As a consequence, on receipt of IP packets the MME needs to send a paging message to the mobile device, which leads to the re-establishment of a radio bearer. In case the mobile device needs to send an IP packet while in RRC idle state, for example because the user has clicked on a link on a Web page after a long time of inactivity, it also has to request the establishment of a new radio bearer before the packet can be transmitted.

Furthermore, the network no longer controls the mobility of a device in RRC idle state and the device can decide on its own to move from one cell to another. Several cells are grouped into a tracking area, which is similar to location and routing areas used in UMTS. The mobile only reports a cell change to the network if it selects a cell which belongs to a different tracking area. This means that the network, or more specifically the MME, has to send a paging message via all the cells that belong to the tracking area when a new packet for the device arrives from the Internet.

2.3.3.17 Treatment of Data Packets in the eNodeB

In addition to radio resource specific tasks, the eNodeB is also responsible for several tasks concerning the data packets themselves before they are transmitted over the air interface. To prevent data modification attacks, also referred to as man-in-the-middle attacks, an integrity checksum is calculated for each data packet before it is sent over the air interface. Input to the integrity checksum algorithm is not only the content of the packet but also an integrity checking key which is calculated from a unique secret key that is shared between the eNodeB and each mobile device. If a message is fraudulently modified on the air interface, it is not possible to append a valid integrity checksum due to the missing key and the message is not accepted by the recipient. Integrity checking applies to IP packets, to radio resource control messages exchanged with the eNodeB and also to mobility and session management messages exchanged with the MME.

In addition to integrity checking, data packets are encrypted before being transmitted over the air interface. Again, the subscriber's individual shared secret key, stored on the SIM card and the HSS, is used to calculate a ciphering key on both sides of the connection. Data intercepted on the air interface can thus not be decoded as the ciphering key is not known to an attacker. Ciphering applies to IP packets, to radio resource control messages and also to mobility and session management messages, the latter two not being based on IP.

A task only performed on user data IP packets before they are transmitted over the air interface is header compression. For LTE networks, this feature is very important, especially for real-time applications such as VoIP. As VoIP is very delay-sensitive, typically only 20 ms of speech data is accumulated in a single IP packet. With a data rate of around 12 kbit/s produced by sophisticated speech codecs, such as an Adaptive Multi-Rate (AMR) codec with a good voice quality, each IP packet carries around 32 bytes of data. In addition, there is an overhead of 40 bytes for an IPv4 header, the UDP header and the RTP header. With IPv6, the overhead is even larger due to the use of 128 bit IP addresses and additional header fields. This means that there is more overhead per packet than speech data. This greatly inflates the required data rate and therefore significantly reduces the potential number of simultaneous calls per base station. As voice calls are likely to be an important feature for LTE networks, it is necessary to compress IP packet headers before transmission. For LTE, the Robust Header Compression (ROHC) algorithm, originally specified in [24], was selected. Its advantages are:

- A very good compression ratio. The 40 bytes overhead of various encapsulated protocols are typically reduced to 6 bytes.
- A built-in feedback mechanism detects compression process corruptions as a result of air interface transmission errors. This allows an immediate restart of the compressor logic instead of letting the error propagate into the compression process of subsequent packets, as was the case with previously used header compression algorithms.
- The ROHC algorithm not only compresses the IP header but analyzes the IP packet and also compresses further encapsulated headers such as the UDP header and the RTP (Real-time Transfer Protocol) header where the data packet contains audio information.
- In order not to focus only on VoIP packets, ROHC is able to detect different header types in a packet and selects an appropriate overall header compression algorithm for each packet. The different compression algorithms are referred to as profiles. For VoIP packets, the RTP profile is used, which compresses the IP header, the UDP header and the RTP header of the packet. Further profiles are the UDP profile, which compresses IP and UDP headers (e.g. of SIP signaling messages, cf. Chapter 4), and the ESP (Encapsulated Security Payload) profile, which is used for compressing headers of IPsec encrypted packets.

Integrity checking, ciphering and compression are all part of the Packet Data Convergence Protocol (PDCP), which sits below the IP layer and thus encapsulates IP packets. The packet size, however, does not usually increase, as the additional PDCP header overhead is more than made up for by the header compression.

2.3.4 Basic Procedures

An important aspect of LTE, in addition to increasing the available bandwidth over the air interface, is to streamline signaling procedures to reduce delay for procedures such as setting up an initial connection and resuming data transfers from idle state. Figure 2.20

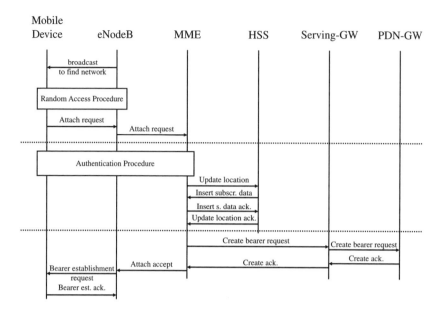

Figure 2.20   Attaching to the LTE network and requesting an IP address.

shows the message exchange of a device attaching to the network after it has been switched on until the point an IP address is assigned. Attaching to the network and getting an IP address is, as mentioned before, a single procedure in LTE as all services are based on IP. It does not make sense, therefore, to attach to the network without requesting an IP address as is the case today in GSM and UMTS networks.

2.3.4.1 Network Search and Broadcasting System Information

The first step in attaching to the network after power on is to find all available networks and select an appropriate network to communicate with. For this, the mobile performs an initial scan in all frequency bands it supports and tries to find downlink synchronization signals. As most LTE-capable devices also support 2G and 3G networks such as GSM and UMTS, the network search procedure also includes such networks. As most bands are not dedicated to a single network technology, the mobile must be able to correlate downlink signals to different radio systems. In case of LTE, the mobile searches for synchronization signals which are placed at regular intervals in the center subchannels (1.25 MHz) of an LTE carrier. Once these are found and properly decoded, the broadcast channel can be read and the mobile downloads the cell's complete system information. In most cases, the mobile device starts its search on the last used channel before it was switched off. If the device has not moved since it has powered off, the network is found very quickly. If a network is found, the registration process continues. If not, the process is repeated until the network is found or all supported frequency bands have been searched. Where the last used network before the device was powered down is not found, the mobile either selects a network on its own based on the preferences stored on the SIM card or presents the list of detected networks to the user who can then select the network of their choice (e.g. in case of roaming).

2.3.4.2 Initial Contact with the Network

After the broadcast information of a cell has been read and the decision has been made to use the network, the mobile device can attempt to establish an initial connection by sending a short message on the random access channel. The channel is referred to as a random access channel as the network cannot control access to this channel. There is thus a chance that several devices attempt to send a message simultaneously which results in a network access collision. If this happens, the base station will not be able to receive any of the messages properly. To minimize this possibility, the message itself is only very short and only contains a 5 bit random number. Furthermore, the network offers many random access slots per second to randomize access requests over time. When the network picks up the random access request, it assigns a C-RNTI (Cell-Radio Network Temporary) identifier to the mobile and answers the message with a Random Access Response message. The message contains the 5 bit random number, so the mobile device can correlate the response to the initial message and the C-RNTI, which is used to identify the mobile device from now until the physical connection to the network is released (e.g. after a longer time of inactivity). In addition, the message contains an initial uplink bandwidth grant, that is a set of resource blocks of the shared uplink channel that the mobile device can use in uplink direction. These resources are then used to send the RRC connection request message that encapsulates an initial attach request.

2.3.4.3 Authentication

When the eNodeB receives the connection request message, it forwards the contained attach request to the MME. The MME extracts the user identity from the message which is either the International Mobile Subscriber Identity or a temporary identity that was assigned to the mobile device during a previous connection with the network. In most cases, a temporary identity is sent which is changed once the user has been authenticated to reduce the number of IMSIs that have to be transmitted before encryption can be activated. If the IMSI of the subscriber is not known to the MME, the Authentication Center in the HSS is queried for authentication information. If a temporary identity is sent that is unknown to the MME, a request is sent to the mobile to send its IMSI instead. Afterwards, the network and mobile authenticate each other using secret private keys which are stored both on the SIM cards and in the authentication center, which is part of the HSS. Once the subscriber is properly authenticated, the eNodeB and the mobile device activate air interface encryption. At the same time, the MME continues the attach process by informing the HSS with an Update Location message that the subscriber is now properly authenticated. The HSS in turn sends the user's subscription data, for example what types of connections and services the user is allowed to use, to the MME in an Insert Subscriber Data Message. The MME confirms the reception of the message which in turn terminates the Update Location procedure with an acknowledgement message from the HSS, as shown in Figure 2.20.

2.3.4.4 Requesting an IP Address

In the next step, the MME then requests an IP address for the subscriber from the PDN-GW via the Serving-GW with a Create Bearer Request message. When the PDN GW receives the message it takes an IP address from its address pool, creates a subscriber tunnel endpoint and returns the IP address to the MME, again via the Serving-GW. Involving the Serving-GW in the process is required, as the user data tunnel is not established between the MME and the PDN-GW but between the Serving-GW and the PDN-GW. Once the MME receives the IP address, it forwards it to the eNodeB in an Attach Accept message, which is the reply to the initial Attach Request message. The eNodeB in turn forwards the Attach Accept message including the IP address as part of a Radio Bearer Establishment Request Message to the mobile device, which answers with a Radio Bearer Establishment Response message containing an Attach Complete message. The Attach Complete Message is then forwarded to the MME and the mobile device can now communicate with the Internet or the network operator's internal IP network (e.g. to connect to the IMS).

Despite the many messages being sent back and forth between the different functions in the network, the number of messages exchanged between the mobile device and the network has been reduced compared with GSM and UMTS by performing several tasks with a single message. This should significantly speed up the overall process. Excluding network detection and reading the broadcast channel, the procedure is likely to take only a few hundred milliseconds.

2.3.5 Summary and Comparison with HSPA

At its introduction, LTE competes with already deployed HSPA networks. It is likely, that some network operators will decide to upgrade their HSPA networks and eventually upgrade the network for the use of higher-order modulation, MIMO and Ethernet-based backhaul. In a 5 MHz band, the performance of LTE and HSPA+ is similar, so other reasons are required for operators to add LTE to their already existing GSM and HSPA infrastructure. One reason for adding LTE to a cell site is to increase the available bandwidth for a certain region by off-loading traffic from HSPA to LTE, once LTE devices become more commonplace. As network vendors are offering base stations capable of supporting several radio technologies simultaneously, the move to LTE could come as part of replacing aging base stations. By 2012, for example, many UMTS base stations will have been in the field for almost 10 years and therefore will have to be replaced anyway. Since it is unlikely that HSPA will be directly replaced by LTE due to many devices still 'only' being HSPA-capable, multiradio technology base stations will become very interesting for mobile network operators. If LTE is used in the same band as the other radio technologies, a single antenna can be used. Therefore, no additional antennas will be required for many base station sites. The existing antenna might be replaced with a new one, however, to enable MIMO transmissions for LTE. When more than a 5 MHz bandwidth is available, LTE can clearly show its advantages over 3G technologies, as LTE radio channels can be easily extended to 10, 15 or even 20 MHz. In addition, the simpler radio and core network with fewer components and new technologies for backhaul transmission from the base station to the rest of the network will lower network operation costs. This will be an interesting driver for network operators, as bandwidth demands keep rising while the revenue per user is flat or even declining. Finally, due to the simplified air interface signaling, LTE is much more suitable for always-on IP connectivity and applications such as instant messaging and push e-mail, which frequently communicate with the network, even while the mobile device is not actively used.

2.3.6 LTE-Advanced

As discussed in Section 1.5, LTE is unlikely to meet the transmission speed requirements of the ITU for 4G wireless systems. 3GPP has therefore started to investigate, how an evolution of LTE could meet these requirements. The following list shows some initial ideas for such a system, currently referred to as LTE-Advanced and LTE Plus, which were presented during a 3GPP workshop in 2008 [25]:

- LTE advanced shall be backwards-compatible to LTE (i.e. like HSPA is backwards-compatible to UMTS).
- The primary focus should be on low-mobility users in order to reach ITU-Advanced data rates.
- Channel bandwidths should be used beyond the 20 MHz currently standardized for LTE (e.g. 50 MHz, 100 MHz).
- The number of antennas for MIMO should be increased beyond what is currently specified in LTE.
- MIMO should be combined with beamforming.
- There should be a further increase in Voice over IP capacity.
- Cell edge data rates should be further improved.
- Self-configuration of the network should be improved.

Details for these and other possible features for LTE-Advanced can be found in [26]. With LTE still in the specification phase and not yet being widely deployed, working on a further evolution presents a significant challenge to 3GPP members for a number of reasons. First, it is likely that significant work on the original LTE and SAE standards will be required beyond 2008. This leaves little room for significant simultaneous developments. Also, there has not yet been much time to allow for adjustments of the standards based on the experience gained from developing and deploying LTE. Finally, some network operators might decide to halt their LTE plans and focus on improving their current HSPA networks to try to bridge the time until LTE-Advanced systems become available.

2.4 802.16 WiMAX

2.4.1 Introduction

Another successor to current 3.5G wireless network technologies is WiMAX, a system based on the IEEE (Institute of Electrical and Electronics Engineers) 802.16 air interface standard. Major infrastructure vendors backing this technology are, for example, Intel, Motorola, Nortel, Alcatel-Lucent and Nokia Siemens Networks. As will be shown in this section, WiMAX shares many basic properties with LTE. From a timing perspective, WiMAX has a head start over LTE, as standards activities were started earlier.

While LTE is mainly attractive for incumbent 2G and 3G operators, WiMAX is very appealing to greenfield network operators, that is operators without an already existing network in place. There are several reasons for this:

- In most countries, frequency bands for UMTS and LTE have already been sold many years ago. WiMAX, however, can be operated in so far unused frequency bands that are still in the process of being auctioned.
- The WiMAX network architecture is fully based on IP, which simplifies the network architecture design and deployment and cuts operating costs compared with current ATM backhaul-based 3G networks.
- New network operators are aiming to offer Internet access and thus compete with fixed line high-speed Internet solutions such as DSL and TV cable. As a consequence, their business model is quite different from that of incumbent wireless network operators, who are still mainly focused on mobile voice services.
- WiMAX network equipment is available today, while LTE equipment will only become commercially available in the 2010–2012 timeframe.

An exception to the rule seems to be the US market, where Sprint, a major incumbent operator, together with Clearwire has decided to use WiMAX due to a lack of a convincing future perspective of its 3G technology.

2.4.2 Network Architecture

2.4.2.1 Small Networks for Stationary Clients

Many early WiMAX network operators have started to deploy small networks based on the early 802.16-2004 standard (previously referred to as 802.16d), which mainly targeted stationary devices with roof-mounted antennas or indoor WiMAX routers with large omnidirectional antennas built in. Such networks require little more than a few base stations and possibly a central server for storing subscription data for subscriber authentication purposes and to control access to the network. Since such networks are designed for stationary devices, no handovers of connections between base stations are required. Each cell can therefore act as a little network of its own.

2.4.2.2 Medium to Large Networks and Mobility

More recently, the air interface standard was enhanced to also support mobility of subscribers, including handovers from one base station to another. This version of the standard is referred to as 802.16e, or 802.16-2005. Since handovers and mobile subscribers require more administration in the network, it became necessary to also specify the network behind the base stations. As the IEEE is only responsible for the air interface standardization, this part was taken over by the WiMAX Forum [27]. Standardizing the radio network and the core network is an important task, since only standardized functionalities and interfaces allow network operators to select compatible components from a wide variety of network vendors. This way, competition between vendors is fostered, which results in lower prices for network equipment. To ensure interoperability of components from different vendors, the WiMAX Forum is also in charge of a certification program for base stations, end user devices and other network equipment.

Figure 2.21 shows how one of the possible WiMAX network infrastructure setups looks like in practice. When compared with the LTE network infrastructure shown in Figure 2.14, there are remarkable similarities. As in LTE, WiMAX base stations communicate with each other for handovers. Furthermore, there is no central element in the radio network, as was the case with UMTS. Instead, WiMAX networks only require gateways between the radio network and the core network, the Access Service Network Gateways (ASN-GW).The ASN-GWs are responsible for user management and mobility.

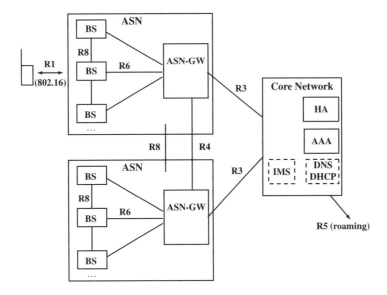

Figure 2.21 WiMAX network architecture. (Reproduced from *Communication Systems for the Mobile Information Society*, Martin Sauter, 2006, John Wiley and Sons.)

The air interface between mobile devices and the base station is referred to as the R1 reference point or interface. In this section, the two words are used interchangeably. The protocol used over this interface is either 802.16-2004 (formerly 802.16d) for stationary wireless installations or 802.16-2005 (formerly 802.16e) for both stationary and mobile clients. Air interface details are discussed in Section 2.4.3.

In the radio network, base stations are connected to the ASN-GW via the R6 reference point. In the first version of the WiMAX standard, the interface is proprietary, which means that an ASN-GW and all base stations connected to it need to be from the same vendor. In the second version of the standard, the R6 reference point will also be standardized to allow mixed configurations. This further increases competition, which results in more competitive pricing. Like in LTE, the R6 interface between the base station and the access gateway is fully based on IP. Consequently, any transport technology that is capable of carrying IP packets can be used. Since multisector WiMAX base stations are capable of air interface data rates of 30 Mbit/s and beyond, as shown in Chapter 3, suitable transport technologies in the last mile to the base station are Ethernet-based microwave systems, VDSL and fiber connections. Because of the prohibitive cost of such connections and their slow speed compared with the capabilities of the air interface, 2 Mbit/s E-1-based connectivity is not likely to be used.

For smooth handovers of connections between base stations, the R8 reference point has been defined as shown in Figure 2.21. Like the R6 reference point, it is also fully based on the IP protocol. In practice, a base station is only connected to the network with a single physical interface. Data between different base stations therefore traverses one or more routers before reaching another base station. In practice, the overhead created by this is likely to be small, as the amount of user data on the R6 interface is likely to be several orders of magnitude higher than the amount of data exchanged for a handover of a connection on the R8 interface.

2.4.2.3 The Access Service Network Gateway

In WiMAX networks, the gateway between the radio network and the core network is referred to as the Access Service Network Gateway. In principle, it is responsible for the same tasks as the Access Gateway (AGW) in LTE. These are:

- subscriber management tasks such as authentication and subscription management;
- mobility management to redirect the connection from one cell to another when the user is moving;
- to actively support the handover procedure in case the R8 reference point between two base stations is missing.

2.4.2.4 Authentication and Ciphering

In the WiMAX standard, user devices are referred to as a Customer Premises Equipment (CPE), a term from the 802.16-2004 standard, which mainly addressed stationary equipment. The term, however, is still being used with the 802.16-2005 standard and mobile devices alongside the term Mobile Subscriber Station (MSS). When a device is powered on, its first task is to search for available networks and to get an IP address from the user's home network, or, in the case of roaming, from another suitable network. Before a device is admitted to the network by the ASN-GW, an authentication procedure is required. Unlike 3GPP networks such as UMTS and LTE, WiMAX does not make use of a secret key that is stored on a SIM card and in the network. Instead, authentication is performed with a public/private key pair in addition to an X.509 certificate.

In theory, the keys and the certificate could be stored on a SIM card. In practice, however, this is not the case today. Instead, the keys and certificate are stored in a safe location in the device itself that cannot be directly accessed to prevent applications from reading the secret private key.

At the beginning of the authentication procedure, the device and the network exchange their public keys with each other, which are then used to derive temporary keys to encrypt further traffic. Data encrypted with a temporary public key can only be decrypted with the corresponding temporary private key which was derived from the secret private key. As private keys are never transmitted over the air interface, it ensures that an attacker cannot decipher or modify the data. Another benefit of having public/private keys is that the private key of the subscriber is only stored in the client device but not in the network, as the network only requires the public key for encrypting the data. Consequently, no sensitive key information has to be stored on any equipment of the network operator.

In addition to the public key exchange, an additional mechanism is used to ensure that the public key sent by a device is tied to its MAC hardware address. This is done by sending a certificate, signed by a certificate authority, in addition to the public key, as shown in Figure 2.22. The certificate authority signs the certificate by encrypting the device's public key and MAC hardware address with its private key. When the client device sends its public key together with the certificate, the network then decrypts the certificate with the public key of the certificate authority. Afterwards, it checks if the subscriber's public key matches the one extracted from the certificate. Furthermore, it is verified that the MAC address, which is part of each data packet, also matches the one given in the certificate. If they match, the client device is authenticated. Tampering with the certificate is not possible since it can only be decrypted by everyone who knows the public key of the certificate authority. However, it cannot be changed and re-encrypted. The certificate also prevents a successful attack by duplicating MAC addresses, as the private key of the original device used in combination with a MAC address is securely stored in the device. Therefore, it cannot be duplicated together with the MAC address.

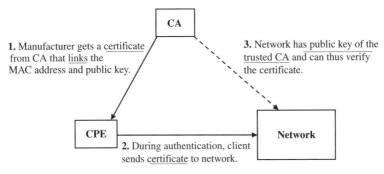

Figure 2.22 WiMAX authentication with the help of a Certificate Authority (CA). (Reproduced from *Communication Systems for the Mobile Information Society*, Martin Sauter, 2006, John Wiley and Sons.)

It should be noted that certificates and public/private keys are also used for authentication and encryption of secure Web sessions, such as for example for online shopping or Web banking. Here, a Secure HTTP (HTTPS) session is established instead of a standard HTTP session and the process is very similar to the one described above. During connection establishment, the Web server sends a certificate signed by a certificate authority to the Web browser. The certificate contains the URL of the Web site and the public key. The Web browser compares the URL the user has typed in with the one in the certificate. If both match, the Web browser can be certain that the connection was not redirected by an attacker. To an attacker, the public key in the certificate is worthless, as they do not have the private key to decrypt the information, which is encrypted by the client with the public key.

Both the WiMAX and the HTTPS authentication processes require certificates to be generated by a trusted certificate authority. Trust is established by storing the certificate authority's public key locally. In the case of WiMAX, the certificate authority's public key is stored in the device itself. In case of HTTPS, the certificate authority's public key is stored in the Web browser. In practice, there are many different certificate authorities that can issue certificates. Verisign, for example is a company issuing both HTTPS and WiMAX certificates [28].

2.4.2.5 Client IP Address Assignments and R6 Tunnels

Once a device is authenticated and air interface ciphering has been activated, the ASN-GW is also in charge of assigning an IP address to a device or to request it from a Home Agent (HA) in the core network, as will be discussed below in Section 2.4.2.7.

As the network between the base stations and ASN-GWs is not necessarily owned by the WiMAX network operator, data traffic on the R6 reference point between gateways and base stations should be encrypted. For this purpose, an encrypted IPSec tunnel could be established between each base station and the ASN-GW. The user's data is thus not only protected on the air interface, but also throughout the radio network up to the ASN-GW.

2.4.2.6 Micro Mobility Management

When a user moves from the coverage area of one base station to another, it is the base station's task to handover the connection. Both the network and the client device can initiate a handover. In the radio network, this means that the current and new base station communicate with each other over the R8 interface or via the ASN-GW during the handover. Part of the handover process is also to inform the ASN-GW that the location of the subscriber has changed, as user data packets now have to be exchanged over a different IPSec tunnel. Figure 2.23 shows how this works in practice. In the example, base stations and ASN-GWs in the radio network use the 10.x.x.x IP subnet. One tunnel is established between the upper base station, which has been assigned 10.0.0.2 as an IP address, and the ASN-GW (10.0.0.1). Another tunnel is established to the base station in the lower part of the figure (10.0.0.3). Before the handover, the user's data, identified by the user's IP address (195.36.219.196), is sent through the tunnel between 10.0.0.1 (ASN-GW) and the upper base station (10.0.0.2). Once the handover

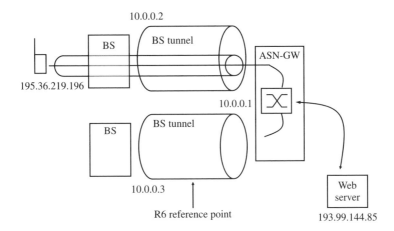

Figure 2.23 Base station and user tunnel for micro mobility management. (Reproduced from *Communication Systems for the Mobile Information Society*, Martin Sauter, 2006, John Wiley and Sons.)

has been performed, the ASN-GW redirects the data flow to the tunnel between itself (10.0.0.1) and the lower base station (10.0.0.3).

2.4.2.7 Macro Mobility Management

Like in larger UMTS and LTE networks, it is required at some point to install several WiMAX ASN-GWs in the network to support a growing number of base stations and users. Once there are several radio network gateways in the network, it is possible that a user will change between cells controlled by different gateways. This means that packets arriving for a user from the Internet can no longer only be routed by default to a single ASN-GW. Instead, it is necessary to introduce core network mobility management as well. In UMTS and LTE, the GPRS Tunneling Protocol (GTP) takes care of this task between the single point of entry to the network (the GGSN in case of UMTS and the PDN in case of LTE) and the radio access network gateway. In WiMAX, it has been decided to use a different approach, as shown in Figure 2.24.

Instead of relying on a proprietary protocol such as GTP, it was decided to use Proxy Mobile IP (Proxy MIP), an already existing IP-based mobility management standard [29]. In principle, Proxy MIP works as follows: when a device requests access to the network, the ASN-GW requests an IP address for the device from the Mobile IP Home Agent. The HA has a pool of IP addresses it is responsible for and all packets arriving from the Internet destined to these IP addresses are always routed to the HA. From this pool, one IP address is assigned to the device and returned to the ASN-GW. The HA notes the IP address of the ASN-GW (64.236.23.28 in Figure 2.24) and begins forwarding all packets arriving from the Internet to the ASN-GW. The ASN-GW in turn forwards packets it receives for this IP address to the base station, to which the subscriber is currently attached via a micro mobility management tunnel as described above.

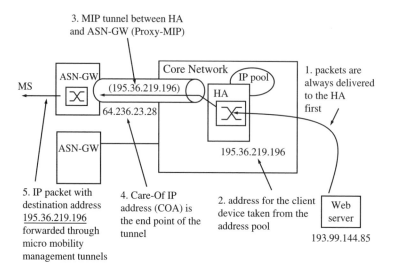

Figure 2.24 ASN-GW mobility management using mobile IP. (Reproduced from *Communication Systems for the Mobile Information Society*, Martin Sauter, 2006, John Wiley and Sons.)

For subscriber devices using IPv4, the ASN-GW terminates the MIP tunnel to make the process transparent for the subscriber device. This is required since it is not desirable to change the protocol stack of the device. Consequently, the ASN-GW becomes a proxy for the subscriber's device, which is why the approach is referred to as Proxy MIP.

For devices using only IPv6 addresses, which is expected to become more commonplace in the future, no proxy is required, as MIP is part of the protocol stack. This means that the mobile device communicates with the HA in the network on its own instead of leaving the task to the ASN-GW.

2.4.3 The 802.16d Air Interface and Radio Network

Like LTE, WiMAX uses Orthogonal Frequency Division Multiplexing (OFDM) to transmit data over the air interface. The systems are therefore very similar on the physical layer and this section assumes the reader is familiar with the LTE air interface described in Section 2.3.3.

2.4.3.1 Fixed WiMAX

The first version of the 802.16 air interface standard, referred to as 802.16d or IEEE 802.16-2004 [30], is currently in use to connect devices such as notebooks and PCs to the Internet via WiMAX modems installed at home or at the office. It is not compatible with the current 802.16e or 802.16-2005 standards, which were developed later and introduce many enhancements required for mobility. The lifetime and use case scenarios of fixed WiMAX deployments are therefore limited, as devices and network equipment are unlikely to be upgradeable to the mobile standard. Also, fixed WiMAX networks do not usually use most of the standardized infrastructure described above, since their network architecture is much simpler and also because no support for mobility is required. Nevertheless, this section takes a look at the fixed WiMAX air interface standard as well, because it is used to some degree in practice and forms the basis for mobile WiMAX, which is discussed later.

The big difference of IEEE 802.16-2004 compared with mobile WiMAX (IEEE 802.16-2005) and LTE is that it uses 256 OFDM subcarriers independent of the bandwidth used for the channel. Out of these, 193 are used for data transmissions. The remaining subcarriers are either unused at the edge of the band or provide pilot signals which are used by devices for channel estimation and filter approximation. Channel bandwidths defined are 1.25, 3, 3.5, 5.5, 7 and 10 MHz. The smaller bandwidths, however, are unlikely to be used in practice, as the resulting throughput is not sufficient to support high-speed Internet access even for a small number of users. Using the same number of subcarriers for all bandwidths means that the symbol transmit time varies depending on the bandwidth. For a bandwidth of 1.25 MHz, for example, the symbol transmit time is 128 μs, while for a 10 MHz deployment, the symbol time is only 22.408 μs. Therefore, the transmission characteristics on the physical layer depend on the bandwidth used for a channel.

Two profiles have been defined by the WiMAX Forum for fixed WiMAX. The wireless Metropolitan Area Network OFDM profile (wirelessMAN-OFDM) is used when a national regulator has officially assigned a frequency band for the use of WiMAX, for example as the outcome of a spectrum auction. Depending on the properties of the assigned band, fixed WiMAX devices are used in either TDD or FDD mode [31].

In TDD mode, the same band is used for uplink and downlink transmissions and the system continuously switches between transmission and reception. The advantage of this approach is that the system can be tuned to reflect the ratio between uplink and downlink traffic. Currently, more bandwidth is required in the downlink direction, which is why more time is allocated for downlink than for uplink transmissions. It should be noted at this point, however, that a 3:1 downlink/uplink ratio on the air interface does not exactly reflect the bandwidth ratio, because uplink transmissions are usually not as efficient due to the limited output power and antenna restrictions of a small device. Furthermore, TDD requires base stations to be tightly synchronized with each other to prevent uplink transmissions of devices in one cell to interfere with downlink transmissions of neighboring cells.

In FDD mode, different frequency bands are used for downlink and uplink transmissions. For licensed frequency bands, this is often the preferred transmission mode. The advantage of FDD is that data can be transmitted in uplink and downlink in parallel. Furthermore, no transmission pause is necessary to give devices the necessary time to switch from transmission to reception mode. In addition, FDD transmission allows more sensitive receivers in mobile devices, which benefits overall data rates.

Figure 2.25 shows what an FDD downlink data transmission looks like in WiMAX with the IEEE 802.16-2004 standard for stationary devices. Downlink transmissions are separated into individual frames with a fixed length between 2.5 and 20 ms. Each frame in turn holds a number of consecutive fields. The first field is the preamble, which has a known bit pattern that mobile devices can use to detect the beginning of a frame. The Frame Control Header (FCH) is next and contains information about the modulation and coding scheme of the first downlink burst that immediately follows.

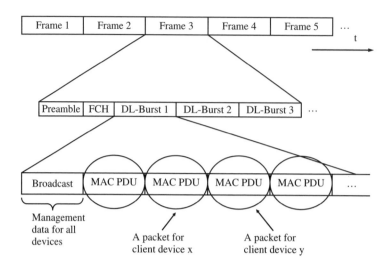

Figure 2.25 WiMAX 802.16-2004 downlink data transmission. (Reproduced from *Communication Systems for the Mobile Information Society*, Martin Sauter, 2006, John Wiley and Sons.)

BPSK modulation and a 1/2 coding rate are used for the FCH to ensure that all devices can receive the information correctly. The first downlink burst of a frame contains downlink broadcast data at the beginning to inform devices if and at what point in the frame data will be transmitted for them. Furthermore, the broadcast zone also contains information for devices when they are allowed to send data in uplink direction. The remainder of the first downlink burst of a frame then contains user data packets for one or more devices. A frame usually contains more than a single downlink burst and each burst can use a different modulation and coding scheme. The location of data for a device thus depends on the radio condition it experiences. Figure 2.25 also shows that data is only sent to one device at a time. This means that subscribers are only multiplexed in time but not in the frequency domain.

As the IEEE 802.16-2004 radio interface was designed for stationary use, devices do not report signal conditions as frequently as is required in systems supporting mobility. Instead, it is the device's responsibility to judge radio conditions and to send a dedicated management message to the network to change the modulation and coding scheme for the uplink and downlink directions when required. Likewise, error detection and correction on the MAC layer is also optional. If used for a connection, a basic Automatic Retransmission Request (ARQ) scheme splits packages into ARQ blocks and each side reports to the other which blocks have been received correctly. Blocks not received correctly are then retransmitted. This mechanism is similar to that used for the Transmission Control Protocol (TCP), which sits above the IP layer. The advantage of additionally checking and retransmitting packets in the MAC layer is that errors can be detected more quickly and that less data has to be retransmitted. This also helps to keep throughput high, as TCP automatically throttles a transmission once errors occur, as it interprets missing packets as congestion.

2.4.4 The 802.16e Air Interface and Radio Network

As fixed wireless applications only address a limited customer base, the IEEE soon decided after the fixed air interface standard was finalized to go one step further and to enhance the air interface with additional functionality for mobility and for power-constrained devices. To support mobility, the air interface management must react quickly to changing signal conditions and it must be able to hand over a connection between base stations when the user is moving. For battery-driven devices, the air interface had to be optimized to be as power-efficient as possible during times when no data is being transferred. In addition, the WiMAX air interface was enhanced to allow higher transmission speeds. The working group responsible for this task is referred to as 802.16e. After finalization of the standard, it is now known as IEEE 802.16-2005 [32]. The mobile WiMAX standard supports FDD and TDD, although initial deployments will only make use of the TDD option.

2.4.4.1 Orthogonal Frequency Division Multiple Access

On the physical layer, the main difference from the fixed WiMAX standard is that the subcarrier spacing is fixed so the number of subcarriers now varies with the bandwidth, as shown in Table 2.2. According to [33], the maximum channel bandwidth supported by the first Intel WiMAX chips for notebooks is 10 MHz. In the future, however, larger

Table 2.2 Bandwidths and subcarriers for WiMAX.

Bandwidth	Number of subcarriers	FFT size
1.25 MHz	85	128
5 MHz	421	512
10 MHz	841	1024
20 MHz	1684	2048

bandwidths are going to be supported as well.

In addition to the parameters shown in Table 2.2, the following physical parameters were selected:

- subcarrier spacing, 10.94 kHz;
- OFDM symbol duration, 91.4 μs;
- cyclic prefix, 11.4 μs.

It is interesting to note that these values are similar but not identical to the values used in LTE. In LTE, the subcarrier spacing is 15 kHz and a shorter OFDM symbol duration of 66 667 μs is used (cf. Table 2.1).

Based on these radio layer parameters, the physical layer throughput of a WiMAX cell can be calculated as follows: The transmission time per symbol is 102.8 μs (91.4 μs for the symbol itself + 11.4 μs for the cyclic prefix), the highest modulation order is 64QAM (6 bits per symbol) and there are 1684 subcarriers in a 20 MHz band:

$$\text{Physical speed} = (1/0.000\ 102\ 8)^*6^*1684 = 98\ 287\ 937\ \text{bit/s (i.e about 100 Mbit/s)}$$

This is almost exactly the same value as calculated for LTE earlier in this chapter and shows that, from this perspective, the two systems provide very similar performances. As was described in the section on LTE, it should be noted that in practice, the throughput of a cell is likely to be only 30–50% of this value. This is because of the overhead for coding, retransmissions of faulty packets, pilot signals and the overhead of the higher protocol layers, and also because of the less than ideal signal conditions for most users in the cell.

Figure 2.26 shows the structure of the downlink subframe. The major difference from the frame shown for the fixed WiMAX air interface is the fact that data transmissions to individual users can now also be multiplexed in both time and frequency, due to the much higher number of available subchannels. For this reason, this form of data transmission is not referred to as OFDM (Orthogonal Frequency Division Multiplexing) but instead as OFDMA (Orthogonal Frequency Division Multiple Access). At the beginning of a frame, the Downlink-MAP (DL-MAP) informs devices when and where data is scheduled for them in the frame. An optional Uplink-MAP (UL-MAP) can also be present in the frame to assign uplink transmission opportunities for this and the following frames.

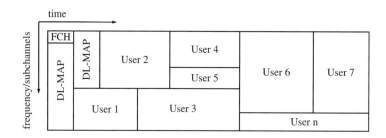

Figure 2.26   A WiMAX OFDMA frame.

2.4.4.2 MIMO

WiMAX also supports MIMO transmissions in the downlink direction with multiple antennas (e.g. two input, two output = 2 × 2) in the same way as already described for LTE. In the uplink direction, mobile devices only transmit a single data stream. Advanced base stations, however, can activate collaborative MIMO and instruct two devices to transmit at the same time. At the base station, the signals are recognized as coming from different devices due to their separate multipath characteristics and are separated accordingly.

Depending on transmission conditions, one of the following two different MIMO transmission modes can be used.

Matrix A: coverage gain. In a 2×2 antenna configuration (two transmitter antennas, two receiver antennas), a single data stream is transmitted in parallel by two separate antennas. A mathematical algorithm known as Space Time Block Codes (STBC) is used to encode the data streams of the two antennas to make them orthogonal to each other. This improves the signal-to-noise ratio at the receiver, which can be used to:

- increase the cell radius;
- provide better throughput for subscribers that are difficult to reach (e.g. in difficult indoor conditions or when moving at higher speeds);
- transmit with higher-order modulation (e.g. 64QAM) while using fewer error correction bits, which in turn increases transmission speeds to that subscriber.

Matrix B: Capacity Increase. This flavor of MIMO, also known as Spatial Multiplexing MIMO (SM-MIMO), sends a completely independent data stream over each antenna, as described in previous sections. Thus, the data rate can be doubled, given that the mobile device is close to the base station and has excellent reception conditions.

2.4.4.3 Adaptive Antenna Systems

Another feature that is already in the 802.16e standard document, but not used in early networks, is AAS (Adaptive Antenna Systems). By using several antennas and connecting them electrically, a beam can be formed towards a client device, thus increasing the signal-to-noise ratio experienced by the client device. To form a beam, the signal is sent over each antenna with a calculated phase shift and amplitude relative to the other antennas. There are no moving parts required for directing the beam in a certain direction, as the beam-forming effect is based on the phase and amplitude differences of the signals. Beamforming can be used in both the uplink and the downlink. For the uplink, beamforming improves the reception of the signal from a device and in the downlink beamforming lowers interference for other devices receiving a transmission from a neighboring cell on the same frequency.

2.4.4.4 Handover Procedures

In 802.16e, both the mobile station and the network can initiate a handover procedure. This is different from UMTS and LTE, in which a handover of a connection is always initiated by the network. In the IEEE specification, a handover is sometimes also referred to as a cell reselection. This is somewhat unfortunate since in other systems cell reselection is the process to change to another cell while no connection is established to the network.

To perform a handover to a new cell, a mobile has to search for neighboring cells when signal conditions deteriorate. During this process, the mobile will not be able to receive data from the current cell. For this reason, the mobile and base station have to agree when such searches can be performed and the base station then buffers all incoming data packets until the mobile device is back and receiving incoming data. When neighboring cells are detected by the mobile. it reports the reception conditions detected to the network via the serving base station. Both the network and the mobile can then initiate a handover procedure, if required. The mobile device on the one hand can initiate the handover if it feels that it would get better service from a neighboring cell. The network on the other hand can initiate a handover for the same reason or for load balancing purposes if it detects that a neighboring cell with less traffic can be received equally well by the device as the current cell.

In the simplest handover variant, the network or the mobile initiates a handover, which causes a short service interruption while the mobile connects to the new cell. If it is already synchronized, the outage will be shorter than if the mobile first has to associate to the cell and the new cell has to request the subscriber's current parameters from the previous cell.

For real-time services such as VoIP, interruptions are undesired and two further options have been standardized to improve the handover behavior. The first optional procedure is referred to as Fast Base Station Switching (FBBS). As in the basic approach above, the base station requests the mobile to frequently scan for the availability of neighboring cells. These are then reported to the serving base station. If signal conditions are strong enough, the serving base station contacts the neighboring base station via the backhaul connection and requests them to set up a context for this subscriber. If the base stations agree, the mobile device is informed that it can select from which base station it wants to receive its downlink data packets. The base stations are kept in a diversity list (the active set) in the mobile which can, by sending a short command, instruct the network to change the cell for downlink transmission. This way, the mobile can quickly react to changing signal conditions. In uplink direction, all base stations that are part of the active set receive the data stream from the mobile and forward correctly received packets to the ASN-GW. This increases the probability that the network receives at least one copy of each packet but has the disadvantage of increasing bandwidth requirements on the backhaul links.

The Macro Diversity Handover (MDHO) is an even smoother form of handover. Here, all base stations in the active set transmit frames on the downlink to the subscriber. The mobile can then combine the received signal and thus increase the chance of successfully receiving a packet. This approach is quite similar to the UMTS soft hand-over, which, however was abandoned again with HSPA and LTE as it was seen as too costly in terms of capacity requirements on the air interface and the backhaul connection. Both FBBS and MDHO require that all cells that are part of the handover procedure use the same frequency, as the mobile only has a single transceiver and can therefore only receive and transmit on a single frequency.

2.4.4.5 Power Saving and Idle Mode

To minimize the power requirements of battery-driven devices, mobile WiMAX intro-duces a number of power-saving modes, referred to in the standard as power-saving classes. With power-saving class 1 the mobile and network agree on a pattern in which the device periodically listens to the downlink and afterwards enters sleep mode for some time, during which it cannot be reached. Over time, the sleep periods are automatically extended as it becomes less likely that data arrives for the device. Should data arrive while a mobile is in the sleep state, it is buffered in the base station and sent to the mobile as soon as it reactivates its transceiver. This automatically ends the power-save mode. Power save mode is also left when the mobile sends data on the uplink.

While this power-save mode is suitable for bursty data traffic with applications such as Web browsing, it is less suitable for VoIP transmissions, which also have long but predictable periods of inactivity between two packets. Here, it would not make sense to enter or exit power-save mode every time a packet has to be sent or received. Power-saving class 2 thus limits data transmissions to certain intervals. Outside these intervals, the device can turn off its transmitter. In practice, this limits the bandwidth available to a device, which is quite acceptable for VoIP applications that require little bandwidth anyway.

Finally, with power-saving class 3, the network and mobile can agree on a single sleep period after which the connection automatically becomes active again.

In practice, several service flows, each with a different IP address and for different applications, can be active per device. Each service flow can be in a different power-saving mode. The transmitter is then only switched off at times in which all service flows have entered a power-saving state.

2.4.4.6 Idle Mode

Even while in a power-save mode, a mobile device is required to wake up periodically and communicate with the network. In cases of long inactivity, this is undesired as it requires resources on the network side to keep a connection active and has a negative impact on the overall standby time of the device. The standard therefore also defines an idle mode state, in which the radio connection to the network is removed, while the device still keeps its IP address(es). Once in idle mode, the mobile can switch off its transceiver and only occasionally check the reception level of the current and neighboring cells and to observe incoming paging messages which could announce waiting packets on the network side. The paging interval is usually in the order of a few seconds. Another advantage of being in idle mode is that the mobile device can roam between different cells of the network that are in the same paging group without reporting the location change to the network. Only if the mobile roams to a cell in a different paging group does it have to inform the network of its new location, that is of its new paging group, so paging messages can be sent to the new paging group in the future. As a paging is sent via several cells, paging coordination is a task of the ASN-Gateway.

2.4.5 Basic Procedures

Figure 2.27 shows the basic procedures to establish a connection with the network after a mobile device has been powered on. In the first step, it will try to find the previously used network, whose parameters it might have saved in nonvolatile memory. Use of this information has the advantage that the mobile can go directly to a certain frequency and, if the user has not moved since the device was switched off, is very likely to receive a signal instantly. If the user has moved and no signal is found, a standard network search procedure is started. At first, downlink transmissions of a network are detected by searching for the preamble of each frame which has a known bit pattern. Once the

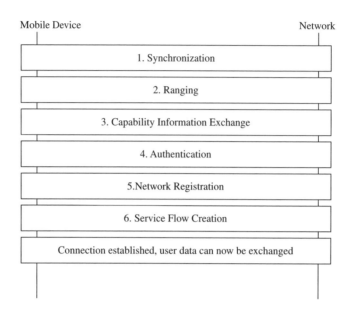

Figure 2.27 Stages required to connect to a WiMAX network.

preamble is detected, the mobile device is synchronized to the frame structure of the downlink transmissions and can start to receive and decode cell information, which is sent after the frame's preamble.

The cell information describes, among other things, where to find the contention-based ranging area, which is used in the second step to get into contact with the network. This area, referred to in other standards as the random access channel, is then used by the device to send a ranging request message with a low power level. The message includes the MAC hardware address of the device and the modulation and coding scheme the device suggests the network use to send an answer. If no answer is received, the message is repeated with a higher transmission power. Once the network has successfully received the ranging request, it answers with a ranging response message. The message contains information on how the device has to adjust its power lever for further communication and its synchronization. Synchronization corrections are required as the mobile station is not aware of the distance to the base station. The more distant it is from the base station, the earlier it has to send its transmissions to arrive in synch with packets of other subscribers that have different distances from the base station. The mobile device applies these values and sends another ranging request to the network. The network verifies that the values have been applied correctly and then returns a ranging response message, confirming the procedure and including Connection IDs (CIDs) that will identify packets to the mobile device in the downlink direction and bandwidth grants in the uplink direction.

In the third step, the mobile device sends a capability request message to the network which contains information about its capabilities such as the supported modulation and coding schemes. The network answers with a capability response message which contains its own capabilities. Capability exchange messages are not transmitted in the contention-based area, but with the basic CID as part of the data area of a frame. This means that, as soon as the network has sent a final range response message, it starts to schedule uplink resources for the mobile device.

In step 4, the device and network authenticate each other as described in Section 2.4.2. Afterwards, the mobile registers to the network. Once registered, the final step in Figure 2.27 consists of establishing a service flow and dedicated CIDs for the user data. This procedure can be initiated by the mobile device or the network, where the service flow is pre-provisioned. Since service flows are agnostic to higher-layer protocols, no IP address is assigned at this point. This is a separate action, which has to be performed by the device once the service flow is active by sending a DHCP (Dynamic Host Configuration Protocol) request to the network in a similar way as is done today in fixed Ethernet and Wi-Fi networks.

2.4.6 Summary and Comparison with HSPA and LTE

When comparing the physical parameters of WiMAX, LTE and HSPA+, it becomes apparent that in a 5 MHz band, performance of the three systems is very similar. Beyond a bandwidth of 5 MHz, LTE and WiMAX perform on a similar level, as both use OFDM modulation and very similar radio parameters. The major difference between the two is that WiMAX will first be deployed in TDD mode, while LTE will mainly be used in FDD mode for historical reasons. Another difference is that LTE uses a different uplink scheme, making it more power-efficient. How much difference this will make in practice remains to be seen. All other differences between the two systems are in the higher layers in the system. While the LTE air interface has inherited a strict channel structure from UMTS, the WiMAX air interface design is much simpler and adheres more to the simple Ethernet-style based MAC layers.

In practice, achievable transmission speed is just one of several important parameters. Equally important is how well a system is able to handle potentially hundreds of always-on devices per cell, each communicating with the system several times a minute, as applications such as VoIP, encrypted VPN (Virtual Private Network) tunnels and instant messengers constantly communicate with their servers in the network to keep their channels open through firewalls and NAT gateways. This requires that the system is not only streamlined for high bandwidths but to also able to handle a significant number of bandwidth requests per second for the keep alive messaging without sacrificing bandwidth and mobile battery power. As can be seen with HSPA, air interfaces are continuously enhanced to also take this issue into account. It therefore remains to be seen how first WiMAX and LTE networks fare in this regard and how their evolution accommodates such device behavior. As a consequence of this continuing evolution, it is impossible to describe one system as better than another in terms of performance.

2.4.6.1 Good Competition Between Network Technologies

Since LTE and HSPA on the one hand and WiMAX on the other are very similar in terms of throughput and usage scenarios, many observers raise the question of whether we will see a similar destructive competition as in the days of the 2G GSM and CDMA networks. Here, users and operators did not benefit greatly from this competition because networks and applications were both in the hands of the operators. This created many incompatibility issues for users. One example is text messaging. While in Europe, text messaging has flourished for a long time, it has only recently become popular in the USA. The main reason for this was that it was not possible for users of different networks to exchange text messages. Thus, the service did not take off until interoperability was finally introduced.

With HSPA, LTE and WiMAX, however, the application landscape is quite different. Here, the networks and applications are separated and do not depend on each other. Applications are based on the Internet Protocol and use whatever network is available. Internet Protocol applications are not and should not be aware of the underlying network technology, which allows people to develop applications independently of the wireless network architecture. Some applications will still be developed by operators but the vast majority will come from Internet-based companies, as will be shown in more detail in Chapter 6. As a result of this split between the application and the network, the competition between different wireless technologies becomes very beneficial because:

- It encourages faster network roll outs, as this is one of the few differentiators between network operators.
- It offers possibilities for new players in the market.
- It creates competition between device manufacturers.
- New applications can be introduced much more easily and quickly, as they are no longer forced into a tight network operator controlled framework.

2.4.7 802.16m: Complying with IMT-Advanced

Like LTE, WiMAX is also set to compete for a place in IMT-Advanced 4G. As the current specification is also not likely to qualify for 4G, several activities have been started to enhance the system. The 802.16m working group has been tasked to specify an air interface with a higher bitrate and the following enhancements are foreseen to improve system performance [34].

2.4.7.1 Use of Several Carriers

Like other standards bodies, the IEEE has recognized that increasing the bandwidth used for data transmission is one of the best ways to increase overall data transfer rates. A multicarrier approach, in which two or even more carriers are used for transferring data, will be used by the future WiMAX air interface. The approach used by WiMAX is backwards-compatible, that is 802.16e and 802.16m mobile devices can be served by the same base station on the same carrier. An 802.16e device, however, does not see the channel bundling and continues to use only one carrier. To be backwards-compatible, high-speed zones are introduced in a frame, which are only available for 802.16m devices. If the carriers used for transmission are adjacent, guard bands that are normally in place to separate the carriers can be used for transferring data.

2.4.7.2 Self Organization and Inter Base Station Coordination

Interference from neighboring base stations and mobile devices is undesirable in wireless systems, as it reduces the overall system throughput. The new version of the standard introduces methods and procedures to request mobile devices to perform interference measurements at their location and send them back to the base station. The base station can then use the information gathered from different devices to adjust its power settings and potentially also to coordinate the frequency use with neighboring base stations.

2.4.7.3 New Frame Structure

In practice, it has been observed that the 802.16e frame structure with frame lengths of up to 20 ms is too inflexible. The downside of such long frames is slow network access and slow repetition of faulty data blocks, as devices only have one transmission opportunity per frame. The standard 802.16 m uses a new frame structure which consists of super-frames (20 ms) which are further divided into frames (5 ms) and again divided into eight subframes (0.617 ms). Within each frame of 5 ms, the transmission direction can be changed once. Since eight subframes fit into a frame, downlink/uplink time allocations of 6/2, 5/3, and so on can be achieved. By switching the transmission direction at least every 5 ms, HARQ retransmission delays are cut by three-quarters, the idle-to-active state transmission time is reduced from above 400 ms to less than 100 ms and the one-way access delay is reduced from almost 20 ms to less than 5 ms [34].

2.4.8 802.16j: Mobile Multihop Relay

In many scenarios, especially in rural areas, there are often only few or no possibilities at all to backhaul high-bandwidth connections via a fixed line copper or fiber links. Consequently, cells need to be connected wirelessly to the network either over high-bandwidth microwave links or via a concept in which the base stations themselves form a mesh-like network to forward traffic between base stations with no dedicated backhaul connection. WiMAX is the first standard to incorporate such a backhaul method and the 802.16j working group specifies how this should work in practice [35].

In addition to rural backhauling, forwarding traffic between wireless network nodes is also an interesting method to fill coverage holes and to improve in-building coverage. At first, it might seem illogical that sending a data packet over the air interface more than once actually increases the data rate. In practice, however, transmitting the packet over two or more links with a high signal-to-noise ratio is better than only transmitting it once but very slowly over a low-quality channel.

The 802.16j amendment to the standard, also referred to as Mobile Multihop Relay (MMR), covers the following points to achieve these goals without increasing the number of base stations with expensive backhaul links:

2.4.8.1 Backwards Compatibility

MMR has been specified in a way that does not require mobile devices to be aware of relay nodes. This is important as introducing relaying would otherwise not be possible in already deployed networks.

2.4.8.2 Multihop Capability

The standard is designed in a way that allows packets to traverse several hops to and from a base station that has a backhaul connection.

2.4.8.3 Relay Station Implementation Options

From the point of view of mobile devices, relays without a dedicated backhaul connection look like a standard base station and have their own base station ID. The specification defines two kinds of Relay Stations (RS). A simple RS behaves almost like a simple repeater and leaves most of the work to a real base station, including even the handling of simple messages such as ranging requests. Such simple relays are also referred to as transparent relays as all links to mobile devices via relay stations are controlled by a base station. More complex relays, referred to as nontransparent relays, are able to locally manage the link to the subscriber and only forward user data packets to a base station and higher layer signaling information.

2.5 802.11 Wi-Fi

2.5.1 Introduction

At the end of the 1990s the first devices appeared on the market using a new wireless local area network technology that is commonly referred to today as Wireless LAN or Wi-Fi. Wi-Fi is specified by the IEEE in the 802.11 standard. It is very similar to the 802.3 fixed line Ethernet standard and reuses all protocol layers down to layer 2, as shown in Figure 2.28. The major difference between the two protocols is on layer 1, where the fixed line medium access has been replaced with several wireless variants. Furthermore, some additional management features were specified that address the specific needs of wireless transmissions that do not exist in fixed line networks, such as network announcements, automatic packet retransmission, authentication procedures and encryption. Over the years, several physical layer standards were added to increase transmission speeds and to introduce additional features. Devices are usually backwards-compatible and support all previous standards to enable newer and older devices to communicate with each other.

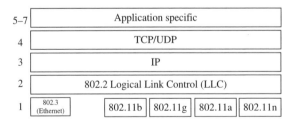

Figure 2.28   The 802.11 protocol stack. (Reproduced from *Communication Systems for the Mobile Information Society*, Martin Sauter, 2006, John Wiley and Sons.)

Initially, Wi-Fi was not very popular or widely known as network interface cards were expensive and transmission speeds ranged between 1 and 2 Mbit/s. Things changed significantly with the introduction of 802.11b, which specified a physical layer for transmission speeds of up to 11 Mbit/s. Network interface cards became cheaper and devices appeared that could be connected to PCs and notebooks over the new high-speed USB (Universal Serial Bus) interface. Prices fell significantly and Intel decided to include Wi-Fi capabilities in their 'Centrino' notebook chipsets. At the same time, the growing popularity of high-speed DSL and TV cable Internet connectivity made wireless networking more interesting to consumers, since the telephone or TV outlet was and still is often not close to where a PC or notebook is located. Wi-Fi was the ideal solution to this problem and Wi-Fi access points were soon integrated into DSL and cable modems. Likewise, Internet access in public places such as cafes, hotels, airports and so on became popular, again enabled by Wi-Fi and cheap high-speed Internet access at the other end of the wireless connection via DSL. Today, Wi-Fi has become ubiquitous in notebooks and many other mobile and portable devices such as game consoles, mobile phones, Internet tablets and Mobile Internet Devices (MID).

Over time, two additional physical layer specifications were added to further increase transmission speeds. The 802.11g standard increased data transfer speeds to up to 54 Mbit/s on the air interface, and the recent 802.11n standard has the potential for up to 300 Mbit/s. It should be noted at this point that these speeds are only theoretical and not measured on the air interface. In practice, protocol overhead reduces the achievable speeds at the application layer to about half those values. This is discussed in more detail in the following sections. Standard 802.11a is another Wi-Fi air interface variant, but has never gained much popularity because it does not use the same standard frequency band as the other 802.11 variants.

The remainder of this chapter is structured as follows: as a first step, the Wi-Fi network infrastructure model is discussed. This is followed by an introduction to the different physical layers and their properties. Like other wireless networking technologies, the network needs to be managed and organized and basic management procedures are discussed next. Wi-Fi security is a very important topic and, as initially encryption algorithms were found to be insecure, it is worth taking a look at this topic and discussing how security was improved over time. Due to the tremendous popularity of Wi-Fi and the growing use of the technology for real-time applications such as VoIP and video streaming, quality of service is becoming an important topic. As a consequence, this chapter then discusses the QoS extension of the Wi-Fi standard and how it can improve reliability for such applications.

2.5.2 Network Architecture

2.5.2.1 The Wireless Network in a Box

Unlike the network technologies described before, Wi-Fi is foremost a local area networking technology and most Wi-Fi networks are deployed as a 'network in a box' as shown in Figure 2.29. In a typical home network, the Wi-Fi network bridges the final meters between the DSL modem and the devices using the fixed line Internet connection. The Wireless LAN Access Point (AP) is usually combined with the DSL modem and also

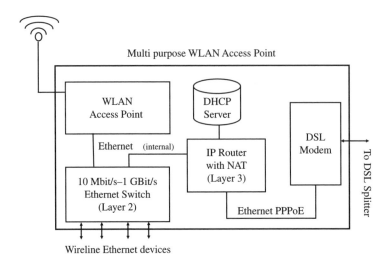

Figure 2.29   A DSL router with a wireless LAN interface. (Reproduced from *Communication Systems for the Mobile Information Society*, Martin Sauter, 2006, John Wiley and Sons.)

serves as the DHCP server, which provides network configuration parameters such as the IP address to notebooks and other wireless devices when they connect to the network. Most multipurpose DSL or cable routers also have a built-in Ethernet switch with several ports to connect PCs and other devices with a twisted-pair Ethernet cable.

2.5.2.2 Network Address Translation

Multipurpose routers such as the WLAN access point can connect many wireless and wired clients to the network; they usually also include NAT functionality, that separates the local network from the DSL connection and translates IP addresses between the LAN and the WAN. This translation is required because the Internet service provider's network usually only assigns a single IP address per DSL connection. By using NAT, local IP addresses and TCP or UDP port numbers are mapped to the external IP address and the same or different TCP or UDP port numbers. This allows all local devices to communicate with servers on the Internet simultaneously via separate connections. From an external network point of view, all devices use the same IP address. While this works well for many applications, there are some for which this translation creates a problem, as incoming packets are discarded if they do not belong to a mapping that was created by an outgoing packet first. This can be solved by configuring static mappings, which forward incoming packets for a server (e.g. a Web server) to a specific internal IP address.

Static mappings, however, are only useful for servers that always use the same TCP or UDP port numbers. SIP, however, which is the dominant protocol for VoIP applications, uses dynamic port numbers. In addition, SIP applications use an IP address they can query from the local protocol stack and include it in application layer messages. As the local network stack is not aware of the external IP address assigned by the DSL network, the wrong IP address is used by the SIP client. Consequently, there are some applications that require more sophisticated solutions for traversing a NAT than static port mapping. Details of these solutions are discussed in Chapter 4.

2.5.2.3 Larger Wi-Fi Networks

Several access points can be used to extend the coverage area of a wireless network. In order that the Wi-Fi network provides a high throughput in each cell, each access point should use a different frequency. All access points broadcast the same network id, referred to as the SSID (Service Set ID) and wireless devices dynamically select which access point to connect to based on signal conditions. Mobile devices can change their association and select a different access point without losing their IP address. For this purpose, the Wireless LAN adapter continuously scans the supported frequency bands to see if it can detect access points with a known SSID and then decides if it would be beneficial to change the association.

Today, most Wi-Fi networks use the Industrial, Scientific and Medical (ISM) frequencies in the 2.4 GHz band. As the band has become very popular, it is becoming more and more crowded. Some 802.11n devices are therefore now also supporting an additional frequency band in the 5 GHz range.

All access points in the same network must be connected to be able to exchange data packets between devices being served by different access points and to have a single gateway to an external network (e.g. the Internet). One possibility is to connect all wireless access points via Ethernet cables to a common backhaul network infrastructure. In home environments, however, this is usually not possible so a better alternative in most cases is therefore to connect them wirelessly. The 802.11e specification contains an extension to the standard referred to as the Wireless Distribution System (WDS). WDS enables an access point to act as a standard access point for client devices and to transmit packets to a neighboring access point also via the air interface. A data packet of a device associated with an access point without a fixed line backhaul connection is therefore transmitted once over the air interface between two WDS access points and then once again from the access point to the client device. As access points usually have only one transmitter, all access points of a WDS have to operate on the same channel. This means that, in practice, devices communicating over one access point create interference for devices using another access point, which further reduces the overall throughput of the network. In home environments, this is often acceptable, as many applications work well even if only a fraction of the total bandwidth is available. Other applications such as high-definition video streaming, however, quickly run into a bandwidth bottleneck if WDS is used or if several geographically overlapping Wi-Fi networks with high traffic loads are operated on the same channel.

2.5.2.4 Campus Networks and Municipal Wi-Fi

Campus and municipal Wi-Fi networks also often use a wireless backhaul channel between base stations. WDS is not used in such networks, however, as it is only for scenarios with only a handful of access points. Instead, proprietary protocol extensions come into play or, in some cases, the access point has two radio modules, one being used to communicate with client devices, and the other one, on a different frequency, to communicate with neighboring access points.

While Wi-Fi has become very popular to cover larger campus areas such as universities, companies, hotels, airports, harbors, and so on, the initial hype around covering even larger areas such as whole cities using Wi-Fi died down once first networks showed the difficulties of using a technology that was designed for short distances and in-house use in outdoor environments over larger distances.

- In outdoor environments, reflections and multipath fading are much more extreme than those Wi-Fi has been designed for. This significantly reduces transmission speeds.
- Wi-Fi is limited to very low transmission power, usually 100 mW or even less, as it is operated in a nonlicensed band. The area that can be covered is therefore only very small. In comparison, cellular base stations usually transmit with 20 or 30 W per sector.
- Due to the limited range, the number of access points required to cover an entire city is immense. To cover an area of 7.5 × 7.5 miles, over 1300 access points are required [36]. The same area is easily covered by less than 20 cellular base stations. Deploying such a high number of access points is both costly and requires significant effort to maintain the network due to the number of network nodes.

2.5.3 The Air Interface – from 802.11b to 802.11n

Over the last 10 years, the IEEE has specified a number of enhancements, also referred to amendments, to the original 802.11 air interface. A number of amendments were made to increase transmission speeds. All changes, however, have been specified in a backwards-compatible manner, which means all devices can communicate with all networks, no matter which version of the standard they support.

2.5.3.1 802.11b – the Breakthrough

The breakthrough for Wi-Fi came after Wi-Fi chips became reasonably affordable and data rates became sufficient for the majority of applications with the 802.11b standard. Prior to the 11b amendment, data transmission rates were limited to 2 Mbit/s on the air interface, or about 1 Mbit/s at the application layer under ideal radio conditions with a channel bandwidth of 25 MHz. The 802.11b standard increased data rates on the air interface to up to 11 Mbit/s and about 7 Mbit/s at the application layer. At the time, this was more than sufficient for DSL and TV cable connections to the Internet, which were mostly in the range between 1 and 2 Mbit/s.

The air interface of both the initial 802.11 standard and the 802.11b amendment is based on a modulation scheme referred to as Direct Sequence Spread Spectrum (DSSS). In DSSS, each data bit is encoded in several chips, which are then transmitted over the air interface. This is similar to the spreading used in UMTS. Wi-Fi, however, always uses 11 chips per bit and only uses two-chip sequences, one for a '0' bit and another one for a '1' bit. This means that Wi-Fi only uses spreading to improve the robustness of the transmission but not for multiple access as in UMTS, which uses separate sequences for each client device.

In the slowest but most robust data transmission mode with 1 Mbit/s, Differential Binary Phase Shift Keying (DBPSK) modulation is used, which encodes one chip per transmission step (symbol). The 2 Mbit/s transmission mode uses Quadrature Phase Shift Keying (QPSK) modulation to transfer two chips per symbol.

To further increase data transfer rates, it was decided to reduce the amount of redundancy. Instead of encoding one bit into 11 chips, the High Rate DSSS (HR-DSSS) physical layer introduced with 802.11b directly translates blocks of 8 bits into different chip sequences which are then transmitted over the air interface. This removes most of the redundancy, which reduces the reliability in less than ideal signal conditions. The amendment therefore also specifies a 5.5 Mbit/s data transfer mode.

To remain backwards-compatible to the original standard, the header of each data frame is transmitted with the original 1 Mbit/s DSSS modulation and Differential Binary Phase Shift Keying. Another reason for using this robust modulation and coding for all headers is that even the most distant devices can decode the headers of all frames and can thus decide if they have to receive and decode the rest of the frame.

The following list gives an overview of 802.11b air interface parameters, which are interesting to compare with those given for LTE in Section 2.3.3 and WiMAX in Section 2.4.4:

- Bandwidth per channel – 20 MHz.
- Frame sizes – 4 – 4095 bytes. Due to IP layer length limits, frames usually do not exceed 1500 bytes.
- Frame transmission time – depends on the modulation and coding used and the size of the frame. A frame with a payload of 1500 bytes requires a transmission time of 12 ms if sent with a speed of 1 Mbit/s. When signal conditions are good and the 11 Mbit/s HR-DSSS modulation is used, the same data packet is transmitted in only 1.1 ms. In addition to those times, each frame is acknowledged by a short ACK frame which, together with the gap between the frames, slightly increases the overall transmission time.
- Retransmissions – when a frame has not been received correctly, it is automatically repeated. Due to the decentralized medium access scheme, which is described in more detail below, a random timer is started before a retransmission occurs. In practice, it can be observed that a frame is usually retransmitted in around 0.5 ms. It is also quite common that a faulty packet is retransmitted more than once since some Wi-Fi implementations do not lower the transmission speed immediately to increase redundancy [37]. Under extreme circumstances in which the selected modulation and coding does not reflect the current signal conditions, more than five retransmissions can be observed before the frame is finally received correctly.

2.5.3.2 802.11g – the Mainstream

The 802.11g amendment to the Wi-Fi standard made a radical break in terms of modulation and coding as the spreading approach was replaced by OFDM.

Orthogonal Frequency Division Multiplexing is also used by LTE and WiMAX. For a basic introduction to OFDM, see Section 2.3.3. It is interesting to note that Wi-Fi was the first popular wireless technology to introduce OFDM. On the air interface, the maximum transmission speed specified is 54 Mbit/s, while at the application layer the highest achievable throughput is around 20–24 Mbit/s. This is, in most cases, sufficient for ADSL2+ and advanced cable modem connections.

To offer the best possible throughput for all signal conditions, 802.11g specifies a number of different modulation and coding modes that result in data transmission speeds between 6 and 54 Mbit/s, as shown in Table 2.3. To reach a transmission speed of 54 Mbit/s, 64QAM is used, which encodes 6 bits per transmission step. To be able to correct transmission errors to a certain degree, redundancy is added and a coding rate of 3/4 is used, that is 1 extra bit is inserted for every 3 user data bits. As Wi-Fi uses 48 OFDM subchannels, 288 bits are transmitted per symbol. When the coding overhead is removed, 216 bits remain for user data.

Table 2.3 The standard 802.11g modulation and coding modes.

Speed (Mbit/s)	Modulation and coding	Coded bits per subcarrier	Coded bits in 48 subcarriers per symbol	User data bits per symbol
6	BPSK, $R = 1/2$	1	48	24
9	BPSK, $R = 3/4$	1	48	36
12	QPSK, $R = 1/2$	2	96	48
18	QPSK, $R = 3/4$	2	96	72
24	16QAM, $R = 1/2$	4	192	96
36	16QAM, $R = 3/4$	4	192	144
48	64QAM, $R = 2/3$	6	288	192
54	64QAM, $R = 3/4$	6	288	216

The slowest transmission mode of 6 Mbit/s is foreseen for harsh signal conditions. Here, Binary Phase Shift Keying (BPSK) modulation is used, which encodes 1 bit per symbol. In addition, the much more robust 1/2 coding is used, which inserts one error correction bit for each user data bit (i.e. 50% overhead). In total, only 48 bits are transmitted using 48 subchannels, out of which only 24 bits are user data.

It is interesting to compare the 48 subchannels used by 802.11g in a 20 MHz band to the 1201 subchannels used by LTE in a similar bandwidth (cf. Table 2.1), as it reveals a lot about the different designs of the two systems. As LTE uses an order of a magnitude more subchannels, each symbol can be transmitted for a much longer time to counter the negative effects of long delay spreads that can appear when the signal travels over larger distances. Wi-Fi on the other hand does not require such equalization, as it is designed for short-range use where delay spread is not as pronounced. It thus uses fewer but broader channels which simplifies system design.

The standard 802.11g has been designed in a fully backwards-compatible manner. This means that older 802.11b devices can co-exist with newer devices in the same network.

Furthermore, 802.11g devices can also communicate with older 802.11b access points. This done as follows:

- Beacon frames, which broadcast system information, are modulated and encoded using the 802.11b standard.
- All frame headers, even those of 802.11g frames, are always modulated and encoded using the 802.11b standard. This means that even old devices can receive the beginning of 802.11g frames and see that they are not the recipient. Consequently, they ignore the rest of the frame, which they would not be able to decode anyway.
- When 802.11g devices detect 802.11b devices in the network, they automatically activate a protection mode and transmit short 802.11b-modulated Ready To Send (RTS) frames, which reserve the air interface for the time required to send the 802.11g-encoded frame. Devices using 802.11b decode the RTS frames and do not attempt to transmit or receive data for time specified in the RTS frame.

As these measures reduce performance, even if no 802.11b devices are in the network, most access points and client devices can be set into an 802.11g only mode.

2.5.3.3 802.11a – the Forgotten Standard

Most of the amendments made by 802.11g to the standard were already included in the earlier 802.11a amendment. The standard 802.11a, however, never became popular, since it was specified for the 5 GHz band. Therefore, it is not backwards-compatible with 802.11b, which exclusively uses the 2.4 GHz band. Devices supporting 802.11a, therefore, had to have two transceivers, one for the 2.4 GHz band and one for the 5 GHz band, which made them more expensive than single transceiver devices. As a result, there were only a few access points and devices available on the market that supported the standard.

2.5.3.4 802.11n – Breaking the Speed Barrier

The latest amendment to the standard is 802.11n, which can potentially raise data transmission speeds by an order of a magnitude compared with the 802.11g standard, if devices that communicate with each other implement all of the options of the standard. In practice, it can be observed today that 802.11n devices are capable of speeds between 100 and 150 Mbit/s at the application layer under good signal conditions. As technology progresses, it can be expected that still higher data rates will be reached with more sensitive receivers and better noise cancellation techniques.

The IEEE 802.11n working group has specified a number of enhancements that all need to be implemented by a device to reach the transmission speeds quoted above:

- Channel bundling –two 20 MHz channels can be bundled to form a 40 MHz channel. This measure alone can more than double transmission speeds, as the guard band that is normally unused between two 20 MHz channels can be used for data transmission as well.

- Support of 2.4 and 5 GHz – since there are only three nonoverlapping channels available in the 2.4 GHz band, it becomes very unlikely that a 40 MHz channel can be used in this band without interfering with other Wi-Fi networks in the same area. As a consequence, the 802.11n standard supports both the 2.4 GHz and the 5 GHz band. In the higher band, up to nine nonoverlapping 40 MHz channels are available.
- Shorter guard time – in many environments, the guard time required to prevent the transmission of a symbol interfering with the next due refraction can be lowered from 800 to 400 ns. This reduces the symbol transmission time from 4 to 3.6 µs.
- New coding schemes – for excellent signal conditions, a 5/6 coding is specified which only inserts one error detection and correction bit for every five user data bits.
- MIMO – like LTE and WiMAX, 802.11n introduces the use of MIMO for Wi-Fi. More details on MIMO can be found in the section on LTE. While LTE and WiMAX are likely to use 2×2 MIMO with two antennas at each side, more antennas are used by Wi-Fi. Dual frequency (2.4 and 5 GHz) capable access points and other devices can have six antennas [38] or even more.
- Frame aggregation – in the default transmission mode, each transmitted frame has to be immediately acknowledged by the receiver. For transmissions of large chunks of data (e.g. a file transfer to or from a server), 802.11n aggregates several frames together. The receiver then only returns a single acknowledgement once all aggregated frames have been received.

In addition to these speed enhancements, the following additional features are also part of 802.11n:

- Options to save cost – since implementing all options described above is costly in terms of hardware and power requirements, most of them are optional. When a device connects to a network for the first time, the access point and the device exchange their capability information and then only use the options that they both support. This way, it is possible to transfer data with some devices with a very high rate, while other devices such as mobile phones, Internet tablets and other small devices that do not require the full data transmission speeds offered by channel bundling, MIMO and so on, operate in a single band and without MIMO to conserve battery power and to reduce hardware costs.
- MIMO power-save mode – if a MIMO-capable device has only a small amount of data to transfer, it can agree with the access point to switch off MIMO to reduce power consumption.
- MIMO beamforming – this option uses several antennas to transmit the same data stream and direct the transmission towards a device as already discussed in the section on WiMAX. This does not increase theoretical transmission speeds beyond the speed possible with a single antenna, but increases the range and the practical throughput for distant devices compared with a standard single stream transmission.
- 5 GHz backwards-compatibility – 802.11n is backwards-compatible to 802.11a. In practice, however, there are only few 802.11a devices that will benefit from this since, by the time 5 GHz-capable 802.11n access points are widely available, it is expected that 5 GHz-capable 802.11n compliant client devices will be as well.

- 2.4 GHz backwards-compatibility – 802.11n is backwards-compatible in the 2.4 GHz band to all previous Wi-Fi standards. This means that 802.11n devices can be operated together with 802.11g and 802.11b devices in the same network. Only 802.11n devices, however, benefit from new features such as MIMO, channel bundling, and so on. As in 802.11g, new devices automatically react to older devices joining the network and start using CTS (Clear To Send) frames before transmitting 802.11n frames that cannot be detected by older devices.

- Overlapping BSS protection – the standard also ensures that channel bundling has no negative effects on other Wi-Fi networks operating in the same area on one of the two 20 MHz channels. For this purpose, access points automatically scan their bands for beacon frames of other access points. If other beacon frames are detected, the use of two channels is discontinued immediately until both bands are clear again. This is required, as older devices cannot properly detect partly overlapping double channel networks and therefore cannot refrain from transmitting frames on the same channel while it is used by the other networks. In practice, it can be observed that some access points offer to deactivate this protection method.

- Greenfield mode – many access point vendors also implement a greenfield mode with no backwards-compatibility and no scans for neighboring networks. This slightly increases performance at the expense of older devices no longer being able to join the network and will potentially cause interference with neighboring networks.

- Quality of service – 802.11n devices should support QoS measures such as giving preference to frames carrying real-time data, as introduced with 802.11e, which is discussed in more detail below.

2.5.4 Air Interface and Resource Management

2.5.4.1 Medium Access

The major difference between Wi-Fi and the cellular wireless systems described earlier in this chapter is the way devices use the air interface. While LTE and WiMAX networks strictly control access to the network to stay in control over quality of service and to prevent network overload, Wi-Fi uses a random medium access scheme which is referred to as the Distributed Coordination Function (DCF).

With DCF, access points do not assign timeslots or transmission opportunities. Instead, client devices autonomously listen to the air interface and transmit frames waiting in their output queue once they detect that no other device is currently transmitting a frame. To avoid simultaneous transmission attempts, each device uses a random backoff time. If after this random time the air interface is still unused, they transmit their frame. While highly unlikely, it is still possible, however, for two devices to start transmitting at the same time. In this case, both frames are lost and both devices have to retransmit their frames. For retransmissions, the time frame for the backoff increases to make it even less likely that devices will interfere with each other a second time.

Each frame also contains a field that informs all other devices of the duration of the transmission. This Network Allocation Vector (NAV) is analyzed by all devices and can be used to switch off the transceiver while the air interface is in use.

To ensure proper delivery of frames, the receiver has to acknowledge the proper reception of each frame, unless frame aggregation is used. Frames are confirmed by immediately returning an ACK frame. The time between a data frame and an acknowledgement frame is shorter than the shortest possible backoff time between two standard frames. This ensures that the ACK frame is always sent before any other device has a chance to send a new data frame.

Another major difference between Wi-Fi and cellular systems is that no logical channels are used, as in LTE, and that a single frame from the access point only contains data for a single device (cf. Figure 2.26). Furthermore, management messages between the access point and a client device are sent in the same way as user data frames. Only the header of the frame marks them as management frames and Wi-Fi chips treat such packets internally instead of forwarding them to higher layers of the protocol stack. This makes the air interface very simple but much less efficient than the air interface of cellular systems.

Figure 2.30 shows how data is transmitted in practice. For each data frame, an acknowledgement is sent by the receiver after a short waiting time (the Short Interframe Space or SIFS), which is required to allow the receiver to decode the frame to check its integrity and for the transmitter to switch back into receive mode. Afterwards, the transmitter has to wait for a random time (the DCF Interframe Space or DIFS) before it can send the next frame. This gives other devices the chance to send their frames in case their random timer was initialized with a lower value. The lower part of Figure 2.30 shows how data is transmitted when frame aggregation is used and only a single ACK frame is sent to acknowledge reception.

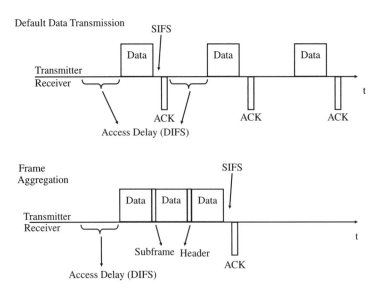

Figure 2.30 Data transmission in a Wi-Fi network.

2.5.4.2 Access Point Centric Operation

The default Wi-Fi operating mode is access point-centric. This means that client devices only communicate with the access point, even if they want to exchange data with each other. This works well for most applications, since the majority of data is exchanged between a wireless device and a server on the Internet. As devices at home or in the office become more and more connected, however, this quickly becomes an issue, as data being streamed from a notebook to a TV screen needs to traverse the air interface twice if both are wirelessly connected. The available bandwidth is thus cut in half. While the Wi-Fi standard also supports an ad-hoc mode in which no access point is required, this mode is not widely used in practice, as most home and office networks require an access point to connect the local network to the Internet.

To improve efficiency when two local wireless devices communicate with each other, the 802.11e amendment introduces the Direct Link Protocol (DLP) that enables two wireless devices to communicate directly with each other. A DLP session is initiated by the two devices exchanging DLP management frames via the access point in the network. If both devices are DLP capable they start to communicate directly with each other once the DLP negotiation is successful. In practice, however, only few devices currently support DLP.

2.5.4.3 An Example Frame

Figure 2.31 shows a typical Wi-Fi user data frame when data was sent from a client device to an access point. The data session to which this frame belongs has been traced with Wireshark (www.wireshark.org/). As Wireshark is freely available at no cost, it is an ideal tool for obtaining hands-on experience with the technology [39]. The upper part of the figure shows a number of frames that have been received by the trace software (frames 778–782). Frame 778, for example is acknowledged by the recipient with frame 779. The trace also shows that the Clear To Send protection frames are sent before a data frame (e.g. frame 780). This indicates that older 802.11b devices are used in an 802.11g network and that the legacy protection mode has been enabled.

The main part of the window shows the header part of frame 778. The frame control field indicates that the frame 'type' is a data frame; that is, it carries user data and not a Wi-Fi control message. Other important pieces of radio layer information contained in the header are the flags that indicate if the frame is a retransmission of a previous frame which was not correctly acknowledged, whether the data part of the frame is encrypted (protected), whether the originator of the frame intends to enter sleep mode after transmitting the frame and the Network Allocation Vector (duration) for the frame.

Unlike fixed line Ethernet frames which only contain the MAC hardware address of the source and the destination of the frame, a Wi-Fi frame additionally contains the MAC hardware address of the Wi-Fi access point. This is necessary, as several independent Wi-Fi networks can be operated in the same area. As an access point receives all frames of all networks, it relies on this information to process only frames of its own clients. At the end of the MAC header, a TKIP (Temporal Key Integrity Protocol) vector is included, which is used as an input parameter for the encryption and decryption algorithm, as discussed below.

Figure 2.31 A typical Wi-Fi user data frame. (Reproduced from *Wireshark*, by courtesy of Gerald Combs, USA.)

2.5.4.4 Sleep Mode

With the increased use of mobile devices it is important to be as power-efficient as possible, especially for battery-driven Wi-Fi devices such as Internet tablets and smartphones. Many of these devices are continuously connected to the network while it is in reach but do not transmit data during most of that time. However, they must be reachable by the network for incoming phone calls, instant messages, and so on. A good balance therefore had to be found between the times during which the receiver is powered down and the time during which the device is reachable. Over the years, a number of different power-saving features have been standardized, but it is still the original sleep mode that is used today by battery-driven devices to reduce energy consumption. This sleep mode works as follows:

• When a device first connects to the network, it agrees the duration of a sleep period with the access point. The access point then assigns the device a bit in the Traffic Indication Map (TIM), which is broadcast in beacon frames. This bit is later set to 1 by the access point whenever frames have been buffered while the device was in sleep mode.
• When a device wants to activate the sleep mode, it sends an empty data frame and sets the Power Management (PWR MGT) bit in the frame header to 1. The access point acknowledges the frame and then buffers all frames that are received for the device.

- After the agreed sleep period has elapsed, the client device's receiver is switched on to receive a beacon frame. If the bit assigned to the device is still 0, no data frames are buffered and the device goes back into sleep mode. If, however, the bit is set to 1, the device usually returns to the fully active mode, powers on the transmitter and sends a request to the access point to forward the buffered frames.
- If at least one client device is in power-save mode, multicast and broadcast are buffered as well.

In practice, it can be observed that battery-driven devices enter sleep mode quickly after all data in the output buffer has been sent. A Nokia N95 smartphone, for example, enters sleep mode after no frame has been sent or received for 100 ms.

2.5.5 Basic Procedures

As in other wireless systems, Wi-Fi devices need to perform a number of management steps before access to the network is granted. This process is a similar but much simpler than the processes shown for LTE in Figure 2.20 and for WiMAX in Figure 2.27. In the first step, the client device scans all possible channels in the 2.4 GHz band and the 5 GHz band (if supported) to detect beacon frames of nearby access points. If beacon frames are received that contain a known SSID (i.e. the name of the network), the device then proceeds to the next step and performs a pseudo-authentication with the network. In practice, this step is only maintained for backwards-compatibility and is no longer used for authentication purposes as the concept was found to be flawed. This is discussed in more detail in Section 2.5.5. Afterwards, the device sends an association management frame to request the access point to accept it as its client device and to inform the access point of its capabilities, such as supported modulation schemes and supported authentication and encryption methods. The access point accepts the request with an association acknowledgement frame in which it in turn informs the client device of its capabilities. Depending on the type of authentication and encryption methods used, the device is then granted access to the network immediately or the access point enforces an authentication and ciphering key exchange message flow before access to the network is finally granted.

From the point of view of the access point, admission to the network means that it forwards Ethernet frames to and from the MAC hardware address of the device. It is therefore possible to use any kind of higher-layer protocol in the network. In practice, however, the IP protocol is dominant and other protocols are rarely used anymore.

Once the device is granted full access to the network, it usually requests an IP address from the DHCP server. This process is identical to a DHCP request in a fixed-line Ethernet network and does not contain any Wi-Fi specific elements.

2.5.6 Wi-Fi Security

2.5.6.1 Early Wi-Fi Security

On the security side, Wi-Fi had a difficult start, as the initial Wired Equivalent Privacy (WEP) authentication and encryption scheme proved easy to break in practice. While first attacks required a considerable amount of time and effort, it is now possible to break into a WEP-secured Wi-Fi network within minutes [40]. As a consequence, WEP has been superseded by more modern encryption techniques.

2.5.6.2 Wi-Fi Security in Home Networks Today

When the vulnerabilities of WEP became apparent, both the IEEE and the Wi-Fi Alliance started programs to improve the situation. These parallel activities resulted in the following authentication and encryption features, which are widely used today:

- Wireless Protected Access (WPA) – this authentication and encryption algorithm builds on a draft IEEE security amendment. The WPA personal mode enforces an authentication procedure and a ciphering key exchange immediately after a device has associated with an access point. During the authentication phase, the client device and access point exchange random values which are used in combination with a secret password known on both sides to authenticate each other and to generate encryption keys on both ends. While there is only one secret password that is used with all client devices, the keys that are generated during this procedure are unique to each connection. This means that devices are not able to decode frames destined for other devices, despite using the same secret password. The algorithm used by WPA for authentication and ciphering is referred to as the Temporal Key Integrity Protocol. It is based on the initial algorithm used by WEP and in addition fixes all known weaknesses. At the time, this implementation was preferred over the more thorough approach proposed by the IEEE to speed up market availability. Nevertheless, WPA is still considered to be highly secure. Today, no attacks are known that could break WPA authentication and encryption, given that the password length is sufficient and that the password used cannot be broken with dictionary attacks.
- Wireless Protected Access 2 (WPA2) – this authentication and encryption algorithm conforms to the IEEE 802.11i security amendment and uses AES (Advanced Encryption Standard) for encrypting the data flow. All new devices offer both WPA and WPA2 authentication and encryption. Many access points can be configured to allow both WPA and WPA2 or only WPA2. To inform client devices which authentication and ciphering method they should use, a number of new information elements were added in the beacon frames.

The only known vulnerability of WPA and WPA2 personal mode is that all devices have to use the same secret password. For home networks, this is usually acceptable and also a pragmatic solution as only a few devices and a limited number of trusted people use the network. For Wi-Fi use in corporate environments, however, using a single password creates a security risk as it is much more difficult to keep the password secret.

2.5.6.3 Security for Large Office Networks

For professional use, WPA and WPA2 also have an enterprise mode that uses a standalone authentication server that is not included in the access point. This is necessary, as companies often deploy several access points, which necessitates the storage of authentication information in a central location. To authenticate devices and to be able to revoke network access for a user, individual certificates are used that need to be installed on each device. The public part of the certificate is also stored in the authentication server. When a device associates with an access point, authentication is initiated by the access point just like in personal mode. Instead of verifying the credentials itself, however, it transparently forwards all authentication frames to the authentication server in the network. Several protocols exist for this purpose and one that is often used and certified for WPA and WPA2 is EAP-TLS (Extensible Authentication Protocol–Transport Layer Security) as specified in [41]. Full network access is only given to the client device once the authentication server authorizes the access point to do so and once it supplies the encryption keys required to encrypt the traffic on the air interface. Encryption is then performed using public/private session keys in a similar way as described for WiMAX above.

2.5.6.4 Wi-Fi Security in Public Hotspots

A major security issue that remains to this day is public Wi-Fi hotspot deployments. For easy access to public Wi-Fi hotspots in hotels, airports and other public places, no authentication and encryption is used on the air interface. This allows a number of different attacks of which two of the most common are described below:

If no encryption is used, data frames can easily be intercepted by anyone in range of a public hotspot. While some Web-based communication over the Internet such as online banking is transported via encrypted HTTPS connections, many other applications such as Web mail, VoIP as well as the standard POP3 and SMTP e-mail are often transported without encryption on the application layer. As a consequence, passwords and HTTP cookies can easily be intercepted and used by an attacker either immediately or later on. A possible countermeasure against such attacks is to use software that encrypts all traffic and sends it through a tunnel to a gateway on the Internet.

An equally serious attack that has been reported from various locations is hackers cloning the start page of public hotspot operators and deploying false access points with network names of public operators. The user is redirected to the clone start page when they first access the net. The clone start page is often secured via HTTPS and looks and acts like the real landing page of the targeted operator. Using this method, credit card information can be stolen and used for other purposes without the user being able to detect the fraud. In practice, it is difficult or even impossible to protect users against such attacks, as only the URL (i.e. the Web address) of the landing page potentially reveals such an attack, as it does not belong to the operator.

2.5.7 Quality of Service: 802.11e

When the network load is low, real-time and streaming applications such as VoIP and video streaming work well over Wi-Fi networks. As soon as the network becomes loaded, however, it is necessary to prioritize the data packets of such applications to ensure a steady stream of data. This is not possible with the default DCF approach, as it treats all frames equally. The 802.11e amendment introduces several features to prioritize packets of real-time and streaming applications. In practice, only the Enhanced Distributed Channel Access (EDCA) is likely to gain widespread acceptance, as it is the only 802.11e feature that has been included by the Wi-Fi Alliance in its Wireless Multi-Media (WMM) certification program. Today, many devices already support WMM and it is likely that in the future the majority of devices will support it, as Microsoft's Windows Vista certification program for wireless network adapters requires support for WMM. In principle, WMM works as follows: when a device wants to send a data frame, it is required to wait for a certain time after which it has to start a random timer. Only once this timer has expired can the frame be sent where no other device that also wanted to send a frame selected a smaller random time and started its transmission earlier. WMM extends this method and defines four priority classes: voice, video, background and best effort. For each priority class, WMM specifies the maximum value of the random timer. The smaller the value, the higher the likelihood that a device will win the race for accessing the air interface. In addition, the standard also defines the max-imum time a frame of a certain class is allowed to block the air interface. A voice frame for example is put into the voice priority class and given the highest priority, which translates into the shortest random timer value. As voice packets are small, WMM also restricts the time on the air interface for this class to prevent misuse. The video priority class gets a slightly higher random timer value, which means it is less likely to gain access to the air interface before a voice packet can be sent by another device.

The different priority queues in a device are also useful to prioritize packets of certain applications over others on a single device. VoIP data frames for example should always be sent before data packets in the best effort queue. This leads to the question of how applications can inform the lower protocol layer of the network stack which priority queue to put the data into. One possibility is the 'diffserv' field in the header of an IP packet, which can be set by an application when it opens a connection.

It should be noted at this point that WMM only ensures quality of service on the air interface. Quality of service on the backhaul link to the Internet via an ADSL or cable modem must be ensured by other means. In the uplink direction, the quality of service can be controlled by the access point/DSL modem which can also analyze the 'diffserv' field of IP packets and expedite the transmission of time critical packets over the back-haul interface. In the downlink direction, the access point/DSL modem has no control over the packet order. It is therefore the network side that should ensure the expedited forwarding of time critical packets. In practice, however, this is rarely done. Despite this lack of quality of service control in the downlink direction, most real-time services such as VoIP still work well even under high network load since the downlink capacity is usually much higher than the uplink capacity. Quality of service control is therefore much more important in the uplink direction, which is controlled by the access point/ DSL modem and not the network.

2.5.8 Summary

This chapter has shown that, from a technical point of view, Wi-Fi does not compete with HSPA, LTE or WiMAX, as these are cellular network technologies designed to cover large geographical areas, while Wi-Fi is a local area network technology for covering hotspot areas, homes and offices. From a commercial point of view there is a slight overlap between the two kinds of technologies since Wi-Fi is not only used in homes and offices but also for hotspot coverage in public places such as hotels, train stations, airports, and so on. Here, Wi-Fi directly competes with cellular network coverage for Internet access.

As will be discussed in the next chapter, the small cell sizes of Wi-Fi and high adoption rates are significant advantages of the technology, as many access points can be operated closely alongside each other compared with the distances required for cellular network base stations. The resulting overall bandwidth is at least one to two orders of a magnitude higher than the bandwidths that can be achieved with cellular networks in the same geographical area. They will therefore be a key element of future converged access network architectures that use cellular technology in combination with personal and business Wi-Fi networks that connect wireless devices to the Internet via a DSL or cable modem connections.

References

1. Drangonwave microwave backhaul solutions (2008) http://www.dragonwaveinc.com/products-whitepapers.asp.
2. Donegan, M. (January 2008) T-Mobile busts the backhaul bottleneck, Unstrung, http://www.unstrung.com/document.asp?doc_id=143211.
3. Sauter, M. (March 2008) The dangers of going SIM-less, WirelessMoves, http://mobilesociety.typepad.com/mobile_life/2008/03/the-danger-of-g.html.
4. 3GPP (19 December 2005) Customized applications for mobile network enhanced logic (CAMEL); Service description; Stage 1, 3GPP TS 22.078, http://www.3gpp.org/ftp/Specs/html-info/22078.htm.
5. Freescale semiconductor. 3G radio network controller (2008) http://www.freescale.com/webapp/sps/site/application.jsp?nodeId=02VS0lyW3P1466.
6. Sauter, M. (May 2007) How file sharing of others drains your battery, WirelessMoves, http://mobilesociety.typepad.com/mobile_life/2007/05/how_file_sharin.html.
7. Holma, H. and Toskala, A. (2006) HSDPA/HSUPA for UMTS: High Speed Radio Access for Mobile Communications. John Wiley and Sons, Ltd, Chichester.
8. Sauter, M. (2006) Communication Systems for the Mobile Information Society, Table 3.6. John Wiley and Sons, Ltd, Chichester.
9. 3GPP (25 June 2008) High speed downlink packet access (HSDPA). Overall Description, TS 25.308.
10. Ferrús, R., Alonso, L., Umbert, A., Reves, X., Perez-Romero, J., Casadevall, F. (2005) Cross layer scheduling strategy for UMTS downlink enhancement, *IEEE Radio Communications*, 43 (6), S24–S28.
11. Caponi, L., Chiti, F. and Fantacci, R. (2004) A dynamic rate allocation technique for wireless communication systems. IEEE International Conference on Communications, Vol. 7, 20–24 June 2004.
12. Rohde and Schwarz (July 2007) HSPA+ technology introduction, Application Note 1MA121.
13. 3GPP (2007)Continuous connectivity for packet data users, 3GPP TR 25.903 version 7.0.0.
14. Rao, A.M. (2007) HSPA+: extending the HSPA roadmap, Alcatel Lucent, http://3gamericas.com/PDFs/Lucent_RAO-MBA-Nov14-2007.pdf.
15. Qualcomm (November 2006) System level analysis for HS-PDSCH with higher order modulation, 3GPP TSG-RAN WG1 #47, R1-063415.
16. 3GPP (2006) One tunnel solution for optimisation of packet data traffic, TR 23.809.
17. 3GPP (2007) General packet radio service (GPRS); service description; Stage, TS 23.060 v.7.6.0.
18. Lescuyer, P. and Lucidarme, T. (2008) Evolved Packet System (EPS). John Wiley and Sons, Ltd, Chichester.
19. 3GPP (2008) Evolved universal terrestrial radio access (E-UTRA) and evolved universal terrestrial radio access network (E-UTRAN); Overall description; Stage 2, 3GPP TS 36.300 version 8.4.0.
20. 3GPP (2008) Evolved universal terrestrial radio access (E-UTRA); physical channels and modulation, TS 36.211, version 8.2.0.

21. Sauter, M. (March 2007) Why IPv6 will be good for mobile battery life, WirelessMoves.com, http://mobilesociety.typepad.com/mobile_life/2008/03/why-ipv6-will-b.html.

22. National Instruments (2006) Addressing the test challenges of MIMO communications systems, http://zone.ni.com/devzone/cda/tut/p/id/5689.

23. Nortel (2008) Nortel MIMO technology provides up to double the existing access network capacity to serve more customers at less cost. *Embedded Technology Journal* 25 October, http://www.embeddedtechjournal.com/news_2006/10/20061025_04.htm.

24. IETF (June 2001) Robust header compression (ROHC): framework and four profiles: RTP, UDP, ESP, and uncompressed, RFC3095, http://www.ietf.org/rfc/rfc3095.txt.

25. The WiMAX Forum (2008) The homepage of the WiMAX forum, http://www.wimaxforum.org.

26. Verisign. (2006) VeriSign custom device certificate service and certificate key ceremony service, http://www.verisign.com/authentication/enterprise-authentication/device-certificate-services/index.html.

27. Perkins, A. (ed.). (1996) IP mobility suport, RFC 2002, IETF, http://www.ietf.org/rfc/rfc2002.txt.

28. The WiMAX Forum. (March 2006) WiMAX forum certifies additional profile for fixed WiMAX products, Press Release.

29. 3GPP (April 2008) Beyond 3G: 'LTE-Advanced' Workshop, Shenzhen, China, http://www.3gpp.org/news/2008_04_LTE_A.htm.

30. 3GPP (2008) Requirements for LTE-Advanced, TS 36.913, http://www.3gpp.org/ftp/specs/html-info/36913.htm.

31. IEEE (2004) IEEE Standard 802.16-2004, Revision of IEEE Std. 802.16-2001, Part 16: Air interface for fixed broadband wireless systems.

32. IEEE (2005) IEEE standard for local and metropolitan area networks – part 16, amendment 2 and corrigendum 1, 802.16-2005.

33. Intel (2006) Intel Centrino mobile technology reference guide for WiMAX networks, Table 2.

34. IEEE (21 April 2008) The Draft IEEE 802.16m system description document, IEEE 802.16m-08/003r1.

35. IEEE (June 2007) Part 16: Air interface for fixed and mobile broadband wireless access systems, , 80216j-06_026r4.

36. Sauter, M. (April 2006) Muni Wifi: How many access points are necessary to cover a city?, http://mobilesociety.typepad.com/mobile_life/2006/04/muni_wifi_how_m.html.

37. Sauter, M. (May 2008) Sniffing Wifi packets and exploring retransmission behavoir, http://mobilesociety.typepad.com/mobile_life/2008/05/sniffing-wifi-p.html.

38. Higgins, T. (October 2007) Slideshow: Linksys WRT600N dual-band wireless-N gigabit router, SmallNetBuilder, http://www.smallnetbuilder.com/content/view/30204/187/1/5.

39. Sauter, M. (April 2008) Wifi tracing with an eeePC, http://mobilesociety.typepad.com/mobile_life/2008/04/wifi-tracing-wi.html.

40. Tews, E., Pychkine, A. and Weinmann, R.-P. (April 2007) http://www.cdc.informatik.tu-darmstadt.de/aircrack-ptw/.

41. Simon, D., Aboba, B. and Hurst, R. (March 2008) The EAP-TLS authentication protocol, RFC 5216, http://tools.ietf.org/html/rfc521.

3

Network Capacity and Usage Scenarios

The way in which mobile networks will be used in the future depends on many factors. This chapter discusses the current state of 2G, 3G and B3G networks and how a rise in capacity could affect usage in the future. Since capacity is limited, this chapter also takes a look at how to steer the use of mobile network resources from a financial point of view and if it is still possible to link profitability with how much a user spends per month for using a network.

3.1 Usage in Developed Markets and Emerging Economies

In developed markets, the use of the Internet to communicate is still mainly bound to specific places where DSL, cable or other broadband connections are available. Wi-Fi has become very popular in recent years due to its ability to un-tether users and allow them to move with their devices through their offices and homes. Small portable devices with built-in wireless connectivity have also become very popular. Wi-Fi has thus created a virtual Internet bubble around people. Anssi Vanjoki of Nokia describes this phenomenon, saying that 'broadband Internet is no longer a socket in the wall'. When the majority of people today leave their homes and offices, however, they leave their personal Internet bubble and instantly lose connectivity. Today, 3.5G networks can already fill this void as enough capacity is available for people using converged Wi-Fi/cellular devices. Future converged 3.5G/Wi-Fi devices will automatically detect this change and switch to a cellular B3G network. B3G networks thus become the natural extension of the personal Internet bubble. Over time, connectivity will get more and more seamless as converged devices will learn which applications can use which networks due to the costs associated with the expected network usage. Music downloads are a good example of such behavior. If a Wi-Fi connection is available, converged 3.5G/Wi-Fi devices will automatically use this type of network as it offers ample capacity, high throughput and cheap connectivity. Users, however, will also want to browse their favorite music store's

Beyond 3G – Bringing Networks, Terminals and the Web Together: LTE, WiMAX, IMS, 4G Devices and the Mobile Web 2.0 Martin Sauter © 2009 John Wiley & Sons, Ltd

catalog. Given adequate pricing for cellular data connectivity and sufficient cellular network capacity, the experience will be almost seamless. How much capacity is available with B3G networks will be discussed throughout this chapter. As fixed-line and wireless networks together provide seamless connectivity for people, cellular network capacity is just one parameter in an overall capacity equation which also takes the fixed broadband access made available via Wi-Fi access points into account.

In emerging economies the picture is quite different. Fixed-line telecommunications infrastructure is not very well developed and it is unlikely that fixed-line DSL or cable connections made available to mobile devices via Wi-Fi will be able to significantly reduce the load on cellular networks. Wi-Fi, however, might turn out to be a great solution for creating wireless mesh networks with projects such as the MIT's One Laptop Per Child initiative. Here, individual computers can be connected to the Internet by using other mobile devices to relay data packets from and to the Internet connectivity hub. The Internet connectivity hub can then use either a fixed-line connection or a cellular wireless connection to route data packets to and from the Internet. While less capacity is available, it is also likely that usage of the Internet will be much lower than in developed markets. This is mainly due to devices that have to be much cheaper than those in developed markets to be affordable. This in turn limits screen resolution, processing power and on-device storage capacity. If music cannot be stored on a device, a music download service is not likely to be appealing to users no matter what the cost. Also, it is unlikely that low-cost phones will be capable of supporting bandwidth-intensive applications such as video streaming. Things are different with notebooks and desktop computers, but it is unlikely that within the next decade use of such devices will become widespread.

3.2 How to Control Mobile Usage

As will be shown later in this chapter, capacity in wireless networks is limited. In fact, capacity on all types of networks is limited and to prevent overload, steering mechanisms are required. This is also important from a financial point of view since operators have a certain amount of capacity in their network which they need to sell for a price that is high enough to recover the initial Capital Expenditure (CAPEX) for acquiring licenses and for buying and installing base stations and the infrastructure behind them. A network also creates recurring costs, the Operational Expenditure (OPEX). These consist among other things of rental costs for properties where equipment such as base stations is installed, costs of leasing transmission lines to backhaul traffic, the monthly power bill, staff for maintaining the network, marketing, customer acquisition and support and so on. When wireless networks were mainly voice-centric, the main instrument to control usage was the price per voice minute. Flat rate packages that include unlimited minutes seem to indicate that the pricing per minute is no longer important. This is not the case, however, since prices of such packages are quite high and network operators estimate the average number of voice minutes spent by users with a flat rate package to calculate the amount to charge for such 'unlimited' use. As it is an average, some flat rate users will use more than the average number of minutes, while some use less. Furthermore, the fine print in many contracts still limits the maximum number of minutes per month. 'Flat rates' and 'unlimited' are thus in most cases anything but flat and unlimited.

3.2.1 Per Minute Charging

For mobile Internet access, charging per minute does not usually make sense. While voice calls create a fixed amount of data that has to be transported through the network per minute, the amount of data generated by accessing the Internet depends greatly on the application. A minute of streaming video produces an amount of data which is an order of a magnitude higher than browsing the Web on a small handheld device. Another application that requires even less throughput per minute is mobile e-mail. Devices specifically tailored for this application are usually always connected to the network or re-connect in short intervals to receive incoming messages. The amount of data transferred over time, however, is very low as most of the time the device is just idle, despite being connected to the network. For packet data, it therefore makes more sense to charge for the amount of data that has been transferred, regardless of the time the device was connected to the network.

3.2.2 Volume Charging

The problem with this approach, from the point of view of many mobile operators, is that charging users for the use of a certain data volume per month does not allow billing based on services. In the example above, a minute of video streaming is likely to generate more data than a push e-mail generates over the course of a full month. If pricing for mobile data is based on push e-mail consumption, downloading videos will not be affordable and most of the capacity of wireless networks will be unused. Vice versa, if data tariffs are based purely on video downloading, the price for transferring mobile e-mail will be almost zero.

3.2.3 Split Charging

In recent years, network operators have started to adopt a dual strategy. On the one hand they are now offering services such as push e-mail by offering dedicated and in many cases preconfigured devices together with a service contract that includes access to the network. On the other hand, operators have also started to sell transparent access to the Internet their subscribers can use with notebooks and other devices. Such offers, even though they are often called 'unlimited', usually come with an upper volume limit to ensure network integrity. Prices for such offers with a monthly usage cap of around 5 Gbytes have initially started at around €50 a month in many countries but have declined over time and are now available for around €25–35. In this regard it is also interesting to see differences in price and usage per country. Especially in Europe, there is a huge price difference when comparing offers in different countries, mostly dependent on how much competition is present between the different network operators.

3.2.4 Small-screen Flat Rates

Some mobile network operators have also started to offer application specific 'flat rates' such as unlimited Web and e-mail access from a mobile device. Based on the fact that mobile devices have smaller screen resolutions and limited storage capacity, these offers are made in the hope that overall consumption per month will be much lower than if the network was used with notebooks. Pricing of such offers is in the range €8–10. The issue with such offers is that it is difficult to ensure that users will only use such offers with small devices. Many schemes have been invented to prevent their use with notebooks, but most of these are easy to circumvent by savy users. Some operators are therefore offering different volume bundles of between 10 and 1000 Mbytes per month at different pricing levels. The price per megabyte of lower-volume bundles is usually significantly higher than that of higher-volume bundles. In this way, mobile device-only Internet access can be sold at a higher price and thus with a higher margin.

3.2.5 Strategies to Inform Users When Their Subscribed Data Volume is Used Up

Today, several strategies exist in practice of what to do once a user exceeds the monthly volume limit. In many cases, operators will start charging a certain amount per megabyte when the limit is exceeded. In most cases this is done without notifying the user, which leads to high customer dissatisfaction. Another approach is to monitor usage and terminate the service when the subscriber exceeds the limit. While this has the advantage that there are no nasty surprises in the next invoice, the approach is equally problematic for the user. Some operators have instead started to introduce soft boundaries by tolerating higher use for some time (e.g. 3 months) and informing their customers in various ways. Extra charges will only be applied if the user does not reduce their use in the following months. Yet another approach is to throttle transmission speed (e.g. to 128 kbit/s) once the limit is exceeded. Operators using this scheme then allow users to remove the throttle by paying for an additional data package. This is done via the Web and the throttle is removed once additional data volume has been bought.

Another way to control mobile usage is to use different pricing strategies for prepaid and post-paid users. Most countries only offer competitive data prices to their post-paid subscribers in combination with a minimum service duration of 12–24 months. In many cases this significantly inhibits uptake of data services. Whether this is the desired effect or an unfortunate side effect is up for discussion.

In some countries, some operators have started to also offer mobile Internet access to prepaid customers. Depending on the country, offers range from high-priced mobile device Web access to prices identical to those offered to post-paid customers. Such offers can usually also be terminated on a monthly basis as operators cannot bind prepaid customers for a longer period of time. The advantage of such an approach is that it attracts young people and students. As this is the main user group for Internet services, such offers have the potential to pave the way for mobilizing the Web.

3.2.6 Mobile Internet Access and Prepaid

Prepaid offers are also interesting for tourists and international business travelers since 3G data roaming prices are still excessive. A Wiki on the Internet has more informationon this topic [1]. At this time there are two notable exceptions. The first is mobile operator '3' with subsidiaries in some European and Asian countries as well as their network in Australia. Postpaid subscribers of '3' are allowed to roam into any other network without paying any roaming charges. The second exception is Vodafone Germany, who offer data roaming via prepaid and post-paid SIM cards [2]. While it is much more expensive than '3's offer, the number of countries in which the offer can be used is much higher. From a technical point of view, it should be noted that mobile Internet access while roaming is only slightly more expensive, the extra expense being incurred in backhauling the data to the home network before forwarding it to the Internet.

In the future it is likely that prices for mobile Internet access will continue to decline to levels that are attractive to more users. As discussed in more detail below, many mobile operators are in the process of acquiring fixed-line assets or are reuniting with their fixed-line divisions to offer mobile Internet access together with fixed-line DSL access at home. This will also be an important instrument to offload data traffic from cellular B3G networks to personal Wi-Fi networks at home or the office that can carry large amounts of data at lower cost.

Future international data roaming scenarios are difficult to predict and will probably vary from country to country. If UMTS operators take a similar path as with voice call roaming, it might be that only regulatory intervention will bring about affordable prices for cellular wireless Internet access while roaming. WiMAX network operators on the other hand might have more interest in picking up additional revenue from tourists and business travelers seeking Internet access while traveling abroad since their business model might be aligned closer to the open Internet rather than to current telecommunication pricing structures. Their lead could open the door to more affordable data roaming prices in the future from other operators.

3.3 Measuring Mobile Usage from a Financial Point of View

Today, mobile operators report their Average Revenue Per User (ARPU) as an indication of the profitability and success of their network operation and marketing. This term is adequate for voice-centric and nonfractured markets in which one user has a single SIM card and only uses voice and SMS. When looking at markets with rising nonvoice use of wireless networks, however, ARPU is quickly becoming an irrelevant key figure for a number of reasons.

First, people in many countries have started using several SIM cards because each SIM card offers an advantage over the others. The average revenue per user is now split between two SIM cards. The mobile network is not run less profitably because of this, but the revenue from such a user is now split over two SIM cards. Business users are a good example of split SIM card use: many business travelers today have one SIM card for their mobile phone and a second SIM card for the 3G data card that connects their notebook to the Internet. The ARPU should consequently contain the sum of both. In practice that is difficult to do because there is usually no way of correlating the two SIM cards to a single user, especially if the SIM cards were bought by a company.

Second, MVNOs (Mobile Virtual Network Operators) in some countries have started to offer cheap voice minutes but sell SIM cards without phones. This raises the question of which of the following two ARPUs would be preferable:

- an ARPU of €30 a month generated with a contract requiring an initial €300 subsidy for an expensive phone which is then spread over 24 months;
- an ARPU of €20 a month generated via a prepaid SIM without subsidies.

On paper the first ARPU value sounds more appealing, but it is likely that the operator will make more money with the prepaid SIM despite the lower ARPU figure.

Third, mobile networks offer a wide range of services today, from voice calls to high-speed Internet access. This raises the question of which of the following two customers is more valuable for a network operator:

- a customer who spends €30 a month on voice calls;
- a customer who spends €30 a month on Internet access.

In most cases the voice ARPU is probably more profitable than the data ARPU. However, prices for voice minutes keep declining so in the end the data customer could become more profitable.

As a consequence the ARPU should be replaced by some other, more meaningful key figure adapted to the continuing changes. The following list shows a number of approaches that could be used in the future to better measure mobile use from a financial point of view:

- average revenue for a voice minute, based on all voice minutes sold in the network over the period of a month;
- average revenue per megabyte for mobile services, that is Web surfing and other Internet activities from mobile phones;
- average revenue per megabyte achieved with high-speed Internet access from notebooks;
- SMS and MMS should also be treated in the same manner as, from a price per megabyte point of view, MMS is no longer more expensive to transport over the network than SMS messages.

3.4 Cell Capacity in Downlink

The data rate in downlink of a cell (network to the user) is often referred to as the capacity of a cell. The theoretical maximum of this value is often used by network manufacturers and network operators to demonstrate the capabilities of network technologies and compare them with each other. The capacity is measured in how many bits a system can transmit per Hertz of bandwidth per second (bits per second per Hertz). Table 3.1 shows the peak data rates, channel bandwidth, frequency re-use and spectral efficiencies for current and future cellular wireless technologies under ideal circumstances [3].

The values given in the table represent the theoretical limit of each technology. Typical speeds experienced in practice are much lower and are discussed

Table 3.1 Theoretical peak data rates, channel bandwidths, frequency reuse and spectral efficiency of different wireless network technologies.

Network type	Theoretical peak data rate	Channel bandwidth	Frequency reuse	Spectral efficiency
GSM	14.4 kbit/s	200 kHz	4	0.032
GPRS	171 kbit/s	200 kHz	4	0.07
EDGE	474 kbit/s	200 kHz	4	0.2
Cdma2000	307 kbit/s	1.25 MHz	1	0.25
1xEV-DO Rev.A	3.1 Mbit/s	1.25 MHz	1	2.4
UMTS	2 Mbit/s	5 MHz	1	0.4
HSDPA	14 Mbit/s	5 MHz	1	2.8
HSPA+ (2 × 2 MIMO)	42 Mbit/s	5 MHz	1	8.4
802.16e WiMAX	74.8 Mbit/s	20 MHz	1	3.7
802.16e 2 × 2	160 Mbit/s	20 MHz	1	8.0
802.16e 4 × 4	300 Mbit/s	20 MHz	1	15.0
LTE	100 Mbit/s	20 MHz	1	5
LTE 2 × 2 MIMO	172.8 Mbit/s	20 MHz	1	8.6
LTE 4 × 4 MIMO	326.4 Mbit/s	20 MHz	1	16.3

further below. The table nevertheless demonstrates that newer technologies make better use of the available spectrum when signal conditions are ideal. This is due to the following reasons:

- Higher order modulation – while GSM, for example, uses GMSK (Gaussian Minimum Shift Keying) modulation that encodes one data bit per transmission step, 64QAM is used in LTE and 802.16e under ideal radio conditions to encode 6 data bits per transmission step.
- Reduced coding – wireless Systems usually protect data transmissions by adding error detection and correction bits to the data stream. The more redundancy is added, the more likely it is that the receiver can reconstruct the original data stream in case of a transmission error. In good radio conditions, however, less error coding is required since the likelihood of transmission errors is lower.
- MIMO – multiple input multiple output techniques exploit the fact that radio signals scatter on their way from the transmitter to the receiver. A 2 × 2 MIMO system uses two antennas and signal processing chains at both transmitter and receiver to transmit two independent streams of data over two independent radio paths. Under ideal circumstances the data rate doubles without using additional spectrum. With 4 × 4 MIMO, the data rate increases fourfold.
- Beamforming – this method exploits the fact that cells usually cover large areas and mobile devices are located at different angles. A cell then forms individual radio beams by using several antennas and transfers an individual data stream over each beam. Again no additional spectrum is required to increase the data rate. Currently, however, MIMO seems to be the preferred way of increasing spectrum efficiency.

- Use of larger frequency bands –using larger frequency bands makes data transmission in the cell faster but does not increase efficiency.

In practice, cell capacity is much lower than these theoretical values, which are only applicable under the most ideal circumstances. Achievable cell capacity depends on the following factors:

- Backhaul connection – for cost reasons, some operators prefer to use a lower-capacity backhaul link to the cell than the capacity supported by the cell over the air interface.
- Inter-cell interference – beginning with 3G radio technologies, all cells of a network use the same frequency band for communication. Thus, each cell interferes with its neighboring cells and the more active users a cell has to handle, the more it interferes with neighboring cells. Another approach is to use different frequencies in neighboring cells. In practice, however, this is no longer feasible with frequency bands of 5 MHz or more as an operator usually has no more than two of these bands available.
- Network capabilities and technology – depending on the radio technology, a given amount of bandwidth is used more or less efficiently as described above.
- Terminal capabilities – similar to networks evolving over time, terminals also undergo changes as radio technology improves. As a consequence, a mix of terminals is used in networks with different capabilities. By improving mobile device capabilities such as antenna performance, sensitivity, signal processing, higher order modulation support, maximum number of simultaneous codes in the case of WCDMA systems, processing power and so on, mobile devices are able to better cope with a given radio environment and receive data more quickly. This increases overall capacity of the cell as the network has more opportunities to use higher-order modulation and coding schemes and thus transport more data during a certain timeframe.
- Terminal locations – networks which are designed for the use with devices using antennas installed on rooftops can have a much higher overall throughput per cell. This is because the average reception conditions are much better than in the case of systems that allow mobile devices with built-in antennas that are used indoors and thus experience worse signal conditions. This in turn reduces the capacity of the cell as the time spent sending data to a mobile which experiences less favorable radio conditions cannot be used to send data much more quickly to devices with better reception conditions.
- Frequency band used – wireless networks that use lower frequencies have a much better in-house coverage than those using higher frequencies. This difference can be observed today between GSM networks that use the 900 MHz band and UMTS networks which use the 2100 MHz band in Europe and Asia. When entering a building, UMTS coverage is lost much sooner than 900 MHz GSM coverage. This is why many wireless network operators are interested in getting permission to re-use lower frequency band allocations currently used for 2G systems for their 3.5G and future B3G networks.

Once all these influences are taken into consideration, the overall achievable spectral efficiency of the data transmission in a cell is much lower than the peak values given in Table 3.1.

While new radio interface technologies are specified with ever higher spectral efficien-cies in mind, it has to be taken into account that a higher spectral efficiency for top performance requires a higher signal-to-noise ratio. The signal-to-noise ratio is deter-mined by the Shannon–Hartley capacity equation:

$$C = B * \log 2(1 + \text{SNR})$$

In this equation C represents the channel capacity, B the channel bandwidth in Hertz and SNR the instantaneous linear signal-to-noise ratio. Table 3.2 shows typical spectral efficiency values for a single channel (i.e. excluding the MIMO entries in Table 3.1) and the corresponding required linear and logarithmic SNR.

The following examples of the evolution of UMTS show that an increasing signal-to-noise ratio is required to reach the theoretical top speed of each new step. As a con-sequence this means that the area in which the specified theoretical top speeds are available is shrinking from step to step.

Table 3.2 Required signal-to-noise ratio for different spectral efficiencies.

Spectral efficiency (bit/s/Hz)	Required SNR (dB)	Required SNR (linear)
10	30	1000
5	15	31.6228
2.9	8	6.30957
2	5	3.16228
1	0	1
0.4	−5	0.316228
0.14	−10	0.1
0.04	−15	0.0316228

Example 1: UMTS

UMTS has a theoretical cell capacity of 2 Mbit/s if all users experience perfect conditions and there is no interference from neighboring cells. With a channel bandwidth of 5 MHz this requires a spectral efficiency of 0.4:

Peak spectral efficiency (UMTS) = 2 Mbit/s per 5 MHz = 0.4 (cf. Table 3.1)

According to the Shannon–Hartley maximum theoretical capacity equation, this requires a signal-to-noise ratio of **−5dB** or **0.316228** (linear), as shown in Table 3.2:

Peak channel capacity (UMTS) = 5 MHz * log2 (1 + 0.316228) = 1.9820 Mbit/s

Example 2: HSDPA

With HSDPA the cell capacity is increased to a theoretical peak of 14 Mbit/s. On the physical layer this is done by using a higher-order modulation and reducing the error coding rate. To reach this theoretical cell capacity the following peak spectral efficiency is required:

Peak spectral efficiency (HSDPA) = 14 Mbit/s per 5 MHz = 2.8 (cf. Table 3.1)

According to the Shannon–Hartley maximum theoretical capacity equation, this requires a much higher signal-to-noise ratio compared with UMTS of about **8 dB** or **6.30957** (linear), as shown in Table 3.2:

Peak channel capacity (HSDPA) = 5 MHz * log2 (1 + 6.30957) = 14.3 Mbit/s

---

Example 3: LTE

3GPP's Long Term Evolution raises the bar once again by specifying a transmission mode with a theoretical peak cell capacity of 100 Mbit/s in a 20 MHz channel. In a 5 MHz channel the peak cell capacity would be 25 Mbit/s, which is 6 Mbit/s faster than HSDPA. To be able to reach this theoretical cell capacity, the following peak spectral efficiency would be required:

Peak spectral efficiency (LTE) = 100 Mbit/s per 20 MHz = 5 (cf. Table 3.1)

Again, according to the Hartley maximum theoretical capacity equation this requires once more a much higher signal-to-noise ratio compared with a UMTS of about **15 dB** or **31.6228** (linear), as shown in Table 3.2:

Peak channel capacity (LTE) = 20 MHz * log2 (1 + 31.6228) = 100.55 Mbit/s

---

As a consequence, this means that the cited theoretical peak cell capacity becomes more and more unlikely since in practice all systems experience the same signal conditions. This means that all 3G, 3.5G and B3G systems discussed in this book will have a similar cell capacity for a given signal-to-noise ratio at a certain location [4].

Thus, other means have to be used to increase cell capacity. B3G systems use the following:

- MIMO technology will counter this effect to a certain extent. The effect, however, is limited by the interference of the independent data streams on the same channel as each other and the fact that the number of antennas in mobile devices and also on rooftops cannot be increased beyond a reasonable limit. This is also reflected in Table 3.1 where 4×4 LTE does not have four times the cell capacity compared with an LTE cell that only uses a single channel. Also, base station costs will increase due to the additional cables between the base station and additional antennas required for MIMO. Cables are quite expensive due to their low signal loss properties and thus already today make up a sizable proportion of the overall base station price.

- Advanced receivers in terminals will also be able to improve signal-to-noise ratios at a given location compared with less capable terminals. Current research indicates cell capacity gains between 40 and 100% for HSDPA [5].
- Increasing the channel bandwidth – this is the only parameter that scales linearly; that is, doubling the bandwidth of a channel also doubles the capacity of the cell. While bandwidth scales linearly, there are two limiting factors for this parameter as well. First, power consumption of battery-driven terminals keeps rising as the channel that has to be received and decoded becomes broader. Increasing the bandwidth thus has a detrimental effect on autonomy on a battery charge. Second, bandwidth for wireless communication systems is in very short supply and it will be very difficult to assign bandwidths of 20 MHz per channel or more.

3.5 Current and Future Frequency Bands for Cellular Wireless

Not every frequency band is suitable for wireless communication with mobile devices. While lower frequencies are better for in-house coverage, they increase the antenna size in mobile devices. A good example are FM radios built into mobile phones. FM radio transmits on frequencies around 100 MHz and typical mobile terminals are too small to have an internal antenna for this frequency range. As a consequence the headset cable is used as an antenna and the FM radio only works if the headset is plugged in. Efficient mobile device antennas are hard to design below 700–800 MHz. While lower frequencies offer better in-house coverage, they also propagate much better in free space and thus reduce overall network capacity due to the larger coverage area of a cell. On the upper end of the spectrum, 6 GHz is around the highest frequency that makes sense for cellular communication. At this end, however, in-house penetration is already quite poor. Thus, this frequency band is also not usable for all purposes. The optimal space for cellular communication is therefore the frequency range between 1 and 3 GHz. Here, however, most bands are already occupied.

Table 3.3 lists current and future frequency bands for terrestrial wireless communication in Europe, as described in the European table of frequency allocations [6] and 3GPP TS 25.101 [7]. There are also some frequency bands reserved for IMT-2000 two-way satellite communications to mobile devices, which are, however, excluded from this discussion.

According to Table 3.3, there are around 540 MHz available for cellular wireless communication in the downlink direction and about the same amount of spectrum for the uplink direction. At the date of publication of this book, the IMT-2000 extension band is not yet in use and only small fractions of the BWA band are used for WiMAX. In the IMT-2000 band there are 12 frequency blocks of 5 MHz each available for UMTS. On average there are four operators per country, each using a single band. Consequently only 30% of this band is actively used at the moment. The 1800 MHz band for GSM is

Table 3.3 Current and future bands assigned to be used for cellular communication in Europe

Frequency band (MHz)	Name	Used for or foreseen to be used for:	Total bandwidth for one direction (uplink or downlink)
880–915 (uplink), 925–960 (downlink)	GSM-900, UMTS-900, LTE-900	GSM, UMTS, LTE	35 MHz
1710–1785 (uplink), 1805–1880 (downlink)	GSM-1800, UMTS-900, LTE-900	GSM, (UMTS, LTE)	75 MHz
1920–1980 (uplink), 2110–2170 (downlink)	IMT-2000 band	UMTS UMTS	60 MHz
2500–2570 (uplink), 2620–2690 (downlink)	IMT-2000 extension band (possibly WiMAX)	UMTS, LTE, WiMAX	70 MHz/25 MHz[a]
3400–3800	Broadband Wireless Access (BWA)	WiMAX	275 MHz[b]

[a] Due to UMTS tx/rx separation of 120 MHz, there is a 50 MHz gap between 2570 and 2620. This could potentially be used by another system. Therefore an extra 25 MHz are counted.
[b] WiMAX is operated in this frequency range in TDD mode. Thus bandwidth in uplink and downlink depends on the ratio selected for uplink and downlink capacity by the operator.

also only partly used at the moment. As a consequence, less than a fifth of the spectrum assigned to terrestrial cellular wireless communication is currently in use.

In other parts of the world, the situation is similar but different frequency ranges are used. In the USA, for example, the 900, 1800 and 2100 MHz ranges are not available to cellular wireless communication. Instead frequency ranges in the 850, 1700, 1900 and 2500 MHz band are used. Furthermore, a frequency range in the 700 MHz band has been auctioned off for cellular wireless communication.

In practice it is getting more difficult for mobile terminals to support all frequency bands for all regions. This means that it is getting more difficult to develop radio chips that will work worldwide, which has far-reaching consequences. Already today, it is difficult to find even high-end mobile devices that can be used all over the world, even if local networks use a cellular standard the device was built for. For mobile device manufacturers the situation is equally undesirable as designing individual chips for different parts of the world increases product prices due to a lower economy of scale.

3.6 Cell Capacity in Uplink

This chapter has so far focused on radio aspects in the downlink direction. In the uplink direction, an equal amount of bandwidth is available to transport the increasing data that users send to the network. Until recently, uplink capacity was not in the spotlight of network designers or the public since a mobile user was mainly seen as a consumer of information who sent little data back to the network. As will be discussed in more detail in Chapter 6, this is changing. One the one hand the use of notebooks that are connected via a cellular network to the Internet is on the rise. Mainly used by business travelers, students who usually have high mobility requirements and by the general public in countries where wireless access and DSL compete heavily [8], the uplink is beginning to carry significant traffic. One main contributor is outgoing e-mail with large file attachments, which is getting more and more common as documents are exchanged between colleagues and friends and presentation documents increase in size. On the other hand, increasing uplink capacity is also required for mobile multimedia devices which are no longer just used to consume content but also for uploading pictures, videos and so on to Internet sharing sites, Web pages, community sites and Blogs. Sending pictures and videos of extraordinary events via MMS and e-mail to public broadcasters also requires uplink resources. It is no longer uncommon to see pictures and videos taken with mobile phones on breaking news [9].

It can also be observed that 3G networks are more often used by semi-professional and professional TV and radio broadcasters. Instead of using specialized radio equipment or fixed-line voice connectivity for event reporting, a 3G network is used for transmitting a live stream to a broadcasting studio, which then broadcasts the content via radio, TV or the Internet to a wider audience interested in the event. Figure 3.1 shows the recording and transmission equipment of a semi-professional radio station reporting from an event and broadcasting their information via FM radio. The mobile phone in the lower part of the picture is connected via a cable to the PC and is used by the software to transmit the audio stream to the Internet. In the future these trends will continue to evolve and the increased uplink bandwidth requirements of voice calls transported over IP will additionally increase the amount of data being sent to the network.

Most B3G systems use two separate frequency bands to transmit data in uplink and downlink direction. This is called Frequency Division Duplex. WiMAX is an exception

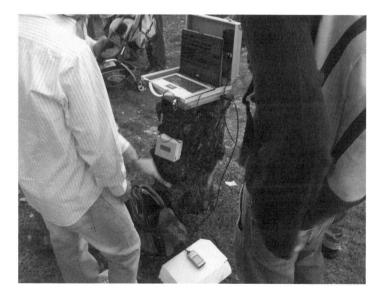

Figure 3.1   Recording equipment of a semi-professional radio station with 3G uplink.

as operating modes (profiles) currently defined to be used in practice are based on a single frequency band to be used for both uplink and downlink. The two transmission directions are separated in time (TDD). In principle uplink data transmission speeds are restricted by the same rules that also apply to the downlink direction. There are, however, a number of uplink-specific limitations to consider for mobile devices:

- Small antennas – while base stations use directional antennas of considerable size to project the available transmission power in certain directions and vertical angles, mobile devices only use small omnidirectional antennas. These antennas are far less efficient than their counterparts at the base station. Transmitting in all directions is necessary since the user can change their location at any time, which means that the direction of the base station keeps changing.
- Limited transmission power – while base stations in a typical urban scenario use a transmission power of 10–20 W per sector, the mobile device is limited for continuous transmissions to 0.25 W or less, depending on the frequency band. To some degree this is counterbalanced by the larger directional antennas of the base stations and more sensitive signal amplifiers in the base station.
- Battery capacity – high-data-rate transmissions require a lot of energy and thus severely impact the operating times of mobile devices on a battery charge. Every mobile device chipset generation tries to reduce the energy required for signal processing and the efficiency of the mobile's power amplifier. It is difficult, however, to keep up with the rising lower layer processing and power requirements.

Comparing the average HSDPA downlink capacity of a cell in use today of 2–3 Mbit/s with the average HSUPA uplink capacity of a cell of around 1.4 Mbit/s [10], uplink efficiency is about 70% of downlink capacity. It is likely that this ratio will not improve with newer technologies. Despite user-generated content requiring more uplink bandwidth, however, it is likely that consumption of information will still require a higher downlink data rate in the future. Thus, it is unlikely that uplink congestion will become a limiting factor.

3.7 Per-user Throughput in Downlink

As has been shown in this chapter, achieving peak data rates of B3G systems by a mobile device is getting less and less likely. This is due to the continuously rising signal-to-noise ratio requirements for reaching these data rates.

In addition, the maximum throughput achieved by a mobile device depends on the following factors:

- The capabilities of the device itself.
- The number of other users and their activity (e.g. voice calls) in the cell.
- Maximum percentage of overall bandwidth that can be assigned to a single user – previously, terminals and networks were designed to use only a fraction of the cell's overall bandwidth for a single connection; B3G networks, however, can assign the majority of the bandwidth to a single mobile device if no other users are currently transmitting or receiving data in a cell.

- Operator-defined bandwidth restrictions – B3G networks can restrict users to a certain bandwidth; many network operators use this functionality charge a premium for higher speeds.
- Amount of traffic on neighboring cells – most 3.5G and B3G networks are based on radio access technologies in which the same frequency band is used by all cells of the network. The overall cell capacity thus also depends on the amount of traffic handled by neighboring cells. High activity in a cell increases interference for neighboring cells and as a consequence the maximum throughput that can be achieved over them.
- Backhaul capacity of the cell – to save cost, cells might be connected with a backhaul link which can carry less traffic than the air interface.

Table 3.4 shows typical per device throughput values in the downlink direction for different types of networks for average to good radio conditions. Values for GPRS, EDGE, UMTS and HSDPA are shown for deployed networks. The first column shows throughput for a lightly loaded cell, that is, only a few subscribers using the cell and a majority using applications such as Web browsing with long inactivity periods, as is typically the case today.

Table 3.4 Typical end users data transfer rates in downlink for cells with light and average load.

Technology	Typical speed with light cell load	Typical speed with average cell load	Download time of a 2 Mbytes file during light cell load
GPRS (5 timeslots)	60 kbit/s	40 kbit/s	266 s
EDGE (5 timeslots)	250 kbit/s	200 kbit/s	64 s
UMTS	384 kbit/s	128 kbit/s	41 s
HSDPA	1 Mbit/s (operator restriction)	800 kbit/s	15 s
HSDPA	2.5 Mbit/s	800 kbit/s	6 s
HSPA + (2×2 MIMO)	5 Mbit/s	1.6 Mbit/s	3 s
WiMAX (10 MHz, 2×2 MIMO)	10 Mbit/s	3.2 Mbit/s	1.5 s
LTE (10 MHz, 2×2 MIMO)	10 Mbit/s	3.2 Mbit/s	1.5 s
LTE (20 MHz, 2×2 MIMO)	20 Mbit/s	6 Mbit/s	0.75 s

The second column shows values during an average cell load, that is, several subscribers are receiving data in a downlink direction. These values are very subjective, since they depend on the number of simultaneous users and their bandwidth requirements. The column is nevertheless included as cells will get more loaded over time as usage picks up and applications require more bandwidth.

Values for HSPA+, WiMAX and LTE are estimated for conditions under which the other networks were tested. For WiMAX, a 10 MHz channel, 2 × 2 MIMO and enhanced signal processing capabilities of the mobile device are assumed. This explains the higher data rate compared with HSDPA as deployed today. For LTE the same assumptions were made. Additionally, LTE is shown twice in the table, once with a channel bandwidth of 10 MHz and again for the maximum LTE channel bandwidth of 20 MHz. This was because LTE will not always be deployed in its widest carrier configuration due to the size of available bands. While the estimates of HSPA+, WiMAX and LTE are estimated based on typical user speeds in networks today and taking the impact of new features such as MIMO and advanced receivers into account, other research, such as the results of simulations described in [11], shows higher data rates.

In the last column of the table, download times for a 2 Mbyte file transfer are shown for the given data rates under light cell load. This file size represents for example a 30 s MPEG-4-encoded video clip with a resolution of 352 × 288 pixels and a frame rate of 20 frames per second.

In a wireless environment data rates are also subject to changing signal conditions. When used while stationary, for example connected to the Internet with a notebook while not moving, throughput during file downloads is stable given that the number of other users in a cell and their behavior do not change. In practice, putting the antenna or the mobile device used for connecting the notebook in a favorable position can have a big impact on the experienced data rate. Figure 3.2 shows an example of a file download in an HSDPA network with a category 12 HSDPA card (1.8 Mbit/s theoretical maximum speed). At the beginning of the file transfer the antenna was in a very unfavorable

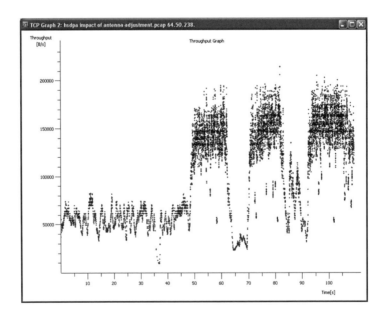

Figure 3.2 Impact of antenna position on transfer speed. (Reproduced from *Wireshark*, by courtesy of Gerald Combs, USA.)

position at a location with below-average network coverage. The data transfer speed was around 60 kbytes per second, which equals 480 kbit/s. At about 50 s into the file transfer the antenna of the network card was slightly redirected while the notebook itself remained in the same place. Immediately data rates increased to about 150 000 bytes per second or around 1.2 Mbit/s. The two throughput drops during the remainder of the file download were caused by moving the antenna back and forth between the two positions to verify that it was the antenna change that was causing this substantial change. From a user point of view, this means that for best performance in stationary use it is best to have an external antenna. In practice, many wireless cards thus have a connector for an additional external antenna connected to the network card via an extension cable. Using a mobile device such as a smart phone connected via a long USB cable to a notebook can also improve reception conditions and thus throughput. Some notebooks are equipped with built-in cellular network cards for HSDPA, WiMAX or other networks. These have the advantage of an internal antenna, which can be much larger than the small antennas of mobile phones and pluggable PC cards. Therefore they might be less susceptible to the effect described above.

Mobility is another factor that can have a great impact on user data rates. Figure 3.3 shows the throughput of a file download while traveling on a train. Again an HSDPA network was used, this time in combination with an HSDPA category 6 mobile device (3.6 Mbit/s theoretical maximum speed). The signal picked up by the HSDPA device was received through the train windows since no repeaters were installed in the train for 3G signals. During the trace the train speed was around 160 km/h. This scenario is one of the most difficult encountered in practice since network coverage was not optimized for track

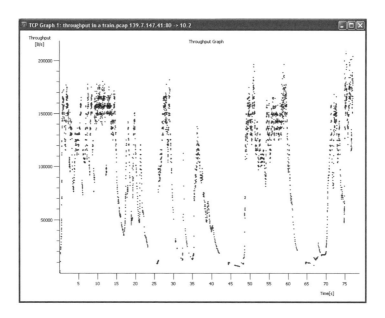

Figure 3.3 Impact of mobility on transfer speed. (Reproduced from *Wireshark*, by courtesy of Gerald Combs, USA.)

coverage. Thus, signal levels varied by a large degree during the file transfer. Also, the train itself and the sun- and heat-insulated windows had a high dampening effect on the radio signal. Handovers were another source of throughput variation.

The impact of mobility, that is the resulting variable data rates, depends on the applications. For file downloads users will notice a lower average speed than while stationary. In the example above, the maximum throughput achieved was around 1.5 Mbit/s while the average throughput of the 6 Mbytes file transfer was around 850 kbit/s.

As long as the connection to the 3G network is not entirely lost, varying throughput has little impact on applications such as Web browsing. This is due to the fact that Web browsing, both in fixed-line and wireless networks, does not usually take advantage of data transfer rates higher than 500 kbit/s, since relatively small amounts of data have to be transferred for a single Web page. This will change over time as Web page content becomes more complex as more pictures and other multimedia elements are added.

On VoIP connections, varying signal conditions such as those shown in Figure 3.3 have a great impact. Firstly, this is due to handovers, which are not yet optimized to hand over high-speed packet-switched connections from one cell to another and thus cause a voice outage that is much longer than the optimized handover in a circuit-switched voice call. Second, lost packets on the air interface are repeatedly sent until they are received correctly. While this approach is favorable for applications such as Web browsing to prevent time-intensive higher layer retransmissions, it is unsuitable for real-time voice or video services. Voice codecs on higher layers have been designed to cope with packet loss to a certain extent since there is not usually time to wait for a repetition of the data. This is why data from circuit-switched connections is not repeated when it is not received correctly but simply ignored. For IP sessions, doing the same is difficult, since a single session usually carries both real-time services such as voice calls and best-effort services such as Web browsing simultaneously. In UMTS evolution networks, mechanisms such as 'Secondary PDP contexts' [12] can be used to separate the real-time data traffic from background or signaling traffic into different streams on the air interface while keeping a single IP address on the mobile device. This is done by an application providing the network with a list of IP addresses in a Traffic Flow Template [13]. The mobile device and the network will then screen all incoming packets and handle packets with the specified IP addresses differently, such as not repeating them on the RLC (Radio Link Control) layer after an air interface transmission error. This is transparent to the IP stack and the applications on both ends of the connection. The IP Multimedia Subsystem makes use of this functionality [14]. External providers of speech services such as Skype, however, do not have access to this functionality.

It is a subject for further study how throughput would look in a train equipped with 3G repeaters in a similar fashion to those used in some trains for 2G and a network coverage optimized along the railway track. The peak throughput values of 1.5 Mbit/s and Figure 3.3 suggest that broadband wireless networks can handle the effects of high-speed mobility quite well.

Data rates of individual users also depend on how many users are active in the same cell. If there are, for example, 10 users in a cell and all are browsing the net, it is highly likely that when one of the 10 users loads a new page all others are reading a page and thus transfer no data. Thus, each user has the full capacity of the cell at their disposal at the moment the Web page is transferred. On the other side of the spectrum is a scenario in which all users are streaming video or downloading large files simultaneously. In this case the users have to share the available bandwidth of the cell while they are transmitting data and not in a statistical way as above. While today practical network use is closer to the first example, it is expected that streaming video will become more popular in a similar fashion to that over fixed-line networks.

Also, file transfers are becoming larger over time and thus also shift usage patterns away from the previous example. Figure 3.4 shows how the data rate during a file download decreases when another user in the cell also starts a file transfer. The data rate of the file transfer is constant at around 120 000 bytes/s (around 1 Mbit/s) until the point where the second user also starts a file transfer. The user's data rate then drops to around 80 000 bytes/s. The combined data rate of the two users in the cell is then at the level expected for an HSDPA-enabled cell which not yet upgraded for 16QAM modulation or a cell which is only connected via a single E-1 (2 Mbit/s) to the network.

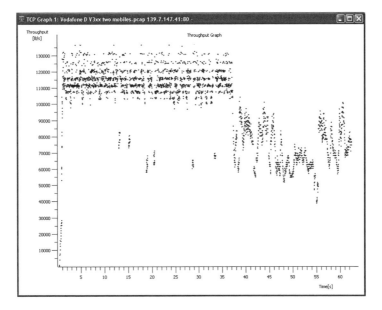

Figure 3.4 Data rate reduction when another user in the cell starts downloading a file. (Reproduced from *Wireshark*, by courtesy of Gerald Combs, USA.)

3.8 Per-user Throughput in the Uplink

While most discussions around performance of beyond 3G networks focus on the downlink, the uplink is mostly neglected. With rising use of the uplink by mobile Web 2.0 applications as well as growing use of notebooks with large file attachments in e-mails and other applications generating large amounts of data to be sent to the network, it is worth taking a look at current and future uplink throughput per user. Table 3.5 gives an overview of typical upload speeds for some network technologies today and expected uplink performance of future systems. To translate throughput into time values, the same file transfer example of a 2 Mbytes file is used as previously for the downlink direction. Typical speeds are not divided into light and average cell load since uplink speeds are usually restricted by the mobile's transmission power and not the overall cell capacity.

Table 3.5   User data rates in uplink direction.

Technology	Typical speed with light cell load	Download time of a 2 Mbytes file during light cell load
GPRS (3 timeslots)	36.6 kbit/s	437 s = 7.3 min
EDGE (3 timeslots)	150 kbit/s	106 s = 1.7 min
UMTS	128 kbit/s	125 s = 2.1 min
HSDPA	384 kbit/s	41 s
HSDPA and HSUPA	1.4 Mbit/s	11 s
WiMAX (10 MHz)	2 Mbit/s	8 s
LTE (10 MHz)	2 Mbit/s	8 s
LTE (20 MHz)	2 Mbit/s	8 s

For GPRS and EDGE (Enhanced Data Rates for GSM Evolution), only three timeslots are assumed in the uplink direction since this is the maximum number of timeslots current mobile phones support for uplink transmissions (GPRS multislot category 32) [15]. For UMTS a 128 kbit/s uplink bearer is assumed in the table. Higher data rates are possible, but in such cases networks are already equipped with HSDPA. For HSDPA, a 384 kbit/s dedicated bearer is assumed that is assigned for average to good radio conditions.

For WiMAX and LTE, standard single-stream uplink transmissions were used for the estimation. While MIMO is also possible in uplink it is unlikely to be used in most situations since uplink transmission speed is usually limited by the mobile's transmission power rather than spectral efficiency. Dividing the signal energy into two traffic independent traffic paths does not therefore make much sense. From a standardization point of view, this has also been considered and mobile stations are predicted to use their transmission power for a fraction of the available OFDM carriers. Thus, more power can be used per carrier than where all carriers are used (cf. Chapter 2). As a consequence, per user data rates do not change for a 10 or 20 MHz network deployment. Since mobiles only use a fraction of the OFDM carriers in power-limited situations, the network nevertheless benefits from wider uplink channels since it allows more mobile devices to transmit their data simultaneously.

It is interesting to directly compare two technologies used in practice today. With GPRS, which was introduced only 8 years ago, the transmission of the 2 Mbyte file takes over 7 min. The recently introduced HSUPA enhancement for UMTS now allows the same file to be transmitted in a mere 11 s. More advanced B3G technologies will decrease transmission times even further.

It should also be noted at this point that file sizes are increasing as well. This counters the trend of rising per user data rates to some extent. The 30 s video file used as an example above was not likely to be sent when GPRS was first introduced. Only a few mobile devices had built-in cameras at the time with no or only very limited video capabilities. As mobile device processing power keeps increasing and camera technology in mobile devices matures, it is very likely that in the future video file sizes will increase due to higher resolution and higher frame rates. It remains to be seen if more advanced video compression algorithms can counter or slow down this trend. One of the best video MPEG-4 compression algorithms available today can encode a TV signal into a 1 Mbit/s data stream (which equals 3 Mbyte for 30 s) in a quality similar to PAL, a color-encoding system used for broadcast television. Starting with HSUPA, such a video stream could thus be sent from a mobile device to the network in real time, given that the device has enough processing power to compress the input signal in real time and enough battery capacity to sustain this operation over a longer time. The sizes of other types of files that users are transferring wirelessly are increasing as well. Thus, the video file example above is not an isolated example but follows a general trend.

In summary, it can be observed that, while both per user uplink and downlink data rates are increasing, the amount of data transferred by users is increasing as well. It is therefore crucial that overall network capacity increases at least as fast as user demand.

3.9 Traffic Estimation Per User

Another factor with a major impact on the capacity required in today's and tomorrow's cellular wireless network is how much data will be transferred per user per day or month. The cost of transferring data over a cellular wireless network will certainly be the main tool for network operators to steer usage to a certain extent, as will be further discussed later on in this chapter. For current pricing strategies the following example shows the amount of data generated by a typical notebook usage of an office worker.

When away from the office, an average office worker generates about 70 Mbytes of traffic in about 10 h, mostly in the downlink direction. This includes e-mail, Web browsing, company database access and VoIP. For this example, it is assumed that about 50 Mbytes are received in the downlink and 20 Mbytes are transmitted to the network. The following discussion focuses on the downlink only. The average data rate over time is 1.39 kbyte/s (50 000 kbytes/s/10 h/60 min/60 s). Compared with the HSDPA cell bandwidth of about 300 kbytes/s (about 2.5 Mbit/s), as shown in Table 3.4, this is not much and more than 200 other subscribers with the same amount of traffic could use the same cell simultaneously. This calculation, however, is just as theoretical as assuming that all users will mostly use high-bandwidth video streaming applications. Thus, a capacity requirement calculation needs to take a number of other variables into account.

• Resource handling – early 3G network technologies such as UMTS Release 99 assigned more bandwidth than required. During Web browsing, for example, the channel is seldom fully used. After the Web page had transferred, the network released the resources on the air interface after some period of inactivity. The channel was not released immediately to ensure a fast reaction to new data packets. This wasted a lot of capacity on the air interface, but was necessary to reduce delays. All 3.5G and beyond technologies use shared channels for data transmission in downlink for which no resource reservation is required. Also, it has been shown in Chapter 2 that systems are continuously optimized to use as little bandwidth as possible for the signaling required for the establishment of a virtual channel between a mobile device and the network and maintaining it. Therefore, no additional overhead is taken into account in this example.

- Busy hour – in most networks, there are certain hours of the day during which users are more active than during others. Let us assume that, during busy hour, usage is three times higher. This reduces the number of users per cell down to 72. On the other side, a single base station site is usually composed of three cells, each covering a 120° sector. Thus, the number of high-speed Internet users per base station increases to about 216. Note that not all 216 users would transfer data simultaneously due to the bursty nature of most of their data traffic.
- Revenue – sers who make use of the network in such a fashion on a daily basis are likely to accept a monthly charge of €30. These users would therefore generate the following revenue per month:

216 subscribers per site * € 30 per month * 12months = € 77.760/year

Over the lifetime of a base station, assumed to be 10 years, this amounts to €777.600. In addition, substantial additional revenue is generated via a base station by services requiring only little bandwidth compared with this intensive usage scenario such as voice calls, multimedia services, mobile Web browsing, e-mail and other mobile device applications.

Holma and Toskala arrive at a similar number of people with a high monthly volume requirement that a base station can serve [16]. Instead of using the amount of data that users transmit per day, they base their calculation on the overall capacity of a cell per month. For a base station with two carriers per sector, compared with a single carrier used in the calculation above, they estimate that a single HSDPA cell can support up to 300 users with a monthly transmission volume between 2 and 4 Gbyte. Furthermore, they come to the conclusion that, when assuming a realistic base station price, a gigabyte of data could be delivered for around €2. This excludes all other costs such as site acquisition, backhaul transmission costs, marketing, customer acquisition and so on.

The values for the calculations used above are certainly open to debate and will even change over time. A few years ago, the amount of data transmitted by the author while on business trips was around 40 Mbytes per day. Lower network charges and faster transmission speeds, however, have encouraged greater use. In the future, it is likely that this trend will continue as prices decline and transmission speeds increase. At the same time, network capacity can increase as required, as shown in Section 3.4. As revenue per user is unlikely to grow beyond the levels outlined in this chapter, it is important that costs in relation to the transferred amount of data keep declining.

3.10 Overall Wireless Network Capacity

So far, this chapter has discussed the current and future per-cell throughput and the per-user throughput. For both, the noise generated by activity in neighboring cells was taken into account. Furthermore, it was discussed how many users can be served simultaneously by a single cell with acceptable quality of service and throughput today and how an evolution path in the future could look. These numbers were calculated based on cell capacity and expected amount of traffic generated per user. Based on these numbers, network operators can then determine how many cells are required for a certain area with a certain population density.

In cities cells are distributed with a site distance of between 500 m and 2 km. Table 3.6 shows the downlink capacity per km^2 for different current and future network types for a single network with an inter-base station distance of 1 km. The first row shows the capacity of a typical initial deployment. Once demand rises, operators usually install additional hardware in base stations to increase capacity. This enhanced throughput per km^2 is shown in a second column. In general the values in the table reflect typical network speeds in deployed networks rather than peak values that are often discussed but which are not achievable in practice for the reasons discussed in this chapter.

Table 3.6 Single network capacity per km^2 with an inter cell distance of 1 km.

Technology	Capacity per km^2 with a low capacity deployment	Capacity per km^2 with a high capacity deployment
GPRS	[a]	256 kbit/s
EDGE	[a]	945 kbit/s
UMTS	2 Mbit/s	[a]
HSDPA	7.5 Mbit/s	15 Mbit/s
HSPA +	15 Mbit/s	30 Mbit/s
WiMAX	30 Mbit/s	75 Mbit/s
LTE	30 Mbit/s	75 Mbit/s

[a]See text.

For GPRS, two carriers per sector and three sectors per cell are assumed. Since the capacity of most GSM networks will no longer increase, no value is given for the low-capacity deployment. This is due to the fact that it is more economical to increase the capacity by deploying B3G networks alongside a 2G network and migrating voice users to 3G. For two carriers per sector, 14 timeslots are available for voice and data traffic. Of these timeslots a varying number are used for voice calls. In Table 3.6, it is assumed that half are available for GPRS traffic. It is further assumed that the maximum GPRS speed per timeslot is 12.2 kbit/s. Higher speeds are possible with better coding schemes. In practice, however, these are only used in a few networks.

Since EDGE is an upgrade for GPRS networks, the same assumptions were used as for GPRS. The maximum speed achieved per timeslot with EDGE is 59.2 kbit/s [17]. To take interference into account and users in areas with weak coverage, a speed per timeslot of 45 kbit/s is used for the estimation.

A single 5 MHz carrier is used per sector for UMTS in an initial low-capacity deployment. Three sectors are assumed per base station. For low capacity deployments, a single E-1 is typically used per base station. This limits the total capacity of a base station to 2 Mbit/s without taking backhaul inefficiencies into account. No value is given for a high-capacity deployment since going to a high-capacity deployment only makes sense in combination with deploying HSDPA.

A typical HSDPA low-capacity deployment consists of cells with a single 5 MHz carrier per sector and three sectors per base station. When upgrading to HSDPA operators are usually also installing additional 2 Mbit/s E-1 connections to the base station. Thus, the overall base station capacity is assumed to be limited by the air interface capacity of 2.5 Mbit/s per sector. For a high-capacity deployment, two carriers per sector are assumed. This in effect doubles the available capacity. It should be noted at this point that some of the base station's capacity is also used for voice calls, which has to be subtracted from the capacity figures in the table. This effect will rise over time as operators migrate users to 3G to avoid further capacity upgrades to their 2G networks due to falling prices and rising use.

For HSPA+, no low-capacity deployment is considered since the use of advanced features only makes sense for further increasing HSPA capacity. The HSPA+ value takes advanced mobile receivers into account as well as the other features mentioned in Chapter 2.

For WiMAX, a 10 MHz carrier, a three-sector configuration and 2×2 MIMO are assumed for initial deployment. As the network expands, operators might increase the bandwidth of the carrier or start using two or even more carriers per sector. Thus, the value given for an initial WiMAX deployment is higher than the value for an initial HSDPA rollout, even though both technologies perform similarly given the same amount of bandwidth. In addition, advanced receivers are assumed in mobile devices for high-capacity deployments since by this time such devices are likely to be used in the network.

For LTE the same assumptions are made as for WiMAX. LTE specifies the use of 4×4 MIMO, but since it still remains to be seen how this can be used in practice, it is not taken into consideration for estimating the values for LTE given in Table 3.6.

As already discussed, the capacity increases shown in Table 3.6 beyond HSPA+ are mainly due to larger channel bandwidths. As capacity increases, achieving such high data rates on the backhaul link from the base station to the network becomes increasingly difficult. This is further discussed in Section 3.17.

The values discussed so far in this section are for a single network. In practice, most areas are typically covered by three or four operators. Thus, the values in Table 3.6 have to be multiplied by that number.

To further increase capacity per km^2 it is possible in theory to decrease the distance between the base stations further by adding additional sites. This is done for example around high traffic areas such as football stadiums, race tracks, downtown city streets and other areas. Such deployments are exceptional and sometimes even only temporary since deployment costs for permanent installations are high. Increasing overall network capacity in this way is not feasible due to the cost involved and the limited number of suitable sites to install base stations. Even if suitable sites exist, the public is often opposed to installing ever more antennas in cities and resistance is likely to grow as network operators are weaving their networks ever denser.

Another option to increase overall network capacity without adding more base station sites and antennas is using more of the available bandwidth. This is what is typically done first when going from a low-capacity to a high-capacity network. As shown in Section 3.4, there is around 500 MHz of bandwidth available for cellular networks between 800 MHz and 4 Hz. In typical deployments in Europe today, the three or four B3G operators per country typically only use a single 5 MHz downlink channel in the 2.1 GHz band. This amounts to a use of 15–20 MHz of spectrum. This means that 66% of the capacity of this band is still unused. Even in the high capacity scenario described in Table 3.6 for UMTS, four operators only use two 5 MHz channels or 40 MHz in total, which still leaves 33% of the capacity of this band unused. In addition, around 50 MHz is used by GSM networks in the lower bands today. In total, 'only' around 70 MHz of the available spectrum is used today.

Table 3.7 shows the amount of capacity per km^2 available today under the following assumptions: four GSM operators have deployed their networks alongside each other. Each network operator uses base stations with three sectors and two carriers per sector. In a two carrier configuration an average of seven timeslots per carrier are used for data transfer (and voice calls). Two timeslots are used for signaling and are thus not counted. It is further assumed that two of the four operators use EDGE with an average data rate per timeslot of 50 kbit/s. The other two operators only use standard GPRS and the data rate per timeslot is 12.2 kbit/s. On average, the data rate per timeslot is thus 31.3 kbit/s. To account for voice calls, it is assumed that half the timeslots are used for this purpose. For UMTS the same number of network operators is assumed to have deployed base stations with three sectors and use a single 5 MHz channel per sector. They all use HSDPA and thus reach about 2.5 Mbit/s per sector per base station. It is further assumed that most voice calls are still handled by the 2G network and thus there is no significant impact on 3G data capacity. As shown in Table 3.7, such a setup amounts to a capacity per km^2 of around 32.6 Mbit/s.

Table 3.7 Downlink capacity per km^2 with four GSM and four UMTS operators today.

Operator	Capacity formula	Capacity per km^2
4 GSM operators, each operating with one base station per km^2, 3 sectors, 2 carriers per sector	4 operators * 3 sectors * 2 carriers/sector * 7 timeslots * 31.1 kbit/s/2	2.6 Mbit/s
4 UMTS/HSDPA operators, each operating with one base station per km^2, 3 sectors and a single 5 MHz carrier per sector	4 operators * 3 sectors * 1 carrier/sector * 2.5 Mbit/s HSDPA throughput	30 Mbit/s
		Total capacity per km^2: 32.6 Mbit/s

Table 3.8 Potential downlink capacity per km² with LTE, WiMAX and several operators.

Operator	Bandwidth and band used	Capacity per km², based on an average spectral efficiency of 0.5, a 3 sector BTS configuration and 2 × 2 MIMO used by all technologies
HSPA/LTE operator 1	3 × 5 MHz in 2 100 MHz 1 × 5 MHz in 900 MHz 1 × 20 MHz in 2 500 MHz	20 Mbit/s
HSPA/LTE operator 2	3 × 5 MHz in 2 100 MHz 1 × 5 MHz in 900 MHz 1 × 20 MHz in 2 500 MHz	120 Mbit/s
HSPA/LTE operator 3	2 × 5 MHz in 2 100 MHz 1 × 5 MHz in 900 MHz	45 Mbit/s
HSPA/LTE operator 4	2 × 5 MHz in 2 100 MHz 1× 20 MHz in 2 500 MHz	90 Mbit/s
WiMAX operator 1	2 × 10 MHz in 3.5 GHz	60 Mbit/s
WiMAX operator 2	2 × 10 MHz in 3.5 GHz	60 Mbit/s
Sum:	Total bandwidth used: 165 MHz	Total capacity per km²: 495 Mbit/s

For a future capacity estimation, WiMAX networks are taken into consideration as well. Their capacity adds up to the UMTS/LTE/CDMA/UMB networks in the same geographical area. Table 3.8 shows how such a possible future combination could look and the resulting total capacity per km². Note that, since the table also considers the use of the 900 MHz band for B3G networks, it is assumed that a major share of voice calls will also be handled by B3G networks either over legacy circuit-switched connections or via VoIP. The impact of this is discussed in more detail in Section 3.10. In the example, 165 MHz out of the 500 MHz available are used. With such a setup, a total capacity per km² of 495 Mbit/s can be reached.

In Section 3.8, two approaches were used to estimate the number of users that could be served with a single network. The more conservative approach of [16] estimated 300 subscribers per base station site for HSDPA using two 5 MHz carriers. Applied to the current network deployment status in many countries as described in Table 3.7, a deployment of four UMTS networks each using only a single 5 MHz carrier results in 600 people who could be served per km² with a monthly data volume between 2 and 4 Gbytes. The second approach discussed in Section 3.8 resulted in 216 users per base station with three sectors and a single carrier per sector. Applied to the example in Table 3.7, 864 people could be served per km² with a daily use of 70 Mbytes.

When the two cell capacity approaches of Section 3.8 are applied to the future scenario described in Table 3.8, the number of people per km² that could be served by HSPA and LTE with a high monthly data transfer volume can be calculated as follows: the more conservative approach estimated 300 subscribers per base station site for two 5 MHz carriers in each of the three sectors. Table 3.8 assumes four operators together using 165 MHz. Furthermore, advanced terminals and 2×2 MIMO are assumed to be in widespread use, which could double spectral efficiency from what was assumed in Section 3.8. The conservative approach would thus result in the following number of people that could be served with a monthly data volume of 2–4 Gbytes:

$$\text{Number of users} = (300 \text{ users}/10 \text{ MHz})^* 165 \text{ MHz}^* 2 = 9900 \text{ users per km}^2$$

The second approach in Section 3.3 resulted in 216 subscribers for an HSDPA system and a bandwidth usage of 5 MHz. Based on this approach the following number of people could be served by the six assumed operators, which together use a bandwidth of 165 MHz with a daily data volume of 70 Mbytes:

$$\text{Number of users} = (216 \text{ users}/5 \text{ MHz})^* 165 \text{ MHz}^* 2 = 14256 \text{ users per km}^2$$

The data volume per user per day also includes data traffic generated by voice calls for this example, either via circuit-switched connections over the HSPA network or VoIP connections via LTE. There is more on this topic in Section 3.11.

When discussing capacity in terms of supported users per km^2, it is interesting to take a look at the population densities of some cities. The population density of Los Angeles is 3168 people per km^2, Munich has 4316 people per km^2, New York is densely populated with 10456 people per km^2 and Manhattan has an astounding 24846 people per km^2. Smaller cities with less than 100000 inhabitants usually have population densities between 500 and 1000 people per km^2.

When comparing the number of users per km^2 calculated above for future network deployments with what today's networks are capable of, it has to be kept in mind that it is likely that the monthly use will rise as well. If the average requirement has doubled by the time networks of this capacity are rolled out, the capacity in terms of number of users per km^2 is cut in half.

As discussed in Section 3.2, there is well over 500 MHz of available bandwidth for downlink transmission available for cellular networks. The total bandwidth used in the example above is well below that value. Thus, overall capacity could be further increased by adding more channels per base station site. Each channel, however, will add to the overall transmission power used at a base station site. At some point, the maximum allowed field strength per site might thus be reached. Today, values for most base station installations are far below the maximum allowed value. In some countries, the national body for telecommunication regulation performs regular field strength measurements. In Germany, for example the federal network agency (Bundesnetzagentur) is responsible for this task and measurement results are published on a Web page [18]. Even in densely populated areas with a high concentration of base stations, the field strength value close to base station sites today is in most cases still below 1% of the maximum allowed value.

3.11 Network Capacity for Train Routes, Highways and Remote Areas

In practice there are some exceptions to the rule of requiring a certain amount of capacity depending on the population density. Areas around overland railway lines and highways, for example, have a much lower population density than city centers or residential areas. Users in such areas, however, might produce a much higher amount of data traffic in cellular networks on average since they are mobile and solely rely on cellular networks for connecting to the Internet. Base stations in those areas on average serve fewer users than base stations in city centers. Despite the smaller number of users, these cells nevertheless generate substantial revenue. Covering such areas can be a competitive advantage since users prefer using networks that cover most if not all areas where they are likely to use the network. In practice it can be observed that highways and to some degree railway lines are specifically covered by 2G networks to ensure coverage along their path to prevent call drops. This is especially important in trains since many people use their phones during train trips. Internet access also benefits from this since many network operators have upgraded their 2G networks with EDGE for higher data rates. In practice it is difficult to predict coverage along railway lines. Some lines are already covered by 3G networks (cf. Figure 3.3) and thus offer excellent connectivity. While this is still the exception today, it is expected that more and more railway lines will be covered by advanced cellular networks. Some train operators are also equipping their trains with Wi-Fi hotspots that use a satellite system and public B3G networks as backhaul connection. An example can be found in [19].

In rural areas, fast high-speed Internet access either via a fixed line or a wireless connection is still rare in many regions. This is due to the fact that the number of potential customers per km^2 is low. Fixed-line high-speed Internet access thus suffers from the fact that the range of DSL for a bitrate of 1 Mbit/s is typically less than 8–10 km. This makes it financially unattractive for telecom companies to install the required equipment. B3G networks might become an interesting solution to this problem. Already today telecommunication companies like Telstra in Australia are using B3G networks to offer high-speed Internet access in remote areas. An article published by Ericsson on this deployment [20] gives further information on the technical and financial background. To make such a venture profitable, a single base station must cover as large an area as possible. One of the main factors in cell range is the frequency used. In the case of Telstra the 850 MHz band was used, which offers substantial increase in coverage range over the 2100 MHz band used for B3G networks in other regions. In addition rural deployments usually require a roof-mounted directional antenna for distant subscribers. In effect, the network is thus used as a mobile network by subscribers close to the base station and as a fixed wireless network rather than a truly mobile network at greater distances. Inside a house, a 3G/Wi-Fi router is used to connect all devices requiring Internet access to the B3G connection. This is similar to initial 802.16 WiMAX deployments. Using a roof-mounted directional antenna greatly improves the link budget (reception conditions) and thus the range over which data can be sent. Ericsson suggests in [20] that, compared with a handheld device used indoors, an indoor window-mounted omnidirectional antenna can improve the link budget by up to 12 dB. An omnidirectional antenna on the rooftop improves the link budget by up to 47 dB due to the greater height and the resulting vertical direction gain. A directional antenna mounted on a rooftop can increase the link budget by as much as 65 dB. In practice, a higher link budget both increases the range of the cell and also the data transfer rate that can be achieved by users and as a consequence overall network capacity. Both network operators and customers thus benefit from rooftop antennas.

The initial version of the UMTS/HSDPA standard allowed cell ranges of up to 60 km. In the case of Telstra, larger cell ranges were required for some regions. The limitation was due to base stations only being able to deal with propagation delays on the random access channel in the order of 768 chips, or a range of about 60 km. The 3GPP standards were thus enhanced and an extended cell range mode was introduced that can extend the cell range from a delay point of view to up to 200 km. While the enhancement requires software changes on the network side, no modifications are required on the mobile device. Further information can be found in the corresponding work item [21].

On the financial side, Ericsson estimates that a base station covering an area of 12 km^2 is financially viable in areas with population densities as low as 15 people km^2. This assumption is based on operator market share of 50%, a mobile penetration of 80% and a fixed mobile broadband penetration of 35% of the subscriber base among other values. The main service revenue was assumed to be €15 for mobile voice telephony and €30 for fixed mobile Internet access.

3.12 When will GSM be Switched Off ?

It has been shown in this chapter that GSM-based GPRS and EDGE have difficulty keeping up with capacity enhancements of B3G networks and subscriber demands. Thus, switching off GSM networks and using the capacity in the 900 and 1800 MHz band for B3G networks would increase total capacity without requiring additional spectrum. It is expected that this will be slowly done over time as the specification of B3G network technologies such as HSPA and LTE now also allows operation in these frequency bands.

Only a few years ago, most industry observers were predicting a rapid decline of GSM networks once 3G networks were in place. GSM had just celebrated its tenth birthday in terms of live network deployments and UMTS was already at the doorstep. Looking at lifetimes of analog wireless systems, it seemed certain that in another 10 years (2012) GSM would be a thing of the past. Today, 2012 is just a few years away and it is quite certain that GSM will be used much beyond then, even in countries where B3G networks have been rolled out. This surprising development has several causes:

- Equipment refresh – in 2002, GSM equipment started to age as network vendors kept selling hardware that had been developed a decade previously. Since then, however, virtually all network vendors have completely refreshed their network equipment from base stations to core network routers. This was a necessity as the parts for aging designs (e.g. 486 processors) were no longer available at reasonable cost. Hardware evolution also meant lower prices. GSM base station controllers sold today, for example, are no less capable than the latest 3G radio network controllers in terms of processing power, memory or storage capacity. GSM base station prices and sizes also keep decreasing, which in turn reduces capital expenditure for network equipment.

- New entrants – another reason for refreshing aging hardware designs is the entry of new Asian companies like Huawei and ZTE into the GSM and 3G markets with new hardware and lower prices. Established vendors could no longer afford to continue selling expensive hardware with the new competition.

- New markets – back in 2002 it was not clear that GSM would have such tremendous success in emerging economies in Asia, India and Africa. Compared with the 2.5 billion GSM subscribers today, the few (hundred million) 3G subscribers in 2008 almost seem like a drop in the ocean. This created economies of scale for GSM beyond anything imagined.
- Continuous evolution – back in 2002, it was assumed that most R&D would be put into the development of 3G networks. This has been true to a certain extent, but instead of being dormant, GSM has continued to evolve. Compared with 2002, GSM hardware is much more efficient due to the technical and economical hardware refresh. New features such as EDGE for higher packet-switched data rates have pushed the GSM standard far beyond the circuit-switched network it was once designed for.
- Network refresh – just like consumer IT equipment such as PCs and notebooks, network equipment such as base stations, controllers, switches and routers have a limited lifetime and require replacement. The cycle is certainly longer than the 2 or 3 years for consumer PCs but after 10–12 years base stations have to be replaced because of aging components or due to their inability to support new features such as EDGE. Also, the power consumption of older systems is much higher than that of new base stations, so at some point the price of replacing a base station is absorbed quickly by reduced operational costs.
- 3G network coverage – even in the most advanced 3G countries such as Italy, Austria, Germany and the UK, 3G network coverage is nowhere near as ubiquitous as GSM coverage. This is different from the 1990s, where GSM coverage quickly approached the coverage levels of the analog networks.
- Roaming – with GSM, international roaming is a major benefit. For the foreseeable future, the majority of roamers will still have a GSM-only phone. Switching off GSM networks makes no sense as revenue from roaming customers is significant.

So where does that leave GSM in Europe and the USA in 2012? In five years, it is likely that the majority of subscribers in Europe and the USA will have 3G-compatible phones that are backwards-compatible to 2G. In urban areas, operators might decide do down-scale their GSM deployments, as most people will now use the 3G network instead of the 2G network for voice calls. Cities will still be covered by GSM, but probably with fewer channels. Such a scenario could happen in combination with yet another equipment refresh, which will be required by some operators for both their 2G and 3G networks. At that time, base station equipment that integrates 2G, 3G and B3G radios such as LTE could become very attractive.

3.13 Cellular Network VoIP Capacity

Even today, voice communication is one of the main applications and revenue generators in wireless networks. As voice communication generates a narrow band data stream, current networks such as GSM and UMTS have an optimized protocol stack for voice transmission in the radio network and on the air interface. As a result, voice transmission in those networks is very efficient in terms of required transmission capacity and hand-over performance. Additionally, a dedicated core network infrastructure is used in such networks for transporting and managing voice calls, as discussed in Chapter 2. B3G networks on the other hand follow a trend that started in fixed-line networks to treat voice calls just as one of many communication applications that can be transported over IP. A detailed discussion on this topic will follow in Chapter 4. B3G networks can thus no longer integrate voice calls as tightly into the protocol stack. As a consequence voice transmissions over the air interface are less efficient in B3G networks. Before looking at B3G voice capacity, the following section discusses the voice capacity of GSM and UMTS base stations to act as a reference for the VoIP discussion that follows.

The Enhanced Full Rate (EFR) codec, used in most GSM networks today, requires a data rate of 12.2 kbit/s. For additional capacity, Adaptive Multi Rate codecs were introduced in the standards some years ago. Due to enhanced processing, almost the same speech quality can be achieved today with the 6.75 kbit/s AMR codec, which requires only half the resources on the air interface. In GSM, an EFR voice call is transported in a single timeslot and requires a bandwidth of about 22.8 kbit/s on the air interface due to channel coding, which adds error detection and correction information. A 6.75 kbit/s AMR voice call is carried in half a timeslot and has a data rate of 11.4 kbit/s after channel coding. An average GSM base station with three carriers, three sectors and 66 timeslots can therefore carry 66 EFR or 133 AMR 6.75 kbit/s voice calls. In practice, however, some timeslots are used for GPRS or EDGE data transmission, which reduces the number of simultaneous calls possible. The number of timeslots reserved for data transmission is an operator-defined value. As a general rule, it can be assumed that about 20–30% of the timeslots are reserved for data transmission. Voice timeslots can be used by GPRS and EDGE transmission while not required for voice calls.

UMTS uses the same codecs as GSM. On the air interface users are separated by spreading codes and the resulting data rate is 30–60 kbit/s depending on the spreading factor. Unlike GSM, where timeslots are used for voice calls, voice capacity in UMTS depends less on the raw data rate but more on the amount of transmit power required for each voice call. Users close to the base station require less transmission power in down-link compared with more distant users. To calculate the number of voice calls per UMTS base station, an assumption has to be made about the distribution of users in the area covered by a cell and their reception conditions. In practice, a UMTS base station can carry 60–80 voice calls [22] per sector. A typical three-sector UMTS base station can thus carry around 240 voice calls. As in the GSM example, a UMTS cell also carries data traffic, which reduces the number of simultaneous voice calls.

In B3G networks such as HSPA, LTE and WiMAX, voice calls are no longer transported over dedicated circuit-switched channels and equipment in the core network. Instead, the voice data stream is packetized and sent over IP. To estimate the number of simultaneous VoIP channels, coding overhead has to be assessed differently since it is highly adaptive. The most important factor in this calculation is the average throughput, as discussed in Section 3.9, since this value includes the average channel coding used in a cell. Furthermore, it has to be taken into account how well a system can transport a high number of simultaneous low-speed connections. Usually it is more efficient to handle few users with high bandwidth requirements since there is a per user overhead for air interface management such as channel access, power control and channel signal estimation. The percentage of this overhead is small for high-speed transmissions but grows for slow data streams such as voice.

When a voice call is transported over IP, the same voice codecs are used by optimized VoIP applications as those mentioned for GSM and UMTS above. In B3G networks voice calls no longer use dedicated and transparent channels but are transported over an IP network together with packets of other applications. One voice packet is usually sent every 20 ms and contains the voice data collected during this time. When the EFR codec with a data rate of 12.4 kbit/s is used, 31 bytes of voice data are sent in each packet. Nonoptimized VoIP applications make use of the G.711 codec, which is used in fixed-line analog voice telephony for compatibility reasons. The data rate of this codec is 64 kbit/s and thus much higher than the data rate of EFR. Sent in 20 ms intervals, a G.711-encoded voice call generates 160 bytes of voice data for each packet.

These 20 ms voice frames are then encapsulated in three protocol layers, each adding its overhead. Figure 3.5 shows the overhead for a 20 ms G.711 voice packet. In the lower part of the figure, the user data carried in the packet is marked in blue. The overhead in

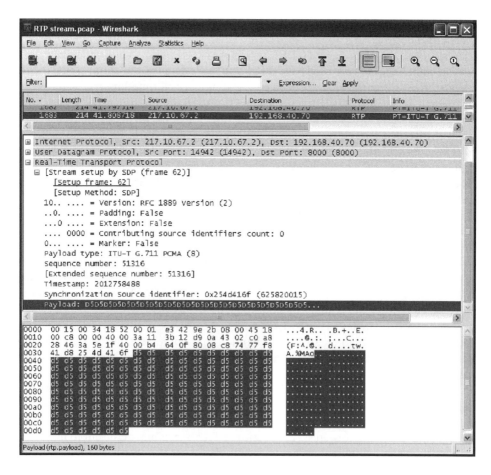

Figure 3.5 Voice transmission over IP, G.711 codec in an RTP packet. (Reproduced from *Wireshark*, by courtesy of Gerald Combs, USA.)

front is shown in white. The IPv4 layer adds 20 bytes to the overhead, the User Datagram Protocol (UDP) adds 8 and the RTP adds another 12 bytes. In total, the overhead for IPv4 is 40 bytes. If IPv6 is used in the network the overhead grows to 60 bytes. For a packet that carries a 20 ms EFR frame, the overhead exceeds the size of the payload. The resulting data rate is thus:

Datarate (EFR) over IPv4 = (31 voice bytes + 40 bytes overhead)/20 ms = 3.55 kbytes/s

In practice the IP overhead would thus reduce the number of simultaneous voice calls per cell at least by two.

Several technical options exist to reduce the overhead. On the application level, more data could be collected before a packet is assembled and sent through the network. From a network perspective this looks appealing as the ratio between over-head and user data could be influenced by the application. This unfortunately comes at a price. Collecting more voice data, that is waiting for a longer time before sending it over the network, would substantially increase the overall delay. Packetization delay is just one part of the overall delay, which additionally includes core and access network delay, air interface delay and jitter buffer delay. A jitter buffer is required for VoIP transmissions at the receiving end since it is not guaranteed that packets arrive in time. As a consequence, the mouth-to-ear delay increases quickly to values beyond 200 ms, which is the limit described in [23] at which the delay becomes noticeable and distract-ing to the user.

Another possibility to reduce the overhead is to use header compression in the network between network elements that do not use the information in the IP header for forwarding the packet. This is possible since most fields of packets exchanged between two specific end points always contain the same values. For UMTS, HSDPA and LTE radio access networks the Robust Header Compression (ROHC) algorithm [24] has been standardized for this purpose. In addition to compressing static fields like the source and destination IP addresses that never change, several profiles have been defined for ROHC so dynamic fields in different combinations of protocol layers used by IP applications can be compressed as well. One of these profiles is used when the ROHC compressor detects a VoIP transmission which uses IP, UDP and RTP. This way an IPv4 or IPv6/UDP/RTP header can be reduced from 40 or 60 bytes down to 3 bytes.

In UMTS and HSDPA, header compression is performed between the mobile device and the RNC, as described in Chapter 2. Thus, both the air interface and the backhaul connection between the base station and the RNC benefit from the compression. It should be noted at this point that header compression is optional and not yet widely used in wireless networks since most voice traffic is still carried over circuit-switched connections. In LTE, header compression is performed between the base station and the mobile device since the RNC has been removed from the architecture and most of its tasks are now performed by the base station.

Due to the almost complete removal of the IP protocol overhead by ROHC, VoIP transmissions over the air interface can be almost as efficient as circuit-switched trans-missions. In [25] it is estimated that over 80 simultaneous voice calls can be transported over a 5 MHz HSDPA channel. The estimation is based on using the EFR voice codec and already includes the radio signaling overhead based on a Release 6 HSDPA implementation with a Fractional Dedicated Physical Control Channel (F-DPCH, cf. Chapter 2). This value is similar to the number of circuit-switched voice calls per UMTS sector given above.

For other B3G technologies, the number of voice calls over IP will be similar given the same amount of bandwidth used per sector. Radio network enhancements such as MIMO and advanced signal processing will further increase the number of simultaneous voice calls. In combination with ROHC, the number of VoIP calls per megahertz of bandwidth can thus exceed the number of circuit-switched voice calls per megahertz of bandwidth today.

3.14 Wi-Fi VoIP Capacity

IP over Wi-Fi shares the same basic technical background as discussed in the previous section for cellular network VoIP capacity. While VoIP capacity in cellular networks plays a significant role due to the amount of the network capacity being used for voice calls, it is likely that in the future only a small amount of capacity will be used for VoIP in private Wi-Fi networks. Wi-Fi VoIP capacity will therefore mainly play a role in office environments where Wi-Fi networks will be deployed for wireless telephony. This could come as part of office environment which relies exclusively on Wi-Fi for networking [26]. While this has not been feasible in the past due to the speed of earlier Wi-Fi standards, 802.11n has the potential to overcome this limitation.

VoIP over Wi-Fi differs from what has been discussed for the cellular world in the previous section:

- Wi-Fi, unlike B3G cellular networks, does not have a centralized scheduler for packets (cf. Chapter 2). With a rising number of clients creating significant load, the number of collisions on the air interface is increasing. Thus, it is not possible to use the full capacity of a Wi-Fi network unlike in a scenario when only few devices are fully utilizing the network.
- The lack of a centralized scheduler makes it difficult to prefer small VoIP data packets to larger packets generated by other applications. The Wireless Multimedia extension discussed in Chapter 2 improves this behavior. Optionally, the Wi-Fi standard also defines a centralized scheduler. It is unlikely, however, that this option will be widely used due to economies of scale in both access point and client devices.
- Most cellular B3G technologies use FDD, that is different frequency bands for uplink and downlink (except WiMAX). Wi-Fi on the other hand uses TDD. Thus, bandwidth requirements of the channel are twice as high as in the FDD system, since the uplink and downlink are sent in the same channel.
- The number of voice calls per access point is ultimately limited by the capacity available on the backhaul connection. For the examples below it is assumed that there is sufficient capacity in the backhaul in both the uplink and downlink direction.
- As in cellular networks, an average bandwidth of a Wi-Fi network has to take into account that some Wi-Fi/VoIP phones will be in unfavorable or distant positions from the access point and thus use a lower transmission speed. The transmission time for

those packets is thus several times higher than the transmission time of packets for VoIP phones close to the access point.

- Unlike in cellular networks, no header compression schemes are currently defined to decrease the IP overhead in Wi-Fi networks.

Table 3.9 shows the number of concurrent voice calls as calculated in [27] in a Wi-Fi network under the following assumptions: all devices in the network are 802.11g capable and no protection schemes are required in the network for older devices. In the first column, the different voice codecs are listed and in the second column their bandwidth requirements. In the third column, the theoretical maximum number of simultaneous calls is given at the maximum speed of the network. In the second column the number of simultaneous VoIP calls is given for a lower average network speed, which is more realistic in practice since not all VoIP devices will be used under ideal network conditions. The table shows that, even when average conditions are assumed and the highest bandwidth codec (G.711) is used, the network supports up to 51 simultaneous VoIP calls. In practice, the number will be somewhat lower due to the negative effects of the decentralized scheduler. Nevertheless, given the short range of a typical Wi-Fi access point, it is unlikely that this limit will be reached in practice. Together with WMM, which prioritizes small and bursty streams, using a combined Wi-Fi network for VoIP and other data is feasible in most environments. Should the overall traffic exceed the limits of a single network, it is also possible to deploy a dedicated Wi-Fi network for voice alongside a Wi-Fi network for general use.

Table 3.9 Number of simultaneous voice calls in a 802.11g network excluding the effect of a decentralized scheduler.

Voice codec	Bandwidth requirement over iP	Number of calls with 54 Mbit/s	Number of calls with 18 Mbit/s (averaged)
G.711	80 bit/s	78	51
GSM Enhanced Full Rate (EFR)	28.4 bit/s	92	71
iLBC 30ms (Skype)	24 bit/s	133	101

3.15 Wi-Fi and Interference

Using Wi-Fi as part of a future converged fixed, nomadic and mobile Internet access network substantially increases overall capacity per km^2. As applications such as TV broadcasting over IP gain in popularity, even Wi-Fi capacity limits are reached quickly in the unlicensed 2.4 GHz ISM (Industrial, Scientific and Medical) band. In countries such as France, TV broadcasting over IP over DSL has already achieved great popularity today. Some DSL access providers have started to offer equipment to wirelessly connect a TV set-top box to the DSL modem/router over Wi-Fi. An example is the Freebox of DSL provider

Free [28]. Due to this and the general popularity of DSL/Wi-Fi routers for PC and notebook connectivity, it is not uncommon to be in the range of more than 10 Wi-Fi networks in a Paris apartment building. Since there is only room for three nonoverlapping 20 MHz channels in the 2.4 GHz ISM band, there is a great partial and full overlap of these networks. While this does not have a big impact on the performance of a Wi-Fi network since the overlapping Wi-Fi networks are only broadcasting beacon frames in idle mode, performance quickly drops when use in overlapping networks increases. In the case of Wi-Fi-connected set-top boxes, it can be observed in practice that a continuous stream of 4–5 Mbit/s is continuously sent independently of whether or not the TV set is turned on. Consequently, this capacity is no longer available in the other overlapping networks.

Figure 3.5 shows a trace taken with a layer 1 tracer [29] under the conditions described. The lower graph in the figure shows the frequency range of the ISM band between 2400 and 2480 MHz. Instead of showing the frequency in MHz, the x-axis shows the 13 available Wi-Fi channels. On the y-axis the amplitude of the signal received over the band is shown. The color of the peak depends on the intensity of the signal received during 60 min. Bright colors indicate high activity level. The graph shows five partially overlapping networks with their center frequency on channel 1 (little traffic so the arch is

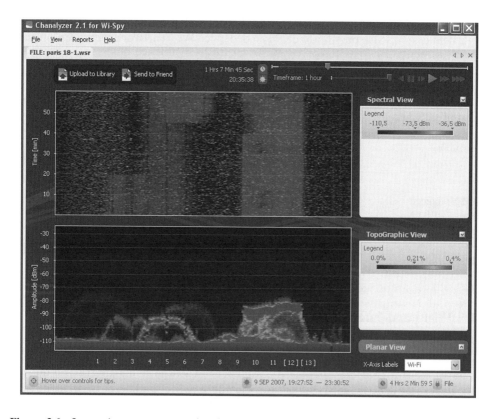

Figure 3.6 Layer 1 trace over overlapping Wi-Fi networks. (Reproduced by permission of MetaGeek LLC, 5465 Terra Linda Way, Nappa, ID 83687, USA.)

not very visible) and channels 3, 5, 6 and 11.The most activity can be observed in the wireless network that is centered around channel 11.

The upper graph in Figure 3.6 shows a time graph over the frequency range. On the y-axis a resolution of 60 min has been chosen to show the activity in the ISM band in the course of 1 h. The Wi-Fi networks on channels 5 and 11 were used for streaming as there was uninterrupted activity throughout the test period. The Wi-Fi networks on channels 3 and 6 were also used for streaming. Streaming was stopped on the Wi-Fi network on channel 6 after about 12 min while streaming was started on the Wi-Fi network on channel 3 about 40 min into the trace.

Table 3.10 shows the impact of partly and fully overlapping networks on the maximum throughput of a Wi-Fi network compared with a situation in which no overlapping occurs. The tests were performed with a Linksys WRT54 802.11g router and Iperf [30], a UDP and TCP throughput measurement tool. One of the notebooks for the test was connected to the Wi-Fi Access Point with an Ethernet cable, while the other one used the Wi-Fi network. With a fully overlapping Wi-Fi network, which is used for TV streaming, the capacity of the Wi-Fi network under test was reduced to 72%. Partial overlapping caused an even bigger speed penalty and performance was reduced to 59%.

Table 3.10   Effect of partly and fully overlapping Wi-Fi networks on throughput.

Situation	Measured throughput
No interference	22.5 Mbit/s
Fully overlapping Wi-Fi network with a streaming client and an estimated continuous stream of 4–5 Mbit/s	16.3 Mbit/s
Fully overlapping Wi-Fi network with a streaming client and an estimated continuous stream of 4–5 Mbit/s	13.4Mbit/s

In the future even more private Wi-Fi networks will be set up and TV and other multimedia streaming over Wi-Fi is likely to become even more popular. As discussed in Chapter 2, the 802.11n standard by default only allows the Wi-Fi channel bandwidth to be increased to 40 MHz in case there are no overlapping networks. As the presented test results have shown, this was a wise choice. In practice, many access points offer an override option. This will increase the problems for the ISM band even further. It is thus likely that equipment sold especially for multimedia streaming purposes will start using the 5 GHz unlicensed band in which there are 18 independent channels for 20 MHz operation available or nine for 40 MHz operation.

3.16 Wi-Fi Capacity in Combination with DSL and Fibre

Wi-Fi in combination with DSL or cable has become very popular for home networking over the past few years and it can be observed that the number of 3G and B3G mobile devices such as PDAs and mobile phones being equipped with a Wi-Fi interface is also rising. In addition, stationary devices such as set-top boxes, multimedia storage devices,

TV and satellite receivers and many other devices are beginning to be equipped with a Wi-Fi interface. As discussed in more detail in Section 3.14, the rising use of Wi-Fi has started to cause interference between networks. This is a trend that is likely to increase in the future as new devices and services are added. As a consequence, many 802.11n products are now additionally operating in the 5 GHz band, which is available for unlicensed use in many countries. Table 3.11 shows the frequency bands currently used by Wireless LAN. The exact bandwidths are country-dependent, but generally in the bands shown in the table. In total, there is around 500 MHz of bandwidth available. It should be noted at this point that the 2.4 GHz band is also used by other systems such as Bluetooth. Their use, however, is limited and narrowband and thus has no significant impact on the use of Wi-Fi, barring some exceptional cases.

Table 3.11 Frequency bands for Wi-Fi.

Frequency range	Total bandwidth available	Number of available 20 MHz channels	Number of available 40 MHz channels
2.410–2.480 MHz	70 MHz	3	1
5.150–5.350 MHz, 5.470–5.725 Mhz	455 MHz	18	9

From a capacity point of view current 802.11g networks provide a maximum practical throughput of around 23 Mbit/s in a channel with a bandwidth of 20 MHz. This requires that there is no interference by other overlapping networks, no legacy 802.11b devices in the network and only a short distance, in the range of several meters, between the Wi-Fi access point and the wireless device. The MIMO functionality of 802.11n, as shown in Chapter 2, is likely to increase this throughput to around 40 Mbit/s in a 20 MHz channel and to over 100 Mbit/s where a 40 MHz channel is used. With increasing distance or obstacles such as walls between the access point and the client device, data rates quickly decrease. In buildings, the range of a Wi-Fi network is thus typically less than 30 m with data rates of 3–5 Mbit/s at the coverage limit. For most applications this range is sufficient. One exception is VoIP over Wi-Fi. Due to the smaller area covered by Wi-Fi networks compared with that of a cordless phone system, Wi-Fi VoIP phones have a significant disadvantage [31].

Several options exist to increase the coverage for the use of VoIP over Wi-Fi or to enlarge a Wi-Fi network in general. Companies mostly prefer to deploy several access points and create a single network by using a fixed Ethernet infrastructure to tie them together. In a home network environment with a single access point connected to the Internet (i.e. no fixed Ethernet infrastructure is available), the Wireless Distribution System (WDS) functionality built into most access points can be used to replace the fixed-line backbone infrastructure with a wireless link. An access point can typically serve clients and forward packets over the WDS link simultaneously. Overall network capacity is significantly reduced, however, since packets traversing a WDS repeater are sent over the air once by the user and once by the repeater of the wireless link.

When making the capacity of fixed-line DSL connections available wirelessly via Wi-Fi, it is important to note that the throughput offered by a Wireless LAN is typically much higher than that of the fixed-line Internet link. For the following estimation it is assumed that 8 Mbit/s DSL links are shared by four people in a household via Wi-Fi. With a population density of 4000 inhabitants per km^2, which was used to set capacity estimations of cellular networks above into perspective, this would result in 1000 DSL connections per km^2, each delivering 8 Mbit/s. With a current DSL subscription rate of around 50% of the households, this would result in about 500 DSL connections per km^2. The line speed, however, is only available up to the DSLAM (DSL Access Multiplexer), which concentrates DSL lines and forwards the aggregated traffic to the wide area optical network. For a concentration rate of 50:1, which is typical for residential areas today, the following capacity per km^2 would result:

$$\text{Wi-Fi-DSL capacity per km}^2 = (4000/4) {}^* 50\% {}^* 8\,\text{Mbit/s}/50 = 80\,\text{Mbit/s per km}^2$$

It is important to note that the estimated value greatly depends on the concentration rate, which is also referred to as the oversubscription. With increasing load on networks due to TV and multimedia streaming, fixed-line operators can quickly change this to much lower values, if the broadcasting server is in their own network. This is done by adding capacity in the optical backbone network with no changes required between the subscriber and the DSLAM. For traffic from and to the Internet via an IP transit point, different rules apply. Here, traffic is typically charged based on required bandwidth. With prices around €20 per Mbit/s [32] in 2007, the use of 80 Mbit/s during peak hours would incur a cost of €1.600 a month. This would be €3 a month per subscriber in the example above. To this cost, the DSL operator has to add the capital and operation expenditure for his own backbone network to the IP transit point, the DSLAMs in the central offices, line rental to customer and so on.

In the future, enhancements of DSL and cable will allow a further increase in fixed-line data rates. ADSL2+, which is already in widespread use in some countries, allows subscriber data rates of up to 25 Mbit/s. VDSL increases subscriber data rates to around 50 Mbit/s. In some regions, optical connections to subscriber homes are currently under deployment, pushing data rates even higher. From a financial perspective, it should be noted that such upgrades are mostly limited to cities, as discussed in Section 3.10. Also, to benefit from the rising speeds in the access for anything but TV and video streaming from a server in the operators network, IP interconnect prices at peering points must fall by an order of a magnitude.

Future bandwidth increases on the last mile to the subscriber come with an additional cost in comparison with today's standard ADSL or ADSL2+ deployments. For ADSL2+ the DSLAM is usually installed in the telephone exchange and the cable length to the subscriber can be up to 8 km for a 1 Mbit/s service. For VDSL, which offers data rates of up to 50 Mbit/s in downlink, the cable length must not exceed 500 m. Thus, DSLAMs can no longer be only installed in central telephone exchanges but equipment has to be installed in street cabinets. This is a challenge since the cabinets are quite large, require power and active cooling and create noise. Also, earthworks are necessary to lay the additional fiber and power cables required to backhaul the data traffic. Figure 3.7 shows a VDSL DSLAM cabinet that has been installed alongside a

Figure 3.7   A VDSL DSLAM cabinet alongside a traditional telecom wiring cabinet.

traditional small telecom cabinet. To connect a new subscriber, a technician is required to manually rewire the customer's line to one of the ports. Different sources currently specify the maximum capacity of such cabinets from about 50–120 VDSL ports [33]. To support 500 VDSL connections per km^2, several cabinets are thus required. If several network operators compete in the same area, the number of required cabinets will grow further.

With VDSL and fiber network deployment to the customer premises, data rates on the Internet link can come close to or even surpass what is currently offered on the wireless link by the 802.11n standard. Using the same oversubscription as in the previous example, the capacity of Internet connectivity via Wi-Fi and DSL could reach the following level:

$$\text{Wi-Fi (VDSL/optical) capacity per } km^2 = (4000/4)^* 100 \, \text{Mbit/s/50} = 2 \, \text{Gbit/}km^2$$

If both the 2.4 GHz band and the 5 GHz band were fully used by Wi-Fi and if it is further assumed that the per network peak throughput in 20 MHz would be 40 Mbit/s, 21 nonoverlapping networks could deliver around 800 Mbit/s in an area with a radius of 30 m. Even if undesired partial network overlapping, no MIMO, reduced cell edge data rates, interference from neighboring Wi-Fi cells and other radio systems using the band are considered, there is still enough capacity available on the wireless link for future fixed-line data rates. It should be noted at this point that some Wi-Fi networks will also carry a substantial amount local traffic, for example between a home media server and an IP-enabled television set, which will generate much more data traffic than what users request from the Internet.

The following summary once more highlights the differences in how capacity can be increased per km^2 in Wi-Fi/DSL networks compared with cellular wireless networks:

The capacity per km^2 of Wi-Fi over DSL is mostly limited at the back end. As the price per Mbit/s capacity at the Internet peering point is expensive, an oversubscription factor is used to limit the maximum bandwidth available to all subscribers at a given time. As prices for interconnection to the Internet decline, the oversubscription can be lowered or the line speed can be increased while keeping the same oversubscription. An advantage of Wi-Fi in combination with DSL compared with cellular wireless networks is that TV and multimedia streaming from servers inside the network can be very cheap as the additional capacity used in the access does not increase the operator's cost.

The capacity per km^2 of cellular networks is limited at the front end. To reach a throughput in the same order of magnitude compared with Wi-Fi over DSL, the air interface resources need to be fully utilized; that is, there is no oversubscription factor between the air interface speed and the speed at the Internet peering point. Thus, mobile operators cannot offer TV and multimedia streaming at the same price as fixed-line operators even if the streaming server is within their own network. To increase capacity in an area they cannot lower an oversubscription factor but have to add more capacity at the base stations and at the Internet peering point.

The following consequences can be deduced from the observations above:

- The capacity of Wi-Fi/DSL networks per km^2 scales with the number of subscriptions per km^2. As each subscriber generates additional revenue, adding more subscribers only requires increased resources in the optical back-end network and increased capacity at the Internet peering point. Both increases are covered by the additional subscriber revenue.
- The capacity of cellular networks per km^2 is lower but in the same order of magnitude as the capacity per km^2 of Wi-Fi/DSL networks for the population density of a mid-size city. Thus, cellular networks can compete for a sizable market share.
- Cellular network operators cannot offer TV and multimedia streaming at a competitive price compared with Wi-Fi/DSL. Their offers are thus limited to Internet access.
- As capacity per base station is increased, the individual backhaul connections to the network must be increased as well. This raises the question of how this can be done economically. This is discussed in more detail in the next section.
- To increase capacity for a Wi-Fi/DSL network, only centralized locations (the DSLAM) must be upgraded.
- Future fixed access technologies such as VDSL and fiber to the home require curb-side installations. The number of subscribers is limited by the number of ports available in the DSLAM at the curb-side.
- Wi-Fi/DSL automatically benefits from falling interconnection peering prices since there is ample capacity between the DSLAM and the subscriber.

3.17 Backhaul for Wireless Networks

It is interesting to note that in reality most links in wireless networks are not wireless. This starts with the connection between the base stations and the rest of the network. This link is also referred to as the base station backhaul or simply backhaul. With increasing air interface throughput, the capacity of backhaul links has to rise as well. This is quite expensive in practice since even current B3G networks use legacy backhaul technology that does not scale well with rising demand. The following section describes the situation today and the technologies used in the backhaul. This is followed by an overview of future technologies.

Currently, two technologies are used to backhaul the traffic from a base station. Wireless operators often choose microwave backhaul connections. This requires extra equipment at the base station and at the other end of the link but frees the operator from monthly fees to a fixed-line operator to lease a wired connection. For this reason, microwave backhaul has become very popular with alternative operators, especially in Europe. Figure 3.8 shows a typical base station setup that uses a microwave connection for the backhaul. The long antennas are used for the connection to the mobile devices while the round antenna creates the directional beam for the backhaul connection and receives the data stream from the other end of the link.

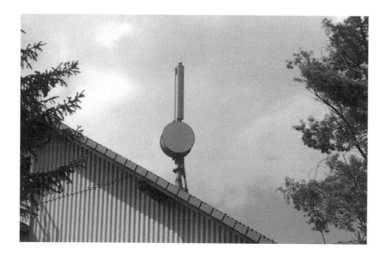

Figure 3.8 Base station antennas with a microwave dish for backhaul. (Reproduced from *Communication Systems for the Mobile Information Society*, Martin Sauter, 2006, John Wiley and Sons.)

Other operators prefer fixed-line connections. This has the advantage that the backhaul network is managed by another company and thus offloads responsibility to an external third-party company.

In North America and Japan a Time Division Multiplexing technology referred to as T-1 is used for the backhaul links with 24 timeslots of 64 kbit/s. In the rest of the world

E-1 links with 30 timeslots of 64 kbit/s are used. The bandwidth of a T-1 is thus 1.5 Mbit/s while the capacity of an E-1 is 2 Mbit/s. From the point of view of B3G networks the use of timeslots on a connection is a relic of voice-centric telephone networks. Here, T-1 and E-1 connections are used to transmit 25 or 31 individual telephone calls over the same line. For GSM and other voice-centric systems it made sense to use the same technology since base stations at the beginning were also exclusively used for voice calls. Due to compression used for voice calls in wireless networks, a single T-1 or E-1 timeslot carries 4 GSM voice calls. Since GSM is also a TDM-based system, one timeslot in a T-1 or E-1 connection carries the content of four air interface timeslots. A base station with three sectors and two carriers per sector each having eight timeslots thus requires 12 of the 24 or 31 timeslots of a T-1 or E-1 connection, respectively. In addition, an additional timeslot is usually used for signaling purposes. A single GSM base station only uses a fraction of the timeslots of a T-1 or E-1. In practice several base stations are thus connected in a chain to the same T-1 or E-1. Note: in the remainder of the text, E-1s are used for further comparisons since T-1 and E-1 connections are essentially the same except for the different number of timeslots.

When the GPRS was first introduced in wireless networks, the same 1:4 mapping between E-1 timeslots and air interface timeslots kept being used. With the introduction of EDGE, which increases packet-switched data transmission speeds, the fixed mapping between air interface timeslots and fractions of E-1 timeslots had to be abandoned. With EDGE the capacity of a single timeslot on the air interface is up to 58.2 kbit/s and thus much higher than the 16 kbit/s of a quarter of an E-1 timeslot. Mapping an EDGE air interface timeslot to a single E-1 timeslot also does not make sense since transmission conditions on the air interface are often not ideal and a lower transmission speed is used for the timeslot. Therefore, starting with EDGE, the timeslot concept which was invented for circuit-switched connections has for the first time become a problem for backhaul connections.

Today, 3G and B3G networks such as HSDPA and CDMA EVDo networks continue to use E-1 connections for backhaul. Since these systems use code division multiple access (CDMA) on the air interface, the time division multiplexing of E-1s is no longer required for transferring data. The decision of using this technology was mainly taken due to the fact that no other technologies were available at the time to transport data in both directions at the required speeds. Furthermore, UMTS/HSPA base stations must be synchronized tightly with each other, which is achieved in practice by locking the base station clock to the E-1 timing which is very precise due to the TDM nature of the technology.

UMTS/HSPA networks use the packet-switched ATM protocol to transfer data between the base station and the RNC. Data exchanged between the base station and the RNC is packetized into ATM packets which have a fixed length of 53 bytes and is then sent over E-1 links. Instead of using timeslots individually the ITU has developed a standard of how ATM packets are logically transported over E-1 links by using all timeslots simultaneously [25]. Thus, while the timeslots still exist on the lower layers of an E-1 connection, higher layers at both ends of the connection are no longer affected by the timeslot structure. Since the timeslot structure on lower layers is maintained, it is not necessary to modify E-1 transmission equipment despite changing to a packet-switched transmission mode.

In practice, the transmission speed of approximately 2 Mbit/s of an E-1 or 1.5 Mbit/s of a T-1 is not sufficient for a three-sector UMTS/HSPA base station. As shown in this chapter such a base station offers a capacity of around 7.5 Mbit/s if one carrier is used per sector or around 15 Mbit/s if two carriers are used. Consequently, a UMTS base station requires four E-1 connections for a single carrier configuration and up to eight E-1 connections for a two-carrier configuration.

Since a single user can achieve speeds on the air interface exceeding the transmission speed of a single E-1 connection, ATM data packets have to be multiplexed over several E-1 links. In UMTS networks, this is done via Inverse Multiplexing over ATM (IMA). In essence IMA sends ATM packets in a round robin fashion over several E-1 links, as shown in Figure 3.9. Vendors usually include IMA multiplexers as part of the base station hardware so no additional hardware is required at the base station site.

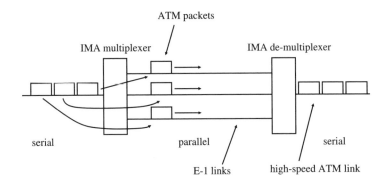

Figure 3.9   Inverse multiplexing over ATM.

As bandwidth requirements keep rising, it is not possible in the medium or long term to increase the number of E-1 links per base station for two reasons. Firstly, there are only a limited number of copper cables available at a base station site. Secondly, network operators are paying line rental fees per link and not per base station site. Depending on the country, E-1 line rental prices per month are currently between €200 and 500. As a consequence, each additional E-1 link significantly increases the monthly operating cost of a base station. Revenue per user on the other hand has reached a ceiling in many countries or is even slightly declining, mostly due to declining prices for voice calls. Data services such as Internet access can compensate for this, but require significantly more bandwidth than voice calls and as a consequence more bandwidth in the backhaul as well. In the future, B3G technologies such as WiMAX and LTE will require backhaul bandwidths of 60 Mbit/s and more per base station. This would require more than 30 E-1 links, which is clearly not practical either from a financial or from a technical perspective.

For the future a number of different technologies already exist or are currently being developed to increase the bandwidth on the wireless backhaul and decrease transmission costs to counter the increasing prices for backhaul connectivity. For the short and medium term some mobile operators may choose to split their backhaul traffic. Data for real-time voice calls will continue to be sent over E-1 connections to ensure quality of service and to have a reliable link for synchronizing the base station with the network. All other types of data flows will also be sent over copper cables but using a different technology. In practice this could be ADSL. Such an approach requires the installation of a device at the base station that can separate the two traffic classes and send them over the different connections. At the other end a similar device is used to combine them again before the next network element is reached. Figure 3.10 shows how such a setup looks for a UMTS/HSDPA network in which the ADSL network of a third-party company is used for the backhaul. The device to be installed at the base station site is usually small enough to fit in the base station cabinet. While this approach reduces the number of links required at the base station site, there are also a number of downsides. Current ADSL deployments are asymmetrical, which means that the bandwidth in the downlink is much higher then in the uplink direction. In a three sector base station configuration in which HSUPA is used there would not be enough uplink capacity in the backhaul to offload potential uplink traffic of all sectors. Another disadvantage is that network operators have to manage and monitor two types of backhaul networks, which creates additional overhead. Finally, this option is mainly interesting for mobile operators that have their own fixed-line ADSL networks, as other ADSL operators might not have a great interest in backhaul wireless traffic via their ADSL network. One of the reasons for this is the oversubscription factor per line discussed before, for which their networks have been dimensioned.

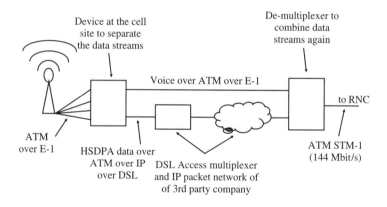

Figure 3.10   Use of DSL and pseudo-wires for backhaul.

A slightly different approach to the one shown in Figure 3.10 is for an operator to install their own DSLAM at a central site and terminate the ADSL links of base stations to their own equipment. In this scenario no IP pseudo-wires are required since ADSL natively transports ATM packets. In this scenario the demultiplexer would receive native ATM packets instead of IP-encapsulated ATM packets.

The use of two networks (E-1 and ADSL) for backhaul is not likely to be a long-term solution as there is a general trend towards IP-based packet backhaul solutions. This trend is driven by wireless technologies such as WiMAX and LTE, which natively use IP over Ethernet instead of ATM. At the time LTE becomes available on the market, many UMTS/HSPA operators are likely to replace their aging GSM or UMTS base stations with equipment supporting multiradio standards. The challenge of this approach is that such a base station requires three different types of backhaul connectivity: the GSM part of the base station requires TDM, GSM/HSPA is based on ATM and the LTE part of the base station is based on IP over Ethernet. As a consequence pseudo-wires will be used to encapsulate TDM and ATM traffic into IP packets, which will then be sent through the backhaul network and an IP metro network to the next node in the wireless hierarchy. This scenario is shown in Figure 3.11. In the metro part of the network, Ethernet over fiber is becoming more popular and metro Ethernet networks will in many cases replace current SDH (Synchronous Digital Hierarchy)-based optical network technology. On the backhaul link which connects the base station site to the optical metro network, several options exist:

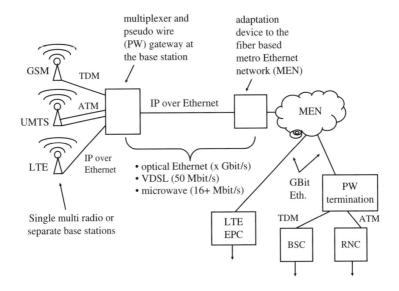

Figure 3.11   Packet-based backhaul options replacing today's E-1/T-1 links.

From a technical perspective using an optical link to connect to the metro network is the best choice. Optical fibers offer very high bandwidths and thus offer scalability for the future. Also, WiMAX and LTE base stations will have native twisted pair copper or fiber gigabit Ethernet interfaces and a connection to an optical metro Ethernet network is therefore straightforward. Unfortunately, only few base stations have fiber connectivity today and deploying new fibers to base station locations is likely to be very expensive.

VDSL is a copper cable-based alternative to fiber deployment. Current VDSL standards allow data rates of up to 50 Mbit/s in downlink and 50 Mbit/s in uplink direction at cable lengths below 1 km. Several VDSL connections per base station site can be used to increase bandwidth if required. At the edge of the metro network the VDSL connection could be terminated by a DSLAM, which in addition to terminating wireless backhaul connections can also be used to terminate consumer or business VDSL connections.

For mobile operators without fixed-line metro network assets, packet-based microwave backhaul solutions are an alternative. Ethernet microwave backhaul solutions support speeds of several hundred megabits/s today and it is likely that even higher bandwidths will be available in the future [34].

3.18 A Hybrid Cellular/Wi-Fi Network for the Future

As has been shown in Section 3.9 for cellular wireless networks and in Section 3.15 for DSL/optical/cable networks in combination with Wi-Fi, there is sufficient wireless capacity available to offer users a broadband connection to the Internet. Each network type, however, has advantages over the other.

DSL/optical/cable Internet connectivity in combination with Wi-Fi will probably be the technology of choice for most households in cities. The Wi-Fi access point is usually built into the DSL modem and thus the devices of all family members can be wirelessly connected within the home. From a financial point of view only one subscription is required to connect all members of the household. Furthermore, DSL/optical/cable network operators can offer TV and multimedia streaming to subscribers from a streaming server in their own networks due to the high capacity available on the last mile to the subscriber. In addition, the Wi-Fi network can also be used to connect devices at home with each other. This becomes more and more important as devices such as network-enabled TV screens, multimedia servers, NAS (Network Attached Storage) servers and PCs within the home communicate with each other. Streaming a recorded movie locally from a multimedia server to a TV screen in HDTV quality requires a large amount of bandwidth, which a Wi-Fi network can support in addition to other simultaneous data traffic such as VoIP, Web browsing and online gaming. The network thus creates a virtual local network bubble around the household and the people living in it. Many devices used in such a network, such as notebooks, are mobile to a certain degree and remain connected to the network even when moved through this bubble. The bubble, however, only has a limited size and once a user leaves it, for example by leaving the house, connectivity is instantly lost.

Cellular networks show their strength outside a local Internet bubble. B3G networks offer an overall capacity that can be sufficient to ensure continued connectivity for people leaving a local Internet bubble. Bandwidth requirements will usually be lower since the storage capacity and display size of mobile devices used on the move are an order of a magnitude smaller than those of devices used at home. For data exchange between personal devices, for example MP3 video streaming between a player and a headset, personal area networking technologies such as Bluetooth are the right choice. Cellular wireless networks are ideal to connect mobile devices back to the network at home, for example via an encrypted IP tunnel. Thus, subscribers can be seamlessly connected to their home network. Little bandwidth is required for home control applications such as checking and changing the status of lights and windows. Bandwidth requirements for streaming of stored content from a multimedia server at home or from a server on the Internet to the mobile device on the other hand requires significant bandwidth and capacity requirements will rise quickly once such applications become a mass market application. For some users, such as students or business travelers, using the cellular network for high-speed Internet connectivity in combination with nomadic devices such as notebooks instead of a Wi-Fi/DSL connection is also appealing. Comparison of the values estimated in Sections 3.9 and 3.15 shows that cellular networks are able to meet these demands for a significant percentage of the population in addition to traffic generated by other applications such as voice calls and mobile access from small mobile devices. In rural areas and at special locations such as in the car or on the train, mobile networks will often be the only cost-efficient way to offer broadband Internet access to subscribers.

Today, there is still little interaction between these two worlds. Users usually leave their Internet bubble behind once they leave their house or office. A change can be observed in practice, however, with mobile e-mail solutions now becoming more popular and mobile network operators starting to offer mobile Web surfing at attractive prices. On the terminal side, as further discussed in Chapter 5, small and portable multimedia devices for both voice and Internet communication become more powerful and user-friendly and usually include both cellular and Wi-Fi interfaces. Voice and video communication is also moving to the IP world, as will be discussed in Chapter 4. Single mobile multimedia devices can thus be used seamlessly in the personal Internet bubble at home and in the much larger Internet bubble created by cellular B3G networks. As will be further discussed in Chapter 6, Web 2.0 and mobile Web 2.0 applications are now taking advantage of both large- and small-screen devices and have thus established a presence in both the personal Internet bubble on notebooks and PCs and on mobile devices in either network depending on the capabilities of the mobile device and the location of the user.

As a consequence, cellular B3G networks are slowly transforming into overlay networks to the private Internet bubble. Many users are likely to spend a significant time in their local Internet bubble. Mobile devices which include a Wi-Fi interface can thus use such private bubbles for a significant amount of time, which reduces the load on high-speed cellular B3G networks. Music downloads are a good example. While high-speed cellular networks deliver a similar experience when downloading music from a central server or via a secure connection from the user's database in their home network, using the local Wi-Fi network for this purpose makes more sense.

A converged use of B3G networks in combination with personal Wi-Fi networks at home, at the office and in hotspot locations also makes sense from a capacity perspective. Even if high-throughput streaming between local devices is taken out of the equation, it is likely that cellular wireless networks in cities will depend on personal Internet bubbles to handle data traffic while users are at home or in the office. This network convergence requires a different approach by network operators. Today, companies offering fixed-line high-speed Internet access and companies operating B3G networks are often separate entities. With the growing trend of using both network types for Internet access by a majority of the population, the former trend of splitting telecom companies into fixed-line and wireless divisions or even distinct companies will revert. Network operators with both fixed-line and wireless assets will have a competitive advantage since they can offer converged network access to their customers. From a backhaul perspective, having both wireline and wireless assets allows a telecom operator to use a single network infrastructure to backhaul both wireless and fixed-line data traffic. It further enables network operators to offer a seamless communication experience to their customers by offering devices that can be used in Wi-Fi networks at home and in cellular B3G networks while on the go. For the mass market it is important to offer and pre-configure services on mobile, nomadic and stationary devices to work in such a converged network environment. Operators with both assets have a further advantage as they can offer Wi-Fi/DSL, B3G cellular access, devices and pre-configured services in a single package. Since the concepts of device and application convergence from a network point of view will form an integral part of tomorrow's communication landscape, they are discussed in more detail in the following chapters in this book. Figure 3.12 shows what such a converged network architecture looks like from the user's and the network operator's point of view.

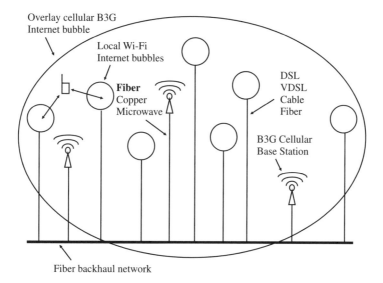

Figure 3.12   Converged cellular B3G and Wi-Fi/fixed-line network infrastructure.

References

1. Prepaid wireless internet access database (2008) http://prepaid-wireless-internet-access.wetpaint.com.
2. Vodafone web sessions (2008) http://www.vodafone.de/privat/tarife-flatrate-vertrag-handy-student/105359.html.
3. Rumney, M. (March 2007) What next for mobile telephony – examining the trend towards high-data-rate networks, *Agilent Measurement Journal*, issue 3, 32.
4. Seymor, M. (2007) Charting the 2007 WiMAX revolution, p. 3, Alcatel Lucent, http://www.ieee-mobile-wimax.org/downloads/M.Seymour.pdf.
5. Holma, H. and Toskala, A. (2006) HSDPA/HSUPA for UMTS, Figure 7.38, John Wiley & Sons, Ltd, Chichester.
6. ECC, CEPT (2007) The European table of frequency allocations and utilisations in the frequency range 9 kHz to 1000 GHz, ERC Report 25, http://www.ero.dk/documentation/docs/docfiles.asp?docid=1650.

7. 3GPP (6 June 2008) User equipment (UE) radio transmission and reception (FDD), TS 25.101, Table 5.0, version 7.8.0.
8. Martin, S. (2007) 3G and 4G wireless is private – DSL is for sharing, http://mobilesociety.typepad.com/mobile_life/2007/06/wireless-is-pri.html.
9. Douglas, T. (2005) Shaping the media with mobiles, BBC News, 4 August 2005, http://news.bbc.co.uk/2/hi/uk_news/4745767.stm.
10. Holma, H. and Toskala, A. (2006)HSDPA/HSUPA for UMTS, Table 8.5, John Wiley & Sons, Ltd, Chichester.
11. Dahlman, E., Parkvall, S., Skold, J. and Beming, P. (2007)3G Evolution; HSPA and LTE for Mobile Broadband, Chapter 19.2. Elsevier, Oxford.
12. 3GPP (11 December 2008) General packet radio service – service description; TS 23.060, stage 2, Release 6, Chapter 9.2.2.1.1.
13. 3GPP (6 June 2008) Mobile radio interface Layer 3 specification; core network protocols; TS 24.008, stage 3, Release 6, Chapter 10.5.6.12.
14. Camarillo, G. and Garcia-Martin, M. (2006) *The 3G IP Multimedia Subsystem*, 2nd edn, Chapter 13.3. John Wiley & Sons, Ltd, Chichester.
15. 3GPP (20 June 2007) Multiplexing and multiple access on the radio path, TS 45.002, Table B.1, version 7.4.0.
16. Holma, H. and Toskala, A. (2006) HSDPA/HSUPA for UMTS, Chapter 7.5. John Wiley & Sons, Ltd, Chichester.
17. Sauter, M. (2006) Communication Systems for the Mobile Information Society. John Wiley & Sons, Ltd, Chichester.
18. Bundesnetzagentur (2008) EMF Datenbank, http://emf.bundesnetzagentur.de.
19. 3G.co.uk (September 2007) Thalys to provide 3G/GPRS wireless on train routes, http://www.3g.co.uk/PR/Sept2007/5120.htm.
20. Dulski, A., Beijner, H. and Herbertsson, H. (February 2006) Rural WCDMA – aiming for nationwide coverage with one network, one technology, and one service offering, Ericsson review.
21. 3GPP (2006) RP-060191, extended WCDMA cell range up to 200 km, http://www.3gpp.org/ftp/tsg_ran/TSG_RAN/TSGR_31/Docs/RP-060191.zip.
22. Holma, H. and Toskala, A. (2007) WCDMA for UMTS: HSPA Evolution and LTE, Chapter 2.8. John Wiley and Sons, Ltd, Chichester.
23. The International Telecommunication Union (2003) One-way transmission time, ITU-T G.114, http://www.itu.int/rec/T-REC-G.114-200305-I/en.
24. Borman, C. (ed.) (June 2001)Robust header compression (ROHC): framework and four profiles: RTP, UDP, ESP, and uncompressed, IETF RFC 3095, http://tools.ietf.org/html/rfc3095.
25. International Telecommunication Union (1 July 2004) ATM cell mapping into PDH, ITU G.804.
26. Roese, J. (30 July 2007) The unwired enterprise, http://blogs.nortel.com/ctoblog/2007/07/30/the-unwired-enterprise.
27. Gast, M. How many voice callers fit on the head of an access point?, http://www.oreilly.com/pub/a/etel/2005/12/13/how-many-voice-callers-fit-on-the-head-of-an-access-point.html.
28. Free (2007) Accédez au monde de la convergence multimédia, http://adsl.free.fr/Wi-Fi.
29. Layer 1 ISM Band tracer (2007) http://www.metageek.com.
30. Board of Trustees of the University of Illinois (2007) Iperf – The TCP/UDP Bandwidth Measurement Tool, http://dast.nlanr.net/Projects/Iperf/.
31. Martin, S. (October 2006) VoIP over Wi-Fi – a field report, http://mobilesociety.typepad.com/mobile_life/2006/10/voip_over_wifi_.html.
32. Goscomb Technologies (2007) IP Transit, Partial Transit & BGP in London, UK, http://www.goscomb.net/services/iptransit.php.
33. Alcatel-Lucent (2008) 7300 advanced services access manager outdoor cabinet, http://www1.alcatel-lucent.com/products/productsummary.jsp?category=&productNumber=a7300oc_etsi&subCategory=DSLAMs.
34. Dragonwave microwave backhaul solutions (2008) http://www.dragonwaveinc.com/products-whitepapers.asp.

4

Voice over Wireless

Despite the growing number of Web 2.0 and mobile Web 2.0 applications being used in B3G networks (cf. Chapter 6), voice telephony continues to be the most important application in a mobile network. As mobile operators have a long history of making voice telephony work over wireless networks, and as they control the mobile infrastructure, they are in a good position to secure a sizable market share of tomorrow's wireless telephony business. Up to 3G UMTS, voice telephony was tightly integrated into the wireless network infrastructure. As a consequence mobile operators enjoyed a voice telephony monopoly in their networks. From the user's point of view the situation changed slightly in many countries in recent years due to fierce competition. MVNOs sprang up who bought buckets of voice minutes from mobile network operators for reselling to customers under their own brand. From a technical point of view, however, the network operator remained in charge, as MVNOs were mere resellers of voice minutes.

In B3G networks the situation has changed. As already discussed in Chapters 2 and 3, B3G networks no longer have a separate core network for voice telephony and voice-optimized protocol stacks in the radio network. As in fixed-line broadband networks, all services and applications are now delivered via the Internet Protocol and a packet-switched connection. This brings both opportunities and challenges. On the positive side, from an innovation point of view, network and services are now decoupled. Thus, mobile services, be they voice or data centric, are no longer solely in the hand of network vendors. Instead, Internet companies are now shaping the service landscape and are working on repeating their success in wireless networks. Since voice telephony has become just one of many services being delivered over IP, fierce competition for next generation voice services has sprung up between network vendors and operators on the one hand and Internet companies on the other. As far as voice telephony is concerned, no clear winner has yet been determined, since for the time being the majority of cellular voice calls are still transported over non-IP connections in 2G and 3G networks.

Both sides have strengths and weaknesses. The following chapter therefore looks at this topic from a number of different angles. First, an introduction is given of how voice telephony works in 2G and 3G networks as well as the benefits of migrating this service to the IP world. Afterwards the chapter takes a look at different telephony over IP services

Beyond 3G – Bringing Networks, Terminals and the Web Together: LTE, WiMAX, IMS, 4G Devices and the Mobile Web 2.0   Martin Sauter   © 2009 John Wiley & Sons, Ltd

from a network operator's point of view and from an Internet company's point of view. Even though there is a general consensus that applications should be access-agnostic, that is that they should not be aware of, or even care, what kind of access network technology (cable, DSL, fiber, wireless, Wi-Fi, etc.) is used, this does not work well in practice for wireless networks. The reason behind this and the implications are also discussed.

4.1 Circuit-switched Mobile Voice Telephony

The main purpose for which 2G wireless systems such as GSM were designed at the beginning was mobile voice telephony. Since it was the only application, each network element was specifically optimized for it. At the time, the state of the art for voice telephony in fixed-line telephony networks was to establish a transparent full duplex channel between two parties. While the connection is established between two parties in such a system, all data sent by the originator was transparently sent to the other side. The only exception are echo cancellation modules which are put into the transmission chain to improve voice quality.

4.1.1 Circuit Switching

Two parties are connected by a switching center which has a switching matrix to connect any telephone line (circuit) with any other. For long-distance calls, several switching centers were daisy-chained. For 2G wireless systems the principle was reused as shown in Figure 4.1. The main difference between a circuit-switched fixed-line telephony system and a wireless circuit-switched telephony system is that a subscriber can no longer be identified by a pair of wires. Instead, a subscriber is now identified with credentials stored on a SIM card. On the network side a database known as the Home Location Register contains a replica of the user credentials and information on which options and

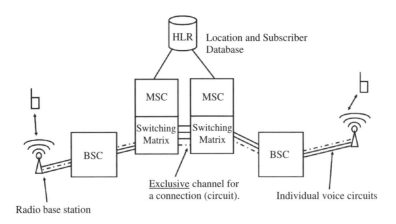

Figure 4.1 High-level GSM network architecture.

supplementary services (e.g. call forwarding) a user is allowed to use. Except for the additional database in the network, most of the hardware and software of a fixed-line switching center could be reused for the design of a mobile switching center. Since the user is no longer identified by a specific pair of wires, a mobility management software component was added to the switching center software.

4.1.2 A Voice-optimized Radio Network

Base Stations Controllers (BSC), as shown in Figure 4.1, are used to decouple switching center from the tasks required to establish and maintain a radio connection. BSCs are, as the name implies, responsible for communicating with base stations, which were initially designed as modems that convert digital information delivered on fixed-line connections into information sent over the air. In addition, BSCs are responsible for handing over a voice call to another base station or another radio cell of the same base station in case of deteriorating radio conditions. A more general term for a voice call in this architecture is 'circuit-switched call' or 'circuit-switched connection', since the switching center uses a switching matrix to connect two transparent circuits together for the duration of the call.

On the user plane the Transcoding and Rate Adaptation Unit (TRAU), which is a logical part of the BSC, converts the speech codec used in the radio network to the speech codec used by the mobile switching center. This adaptation is necessary as GSM reuses the hardware of fixed-line switching centers, which is based on 64 kbit/s circuit-switched channels while on the radio network side it was necessary to compress the voice signal to squeeze as many voice calls as possible through the narrow radio channel. Furthermore, the adaptation is necessary as mobile networks are interconnected with fixed-line networks which also use 64 kbit/s circuit-switched channels for voice telephony.

4.1.3 The Pros of Circuit Switching

As each voice call is transported in a circuit-switched channel, the behavior of the system when setting up a new call is deterministic. If there is a free circuit between the originator of the call, the base station, the base station controller, the mobile switching center, and from there to the terminator of the call, the call is established. If no circuit is available on one of these links, the call request is rejected by the MSC. During a call, system behavior is also deterministic. As circuits are uniquely assigned to a call, each call is independent and thus they cannot influence each other.

For further details on the design of 2G GSM circuit-switched wireless networks, the 2G radio network and call establishment, see chapter 1 of [1]. It should further be noted at this point that, over the years, the GSM network as presented above was enhanced by packetizing the connection between the MSCs with an approach known as the Bearer Independent Core Network (BICN). For an overview of BICN see Section 1.6.

4.2 Packet-switched Voice Telephony

Designing a network for circuit-switched connections is ideal for voice telephony, fax and narrowband circuit-switched data calls. Unfortunately, this also limits the use of the network to a narrow set of applications. As networks were designed with these applications in mind, there is no separation between the network and the applications, which ultimately prevents evolution. The SMS service is a good example. Adding SMS to the GSM network meant misusing signaling channels originally designed to carry messages required for voice call establishment. Again the application (SMS) was tightly integrated into the network design and only worked in this specific type of network. As a result, SMS messages could not be sent between wireless networks in the USA that used different kinds of 2G network technology for many years. Thus, tight integration of network and service prevented the take-up of a new service for many years until at last gateways were put in place between networks that convert the SMS signaling on one network to the signaling standard used in another.

Tight integration of applications and networks also prevents the evolution of an application. This is because changing an application also requires changes to the network itself. This is a process which network operators are only doing with great reluctance since changing the network structure is difficult, expensive and bears the risk of unforeseen side effects on other parts of the network.

4.2.1 Network and Applications are Separate in Packet-switched Networks

The Internet, on the other hand, follows an entirely different approach. Here, network and applications are independent of each other. This is achieved by creating a neutral transport layer that carries packets. Each packet has a source and destination address, the IP address. Nodes in the network then use the destination IP address to decide on which link to forward a packet to. Packets can be of variable lengths and can be concatenated on higher layers of the protocol stack. Thus any kind of application can efficiently send high and low volumes of data through the network. The network and the applications that run over the network are decoupled since the network does not see applications, just IP packets. At the top of the protocol stack applications do not see IP packets, just streams of data. This separation has worked very well in practice, as can be seen for example by the birth of the HTTP protocol and Web browsing, which were only invented many years after the invention of the IP protocol and the launch of first networks carrying IP packets. Since then, the Internet, or rather the accumulation of IP networks that form the Internet, has mainly evolved to offer ever higher speeds to the end user and to make access to the network cheaper. Applications have evolved almost independently of the network except for the fact that new applications such as video streaming and IPTV require much higher bandwidths.

4.2.2 Wireless Network Architecture for Transporting IP packets

The support for transporting IP packets in wireless networks was added in several stages. The first stage of the process was to add a packet-switched core network domain alongside the already known circuit-switched part of the network. In addition, the Radio Access Network (RAN) was enhanced to support both circuit-switched and packet-switched services simultaneously. This is shown in Figure 4.2. When a mobile device establishes a connection in a UMTS network, it informs the radio network or the RNC if it wants to establish a circuit-switched connection for a traditional service such as voice

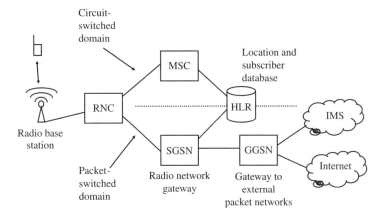

Figure 4.2   The UMTS network architecture with a circuit-switched and a packet-switched domain.

telephony or SMS or a packet-switched connection. The RNC then forwards the request to the MSC in case a circuit-switched connection is to be established or to the gateway node of the packet-switched core network (SGSN). Where a packet-switched connection is requested, the mobile device and the SGSN then perform authentication and activate ciphering for the radio link. Afterwards the mobile device requests an IP address in a procedure referred to as the Packet Data Protocol context activation. The SGSN then communicates with the gateway to the Internet (GGSN) which selects an IP address and returns it to the SGSN. The SGSN then returns the IP address to the mobile device and the connection is established. During the PDP context activation the mobile device and the network can also negotiate the quality of service to be used for the connection. An important property of a certain QoS level is, for example, the maximum bandwidth the network will grant to the mobile device and the minimum bandwidth it will try to enforce in case of congestion.

For applications, this process is usually transparent, as the establishment of an Internet connection is usually the task of the device's operating system. On a notebook, for example, the process is usually invoked manually by the user either with a special program delivered together with the wireless network card or by using the dial-up application of the operating system. Applications such as a Web browser, instant messaging program or VoIP client will just notice that the notebook is connected to an IP network and start using the connection.

It is important to note at this point that the initially negotiated QoS level of a wireless connection applies to all applications using the connection; that is, it applies for all packets being sent and received. In practice such a QoS negotiation is mainly used to limit the maximum allowed bandwidth per user. Such a general assignment is unsuitable, however, to ensure the priority of IP packets of a voice telephony application over IP packets exchanged between a Web browser and a Web server in case of congestion. Depending on the type of network, it is thus possible to negotiate a separate quality of service level for individual applications or traffic flows. Since this requires the intervention of the application, it breaks the separation of network and application. As a consequence the application must become network-aware. Also, many network operators only allow their own applications to request a certain quality of service for their individual data stream.

While the connection is established, the RNC is responsible on the network side for maintaining the physical connection, enforcing the requested quality of service and handing the connection over to a different cell if the user is mobile. For further details refer to Chapter 2, which discusses these processes for different B3G network types.

4.2.3 Benefits of Migrating Voice Telephony to IP

Given the increase and abundance of bandwidth, it seems like a natural step to also transport voice calls over the Internet instead of over a separate and dedicated network. There are many different angles from which to view this shift. From a network operator's point of view, transporting voice calls over IP reduces cost in the long term since only a single network needs to be maintained. From an enterprise point of view, using the IP network inside the enterprise for voice telephony reduces the number of cables in the building since a desk no longer requires a separate cable for the telephone. From a user's point of view, migrating the circuit-switched voice service to the packet-switched IP domain potentially reduces cost as they no longer have to pay for analog telephony and Internet access separately. Instead of connecting the analog telephone to a traditional telephone jack in the wall, it is now connected to a telephone jack in a DSL router or cable modem. The router contains the required hardware and software to enable call establishment and to digitize the analog signal received from the analog telephone, packetize it and send it over the Internet. If the network is well designed, the change is transparent to the user. Except for the potentially lower price, however, the user has no incentive to migrate to VoIP this way. This applies for fixed-line as well as for wireless networks.

4.2.4 Voice Telephony Evolution and Service Integration

Fortunately there is another good reason for migrating voice telephony to the IP world, both in wireline and wireless networks: service integration. Service integration happens when a communication device is capable of running several services simultaneously and is able to combine them in an intelligent way. Examples of this include telephony plug-ins in Web pages. When a user searches for a suitable hotel they might do this by accessing the Internet and querying a search engine. The search engine will return a number of Web pages of hotels in a city and the user will then go through those pages to see if some of those hotels are suitable. Several hotels are interesting but she wants to be sure that there will be Wi-Fi access and that the costs are reasonable. She therefore decides to call the hotel instead of ordering online. In a circuit-switched telephony world the user now has to pick up the telephone and dial a telephone number. This takes some time and is potentially expensive if this involves making an international call. In a world where voice telephony is just another application running over the same network, the Web page can have an embedded button which will launch a VoIP program on the notebook which automatically calls the hotel. The user is connected instantly, does not have to type in a telephone number and the call is likely to be free no matter what part of the world the hotel is located in.

Another example of service integration is a document with embedded information about those who created it. When downloading an interesting document it is usually difficult to find out who created the document and how to contact them, and then it is necessary to pick up a telephone and dial a telephone number. This means that there is a gap between the desire to communicate and the ability to do so. Once voice telephony has just become another application running over a single network, the information in the document can be used by the voice telephony application to find the user and call them. This significantly shortens the gap between the desire and the ability to communicate.

Service Integration also means making the voice telephony service more flexible and grouping new types of communication applications around it. In the IP world, voice telephony can be extended with presence capabilities so a person can see if another person is available before making a call. Instant messaging in combination with presence is also an important addition to voice telephony as users often prefer a text-based message to a call.

Voice telephony evolution also means separating the service from the device. Since the network and the application are separate, it is also no longer important which device is used with a certain voice telephony user account. A user can have several devices for voice communication, like a cordless telephone, PCs and notebooks, B3G cellular telephones, a PDA, a game console and so on that all use the same voice telephony account. Some or even all devices can be attached to the network simultaneously and an incoming call will be delivered to all active devices or a user-defined subset instead of only to a single device, as in the past. The user then accepts the call with the most suitable device in the current situation.

4.2.5 Voice Telephony over IP: the End of the Operator Monopoly

Another important element in the transition of the voice telephony service into the IP world is that it ends the operator monopoly on this service in both fixed-line and wireless networks. This has allowed Internet companies such as Skype, Yahoo, Microsoft, Sipgate and many others to offer telephony services either with proprietary or standardized protocols. This has sparked competition and innovation.

The remainder of this chapter now looks at a number of different approaches to offering voice telephony in different standardized fashions over wireless B3G networks. Section 4.3 introduces the SIP which is the most popular protocol and service platform for establishing voice calls over IP networks. Since it is completely network-agnostic, it will run over both wireline and wireless IP networks and allows Internet companies to offer voice telephony services.

Section 4.4 takes a closer look at the IMS, a telephony and general application framework favored by current fixed-line and mobile operators as their next-generation telephony and services platform. The IMS itself is based on the SIP architecture, which was enhanced with standardized procedures to act as a general multimedia service delivery platform over both fixed-line and wireless networks. The IMS system also contains additions to communicate with the network itself to ensure quality of service, which is something that 'naked SIP' (described in Section 4.3) is not capable of doing. Why this might be required in the future and the pros and cons of this are then discussed in Section 4.5.

4.3 SIP Telephony over Fixed and Wireless Networks

The most popular standardized system to establish voice calls over IP networks is SIP. It is standardized in RFC 3261 [2] and is the abbreviation for Session Initiation Protocol. As the name suggests, the protocol is intended for establishing not only a voice connection between two parties, but also a general messaging protocol to establish a 'session' between two or more parties. The term 'session' is quite generic and in practice SIP can be used to establish voice and video calls and instant message exchange between two parties; it can carry presence information so users can see when their friends are online and many other things. In practice, however, SIP is mostly used for establishing voice calls today, despite its general nature.

In the wireless world, SIP telephony is used today in various ways. Cable and DSL modem routers often include a SIP application which converts the signals received from an analog voice telephone plugged into a telephone jack at the back of the router. Since from the telephone's point of view the telephone jack behaves just as the standard analog telephone network jack, it can be used with traditional wired and cordless telephones. There are also a number of Wi-Fi telephones available on the market today which use the SIP protocol and a Wi-Fi connection at home or at the office to connect to the Internet. Most importantly, however, there are cellular telephones which in addition to 3G or B3G are also Wi-Fi capable. Many of them also have a built-in SIP client software and can thus be used as cellular telephones and SIP telephones over either Wi-Fi or the B3G network. Last but not least, there is a great variety of SIP software available for PCs and notebooks.

4.3.1 SIP Registration

Figure 4.3 shows the network components required for voice telephony over IP with SIP. In SIP terminology the VoIP software running on the client device is referred to as the User Agent. This abstraction is a good fit for the B3G world since mobile devices are

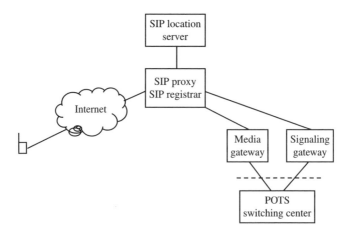

Figure 4.3   The SIP network architecture.

becoming ever more versatile and, although mobile voice telephony is surely an important application, it is nevertheless just one of several applications on a mobile device.

When the User Agent is started, for example when a mobile device is switched on, it will register its availability with the SIP registrar in the network in order to be reachable and to allow the user to make outgoing calls. For the following description it is assumed that the mobile is already attached to a wireless network and an IP address has been assigned. This procedure is not part of the SIP specification since SIP voice telephony is an application running on an IP network. How a device attaches to the network and how an IP address is requested in Wi-Fi and cellular B3G networks has been shown in Chapter 2.

If an IP address is used to identify the registrar, the User Agent can send a registration message right away. If a domain name is used instead, a DNS server has to be queried first to convert the domain name into an IP address.

The most important information elements of the SIP registration message are the IP address and UDP port of the User Agent and the SIP identity of the user. The format of the SIP identity, also referred to as the user's SIP Universal Resource Identifier (URI) [3], is similar to that of an e-mail address and contains the user ID and the SIP domain (also known as 'realm'). The realm identifies the SIP provider with whom the user has a subscription. Examples of valid SIP addresses are '5415468@sipgate.de' and 'martin.sauter@mydomain.com'. Both the ID ('5415468' or 'martin.sauter') and the realm (sipgate.de or mydomain.com) are required to be reachable by users of a different SIP network provider, as will be shown below.

When the registrar receives a 'register' message, it searches the database for the corresponding account information and then attempts to authenticate the user. This is done with a password which is shared between the User Agent and the registrar. Unlike cellular wireless systems such as GSM and UMTS, where the common secret is stored on the SIM card of the mobile device, the password for SIP telephony is usually stored in the mobile device itself and can be changed via a menu in the User Agent software as required. Verifying the password is done by the SIP registrar rejecting the first registration request with an 'unauthorized' response message which contains a random value for the User Agent, which is referred to as a 'nonce'. The User Agent then takes the nonce and the password as an input for a shared encryption algorithm to create an authentication response value. This value is then sent back to the registrar in another register message. When receiving the second register message the registrar compares the authentication response value to the value it has computed itself. If the values match, the subscriber is authenticated and the registrar answers with an 'ok' message. Afterwards the User Agent is available and the user can initiate and receive calls. As a final step the registrar stores the subscriber context (IP address, UDP port number, etc.) in the SIP location server database. Figure 4.4 shows the message flow in a graph.

Associating the subscriber's context with the SIP identity is also referred to as a binding since the registration process binds the user's identity (the SIP URI) to the information on which User Agent on which device the user can be reached. The user can register several User Agents, that is several devices, to their SIP identity. Incoming calls will then be signaled to all User Agents/devices that are bound to a SIP identity. In the circuit-switched telephony world this feature is known as simultaneous ringing or 'simring'.

Figure 4.5 shows the message content of the second SIP register message. All information in the SIP message is in human readable format. This makes the message

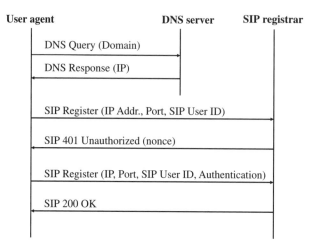

Figure 4.4   SIP registration message flow.

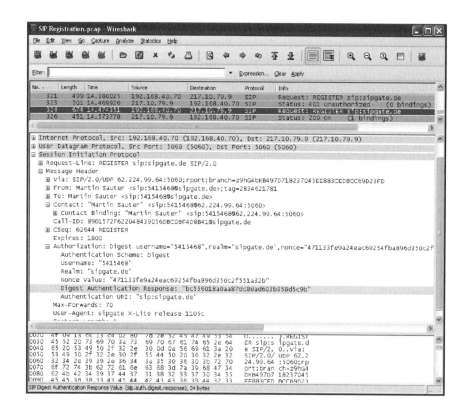

Figure 4.5 A SIP register message with authentication information. (Reproduced from *Wireshark*, by courtesy of Gerald Combs, USA.)

quite large (674 bytes in the example), but has the advantage that it allows easier development of new functionality and debugging of a system. The syntax of SIP messages is similar to that of the request and response messaging of the Hypertext Transport Protocol (HTTP), which is used by Web browsers to request a Web page. Each SIP request message starts with a request line which announces the type and the intention of message. The message header then contains the information required, depending on the message type. The register message in Figure 2.6, for example, contains the identity of the user and the authorization information. All requests also contain a command sequence number (CSeq) to be able to correlate all messages belonging to the same request/response dialogue. This allows simultaneous dialogues for different purposes, like receiving an instant message while at the same time establishing an outgoing call.

Responses to a SIP request contain a status line at the beginning of the message and the message header. The status line contains a numeric status ID and the message name in clear text. The header then contains further response information.

4.3.2 Establishing a SIP Call Between Two SIP Subscribers

Figure 4.6 shows the message flow between two User Agents via two SIP proxies for establishing a voice call. A SIP proxy is usually physically implemented in the same server as the SIP registrar, as shown in Figure 4.3. From a logical point of view, however, the SIP proxy is independent. The proxy gets its name from the fact that the User Agent does usually not know the IP address of the other party the user wants to get in contact with and thus sends the message to the SIP proxy. The SIP proxy then locates the other party and forwards the message on behalf of the user.

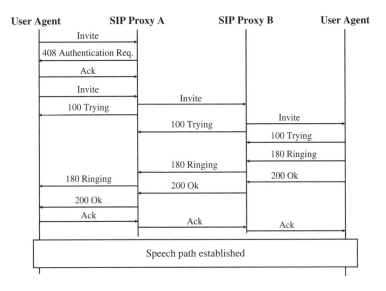

Figure 4.6   SIP session establishment message flow.

The first message sent by the originator to the SIP proxy is a SIP 'invite' message. The most important parameter of this message is the destination identity. The identity can have two different formats. SIP subscribers can be identified either with a username (e.g. sip:martin.sauter@sipgate.de) or a standard telephone number to be reachable from external fixed and mobile networks. Where the user dials a standard telephone number the User Agent automatically appends the local realm (e.g. sip:00497544968888@ sipgate.de).

When the SIP proxy receives the message it responds with a '407 Proxy Authentication Required' message to authenticate the User Agent. The User Agent then terminates this invite dialogue with an acknowledge message. Afterwards, the User Agent calculates an answer for the random value given back by the SIP proxy and sends another invite message, this time including the security information.

In the next step the SIP proxy verifies the authentication response in the invite message as described in the register dialogue above and proceeds with the session establishment by analyzing the destination party identity in the 'To:' field of the message header. Depending on the realm part of the identity, the SIP proxy searches either its own subscriber database or a database containing the IP addresses of proxy servers of other SIP networks. In the case shown in Figure 4.6, the realm belongs to a different SIP network and the message is thus forwarded to the SIP proxy server of the other network. To show that the message has been forwarded, the SIP proxy then returns a '100 trying' message back to the User Agent.

When the SIP server of the other SIP network receives the incoming message it also looks at the 'To:' field of the 'invite' message. Since the realm part of the ID is its own it will query the local user database to retrieve the location (IP address and UDP port number) of the subscriber. It then forwards the 'invite' message to the destination subscriber and returns a '100 trying' message back to the SIP server in the other SIP network.

When the destination User Agent receives the 'invite' message it also returns a '100 trying' message to the SIP proxy to indicate that it has successfully received the invitation. The User Agent then informs the user of the incoming call and returns a '180 ringing' message to the SIP proxy. This message is then sent back via the SIP proxy in the other SIP network to the originating User Agent. When the user accepts the call, the User Agent sends a '200 ok' message to the originating User Agent via the two SIP proxies. At the same time the audio channel is established between the two User Agents. How this is done is described in the next section. It is important to note that the audio data is exchanged directly between the two User Agents and not via the SIP proxies, as they are only used for establishing a session.

Each proxy that forwards a message to another SIP proxy adds its IP address to the header part of the message. The recipient of the message is thus aware of all the SIP proxies the message has traversed. When returning an answer the User Agent includes these IP addresses in the SIP header of the message again. This way a response can efficiently traverse the SIP network without requiring a decision at each SIP proxy as to where to forward the message.

SIP proxies are not only allowed to forward messages but can also change their content or react to certain responses. If the terminating User Agent, for example, sends back a 'busy' response, the SIP proxy can either return this message to the originating User

Agent or decide to discard the busy message and forward the call to a SIP voice mail system. Proxies can also fork a message, which is required, for example, if a user has registered several devices to the same SIP identity. In this scenario a single incoming SIP 'invite' message is forked and sent to each device registered to the identity. To do this the SIP proxy needs to be 'stateful' as it needs to remember how often it has forked a request to appropriately react to incoming SIP messages of the different destinations.

It should be noted at this point that the example in Figure 4.6, which reflects what is done in practice, does not contain messages to authenticate the destination User Agent. This allows a potential attacker who has managed to take over the IP address of the terminating User Agent to accept the call.

4.3.3 Session Description

As SIP is a generic session establishment protocol for many different kinds of media streams, the originator of a session has to explicitly inform the other party what kind of session (voice, voice + video, etc.) is to be established. This is done by describing the types of media streams and their properties in the body of the 'invite' message. The protocol used for this purpose is the Session Description Protocol (SDP), standardized in RFC 4556 [4].

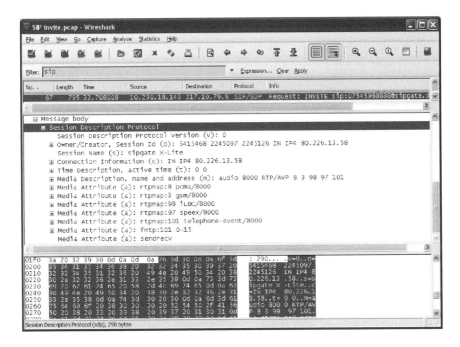

Figure 4.7   Session description in a SIP 'invite' or '200 ok' message. (Reproduced from *Wireshark*, by courtesy of Gerald Combs, USA.)

Figure 4.7 shows the SDP media description in the message body of the SIP 'invite' message for a voice call. The most important parameters are:

1. The connection information parameter – the originator uses this parameter to inform the other User Agent of the IP address from which the media information for the session will be sent and received.
2. The media description parameter – contains information on the type of media stream the originator wants to send:

   • The first subparameter specifies the type of media stream (e.g. audio).
   • The second subparameter represents the local UDP port number from which the stream will be sent and where the return stream is expected if a full duplex connection is to be established. This will be the case for most types of communication sessions.
   • The next subparameter is the description of the transmission protocol to be used for the media stream. Most audio and video applications use the RTP/AVP (Audio Video Protocol) for this purpose. Further detail on the streaming protocol is given below.
   • The remaining numeric parameters are RTP payload numbers, which indicate the supported voice and video codecs of the User Agent.

3. The media attribute parameters – these follow in subsequent lines and give further media details:

   • Rtpmap – this media attribute gives further details on the payload numbers. Payload number 8, for example, corresponds to the A-Law Pulse Code Modulation (PCMA) codec, which is also used in circuit-switched telephone networks to encode the voice signal for a 64 kbit/s circuit-switched channel. The line also informs the terminating User Agent that the input signal was digitized with a sampling rate of 8000 Hz.
   • Sendrecv – this media attribute tells the remote party that the originator will send a media stream and expects to receive a similar media stream from the remote party as well.

If several media streams are required for a session (e.g. voice + video), another media descriptor parameter and corresponding media attribute parameters are appended to the SDP message.

The remote User Agent sends their session description information as part of the '200 ok' message once the user has accepted the session. The same parameters are used in this SDP message as in the SDP message of the originator. The values transported in the parameters, however, might be different. This is the case, for example, if the remote User Agent supports a different set of media codecs than the originator. If they share at least one codec for a type of media stream the connection can be established. Otherwise the session setup will fail.

Once the '200 ok' message is received by the originator both ends of the connection will start sending their media streams. The media stream of the originator is sent from the UDP port described in the media descriptor parameter of the 'invite' message to the UDP port of the remote User Agent given in the '200 ok' message. The media stream of the remote User Agent uses the reverse port combination. The media codec chosen by both ends depends on the local list of supported codecs and the remote list of supported codecs received from the other end.

4.3.4 The Real-time Transfer Protocol

In circuit-switched networks a media stream can be transparently exchanged between the two parties through a circuit-switched connection. In IP networks, however, transmission is packet-switched and the media stream is encapsulated in IP packets which are then routed through the network. As each IP router between the two parties has to make a routing decision, each packet has to contain the IP addresses of the two parties and the UDP port numbers between which the media is exchanged. On top of IP and UDP, the RTP is used to encapsulate the media stream and to give the destination further information about how to interpret the incoming data. RTP is standardized in RFC 3550 [5] and Figure 4.8 shows the contents of an RTP header. As RTP is a generic media stream protocol, one of the first parameters of the header is the payload type, which informs the receiver of the protocol and the RTP profile to be used for decoding the payload information. RTP profiles for audio and video transmissions with different codecs are defined in RFC 3551 [6].

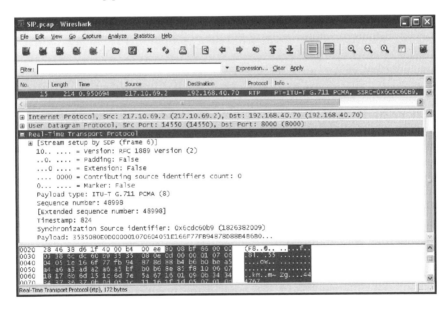

Figure 4.8 Real-time transport protocol header. (Reproduced from *Wireshark*, by courtesy of Gerald Combs, USA.)

In IP networks, the order and timely delivery of UDP packets is not guaranteed. Furthermore, it is possible for UDP packets to be dropped during times of network congestion. For real-time voice and video transmissions this is challenging. To compensate, the receiver uses a jitter buffer in which incoming data packets are stored for a short time (e.g. 50–100 ms) before they are played to the user. If a packet is late, the time it remains in the jitter buffer is reduced. If packets arrive in the wrong order, but still within the limits of jitter buffer, they can be reordered and the event is not noticeable to the user. The longer the jitter buffer stores packets, the higher the chances are that temporary network congestion will have no impact on the voice or video quality. The time for which the jitter can store packets in practice is very limited, however, since this adds to the overall delay of the connection. End-to-end delays including delays caused by the codec, by transmission and by the jitter buffer exceeding 150 ms are already noticeable to the user.

To control the jitter buffer the RTP header contains two parameters. The first one is the sequence number field which is incremented in each packet. This allows the receiver to detect missing or reordered packets. Constant bitrate codecs can then use the payload size and their knowledge of the sampling rate to determine if a packet is late, early or on time. Another way to determine the correct arrival time is the timestamp parameter. The use of this parameter depends on the type of audio or video codec used. The PCM codec, for example, samples an audio signal 8000 times a second. With 8 bits per sample the resulting codec rate is 64 kbit/s. The User Agent then typically puts samples of a 20 ms interval into one RTP packet. For each packet the timestamp value is thus incremented by $8000/0.02 = 160$.

Every RTP session is accompanied by a Real Time Control Protocol (RTCP) [5] session that uses the next higher UDP port number. Here, messages are exchanged periodically, in the order one every 5 s, in which the members of a session inform each other about the number of lost packets and the overall jitter experienced.

4.3.5 Establishing a SIP Call Between a SIP and a PSTN Subscriber

Many SIP networks enable their users to call subscribers of a Public (Circuit) Switched Telephone Network (PSTN) or of 2G/3G cellular networks. Since these subscribers can only be reached via a circuit-switched network, a gateway is required as shown in Figure 4.3 above. The gateway has two logically separate components which can be implemented either in a single device or separately if this is required for scalability reasons. The first logical component is the signaling gateway, which translates the SIP messages shown in Figure 4.7 into Signaling System Number 7 (SS-7) messages, which are used in circuit-switched networks. The second component is the media gateway that takes the digitized voice information from incoming IP packets and puts it into a circuit-switched connection and vice versa. Usually the SIP network and the PSTN network use the same voice codec. In this case the data on the application layer can thus be forwarded transparently. If different codecs are used in the two networks the media gateway additionally transcodes the media stream. Large networks might require several media gateways due to the high number of simultaneous calls or for redundancy reasons. Also, networks usually have at least two signaling gateways to be able to continue service in case one of the gateways malfunctions. The SIP proxy usually contains the functionality to detect that a gateway has failed and automatically redirects new sessions to the alternative gateway.

In the circuit-switched telephony world the terminating switching center is responsible for generating the alerting tone that is returned to the originator of the call until the terminating party has accepted the call. With the message flow as shown in Figure 4.6, the gateway is not able to forward the alerting tone to the SIP originator since the originator first requires the UDP session description from the SIP gateway which contains the UDP port number from which the audio stream originates. The standard therefore already allows the sending of the session description information in the '180 ringing' message or in a '183 session progress' message that is sometimes used as an alternative. This is the preferred solution even for SIP-to-SIP calls as the media streams can be established in the background before the terminator has accepted the call. Thus, no time is lost after the terminator has accepted the call.

Figure 4.9 shows the session establishment message flow from a SIP subscriber to a PSTN subscriber. The message flow up to the Signalling Gateway is identical to the message flow for the SIP-to-SIP call shown in Figure 4.6 except for the following: instead of using a '180 ringing' message, the alternative '183 session progress' is used in Figure 4.8, which contains the early session description of the signaling gateway. Also, additional signaling is required to inform the media gateway of the call.

The signaling gateway converts the SIP messages for the POTS, GSM or UMTS switching center as follows: the parameters of the SIP 'invite' message are used to create an SS-7 'Initial Address Message' (IAM), which requests the switching center to establish a circuit-switched connection to a subscriber.

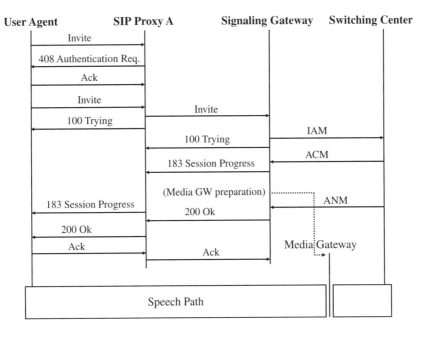

Figure 4.9   Messaging flow for a SIP to PSTN call.

The switching center responds with an Address Complete Message (ACM) which tells the SIP signaling gateway that the call is proceeding and which timeslot on which circuit-switched link between itself and the media gateway will be used for the voice data. This information is then used by the SIP network to configure the media gateway for the session. This is done by informing the media gateway of the link and timeslot number to be used for the session on the circuit-switched side and the IP address and UDP port number of the User Agent on the SIP side. Once the media gateway is ready, a '183 session progress' message is created and sent back to the SIP proxy, which in turn forwards the information to the User Agent. Since the session progress message contains a session description, the User Agent can now already activate the speech path to forward the alerting tone generated by the switching center to the user. Note that in Figure 4.8 the activation of the speech path is shown in the lower part for clarity reasons, even though it can already be active after the session progress message.

When the circuit-switched subscriber accepts the call, the switching center generates an Answer Message (ANM) and sends it to the signaling gateway. The signaling gateway converts this SS-7 message to a '200 ok' SIP message and forwards it to the User Agent via the SIP proxy.

Finally, it should be noted at this point that, from the switching center's point of view, the signaling gateway behaves like another switching center. Thus, no modifications are necessary in the SS-7 messaging. From the SIP proxy's point of view the signaling gateway behaves just like a User Agent. Thus, no changes are required in the SIP messaging either.

4.3.6 Proprietary Components of a SIP System

The functions and entities of a SIP network discussed up to now are only related to establishing a session. In practice, however, a SIP network usually has a number of additional functions:

- Some operators use in-band announcements to inform users about the cost per minute of a call before the connection to the terminating party is established.
- While many operators offer SIP-to-SIP calls for free, calls to PSTN or mobile subscribers are not free. Therefore a billing solution is required. In case of post-paid billing, which means that the subscriber receives a monthly bill, the SIP Proxy or the PSTN gateway needs to forward billing information to a billing system.
- It is also common for network operators to offer prepaid billing where users transfer a certain amount of money to the operator that can then be used for calls. Transferring funds usually requires a Web-based interface. To be able to bill for calls in real time, the SIP proxy or gateway needs to have an interface to a prepaid billing server. Alternatively the prepaid billing solution can also be a part of the proxy or gateway.
- Many SIP networks also offer a Web-based configuration and customer care interfaces with many functionalities. These include user self-configuration of call forwarding, prepaid top-up, review of billing information and so on.

How and where these functions are implemented is not standardized. From a technical point of view this is not necessary since their implementation has no impact on the SIP signaling between a User Agent and an SIP proxy. Lack of standardization has the advantage that companies can be innovative and offer additional functionality to network operators to make their service more attractive to end users which, in turn, can lead to a competitive advantage. The disadvantage of a lack of standardized interfaces for these functions to SIP components, however, also means that the market is fractured. The availability and evolution therefore depends on the vendor of the SIP equipment. This in turn has the disadvantage for network operators that there is no direct competition in this area, which usually results in higher prices for any additional functionality network operators want to buy once they have bought SIP equipment from a vendor.

4.3.7 Network Address Translation and SIP

Most Wi-Fi home networks connected to the Internet via DSL or cable are only assigned a single IP address by their network operator. To be able to use several devices within the home network the DSL or cable router has to map the local IP addresses to a single external IP address. Since two computers in the private network can use the same UDP or TCP port numbers, these have to be mapped between the private and the public network as well. This process is referred to as Network Address Translation (NAT). Figure 4.10 shows a setup which requires NAT. For applications such as Web browsing this mapping process is completely transparent as they do not use their own IP address and port number on the application layer.

The SIP and SDP protocols, however, use the local IP address and port number on the application layer, as shown in Figures 4.5 and 4.7, to signal to a SIP proxy and to a remote User Agent where to send messages and media streams. As the User Agent on a device is only aware of the private IP address and port number, an additional step is required to determine the external IP address and port used for a request before a SIP operation. This is usually done with 'Simple Traversal of User Datagram Protocol Through Network Address Translators', or STUN, which is standardized in RFC 3489 [7].

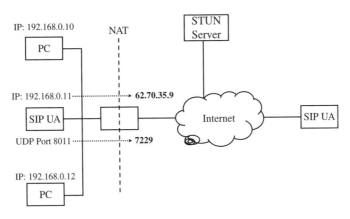

Figure 4.10   Network address translation in home networks.

The principle of STUN is as follows: before a User Agent contacts the SIP registrar to register, it first contacts a STUN server in the public Internet with a 'binding request' message. When the 'binding request' packet arrives at the DSL or cable router the NAT algorithm changes the IP address of the packet from the private IP address to the public IP address. If the local UDP port number used by the packet is already in use for a conversation of a different client device, it is also changed. The STUN server thus receives the packet with the external identification. The STUN server then sends a 'binding response' packet back to the User Agent. At the NAT device the destination IP address (and UDP port) of the packet is changed to the private IP address (and private UDP port). The 'binding response' packet, however, also contains the real IP address (and UDP port) in the payload of the packet, which is not changed by the NAT router. This way the SIP User Agent can determine which external IP address was used and if the UDP port number was changed as well. As a number of different NAT implementations exist, the User Agent will send additional 'binding request' messages to probe the behavior of the NAT device when the STUN server replies to the request from a different IP address and port number. If the NAT device delivers these replies from the STUN server to those requests, the User Agent is aware that it does not need to start sending a media stream to open the NAT port. If no answer is received for such additional binding requests, the User Agent takes this into account later on and does not wait for an incoming media stream from the other end before starting its own transmission.

Figure 4.10 shows one User Agent behind a NAT device and another User Agent directly connected to the Internet. STUN, however, also works for scenarios in which both User Agents are behind network address translators.

It should be noted at this point that there is one network address translation scheme which STUN is not able to overcome. Most NAT implementations will always use the same mapping from internal to external UDP port number regardless of the destination IP address. If, however, the NAT implementation uses a new port mapping for each external IP address, it is not predictable which UDP port will be used for the media stream since it will be sent to an IP address that is different from that of the STUN server. In this case, no media connection can be established and session establishment from User Agents behind such a NAT device fail. Fortunately, most routers used in home environments do not use this kind of address translation.

4.4 Voice and Related Applications over IMS

Section 4.3 has shown how SIP is used to establish Voice over IP sessions. SIP, however, is a general protocol to establish any kind of session and is thus capable of much more than just to establish voice sessions. In practice, however, voice telephony is the main application for SIP. As SIP and VoIP are based on the IP protocol, it can also be used in the packet-switched part of 3G networks and of course in B3G networks. From a wireless network operator point of view, however, a number of critical elements are missing in today's SIP specifications to support millions of subscribers. These include:

Mobility aspects:

- General SIP implementations are network-agnostic and cannot signal their quality of service requirements to a wireless access network. Thus, VoIP data packets cannot be preferred by the system in times of congestion.
- Handling of transmission errors on the air interface cannot be optimized for SIP calls. While Web browsing and similar applications benefit from automatic retransmissions

in case of transmission errors, VoIP connections would prefer erroneous packets to be dropped rather than be repeated at a later time since such packets are likely to come too late (cf. jitter discussed in Section 4.3.4).

- SIP VoIP calls cannot be handed over to the 2G network such as when the user roams out of the coverage area of B3G networks.
- VoIP over SIP does not work in 2G networks.
- Most SIP implementations today use the 64 kbit/s PCM codec for VoIP calls. Compared with optimized GSM and UMTS codecs, which only require about 12 kbit/s, this significantly decreases the number of VoIP calls that can be delivered via a base station. Furthermore, mobile network optimized voice codecs have built in functionality to deal with missing or erroneous data packets. While this is not required for fixed networks due to the lower error rates, it is very beneficial for connections over wireless networks.
- Emergency calls (112, 911) cannot be routed to the correct emergency center since the subscriber could be anywhere in the world.

Functionality aspects:

- There is no billing flexibility. Since SIP implementations are mostly used for voice sessions, billing is usually built into the SIP proxy and no standardized interfaces exist to collect billing data for online and offline charging.
- Additional applications such video calls, presence, instant messaging and so on are usually not integrated in SIP clients and networks.
- It is difficult to add new features and applications since no standardized interfaces exist to add these to a SIP implementation. Thus, adding new features to User Agents and the SIP network such as a video mailbox, picture sharing, adding a video session to an ongoing voice session, push-to-talk functionality, transferring a session to another device with different properties and so on is proprietary on both the terminal and the network components. This is costly and the use of these functionalities between sub-scribers of different SIP networks is not assured.
- Insufficient security:
  - Voice is usually sent unencrypted from end to end, which makes it easy to eavesdrop on a connection.
  - Signaling can be intercepted since it is not encrypted. Man-in-the-middle attacks are possible.
  - No standards exist on how to securely and confidentially store user data (e.g. username/password) on a mobile device.
- Scalability – mobile networks today can easily have 50 million subscribers or more. This is very challenging in terms of scalability since a single SIP proxy in a network cannot handle such a high number of subscribers. A SIP network handling such a high number of subscribers must be distributed over many SIP proxies/registrars.
- There is no standardized way to store user profiles in the network today. Also, no standardized means exist to distribute user data over the several databases which are required in large networks (see scalability above).

To address these missing pieces the wireless industry has launched a number of projects. The most comprehensive is that of the Third Generation Partnership Project, which is also responsible for GSM, UMTS and LTE standardization. The result of this ongoing activity is the IP Multimedia Subsystem, which is based on the SIP as the core protocol for a next generation service delivery platform. Additional to the core SIP standards discussed above, many new standard documents were created to describe additions required for an IMS system.

Other standards bodies have also started to define next-generation IP-based voice and multimedia architectures. The most noteworthy of those are TISPAN, which has specified an IMS-like architecture for fixed-line IP networks, and 3GPP2, which has defined an IMS for CDMA-based networks. With Release 7 of the 3GPP standard, 3GPP and TISPAN have joined forces to create a single next-generation network architecture that will work in both fixed and 3G, B3G and Wi-Fi networks. Furthermore, interworking between the 3GPP IMS and the 3GPP2 IMS is assured. This is good news from a network convergence point of view since operators with fixed and wireless assets are very interested in having a single platform with a single service offering and to allow subscribers to use one subscription with all types of devices and access networks.

In the following sections an overview is given of how the IMS uses the SIP protocol and how it addresses the missing elements described above to become a universal session-based communication platform. Because of its centralized design, the IMS is likely to be successful with applications grouped around voice and instant messaging services. The following list shows applications which IMS networks are likely to offer to users in the short to mid term:

- Voice telephony as the main application – this includes handing over voice calls between networks as the user roams out of coverage of a network. Furthermore, advanced IMS solutions will enable handovers of voice calls to a 2G network when B3G coverage is lost.
- The IMS enables video calls with the advantage over current 3G circuit-switched mobile video calls that the video stream can be added or dropped at any time during the session.
- Presence and instant messaging.
- Voice and video session conferencing with three or more parties.
- Push message and video services such as sending subscribers messages when their favorite football team has scored a goal, when something exciting has happened during a Formula 1 race and so on.
- Calendar synchronization among all IMS devices.
- Notification of important events (birthdays, etc.).
- Wakeup service with auto answer and the user's preferred music or news.
- Live audio and videocasts of events. The difference between this and current solutions is the integrated adaptation of capabilities on the device.
- Peer-to-peer document push.
- Unified voice and video mail from all devices used by a person that are subscribed to the same IMS account.
- One identity/telephone number for all devices of a user. A session is delivered to all or some devices based on their capabilities. A video call would only be delivered to

registered devices capable of receiving video. Sessions can also be automatically modified if devices do not support video.

- A session can be moved from one device to another while it is ongoing. A video call, for example, might be accepted on a mobile device but transferred to the home entertainment system when the user arrives at home. Transferring the session also implies a modification of the session parameters. While a low-resolution video stream is used for a mobile device, the resolution can be increased for the big screen of the home entertainment system if this is supported by the device at the other end.
- Use of several user identities per device – this allows the use of a single device or a single set of devices to be reached by friends and business partners alike. With user profiles in the network, incoming session requests can be managed on a per user identity basis. This way, business calls could be automatically redirected to the voice mail system at certain times, to an announcement or to a colleague while the user is on vacation while private session requests are still connected.

Which of these applications are offered in an IMS network depends on the individual network operator. It is likely that the applications and services listed above will not remain the only ones for IMS and more will be added in the future. The Daidalos project [8], which was part of the 6th EU Framework Program, has developed an interesting vision of how these services could be used in practice. A video, available on their Web site, impressively shows the high-level results of their research.

IMS systems can also be the basis for many of the Web 2.0 services described in Chapter 6. In practice, however, such services are mostly developed outside the IMS for a number of reasons. Firstly, only a few IMS networks are deployed today. Secondly, Web 2.0 applications are not designed by network operators but by Internet companies. These companies are not keen to integrate their applications with potentially hundreds of different IMS networks since this requires negotiations with potentially hundreds of IMS operators for a global rollout. Finally, it is also quite difficult to integrate IMS applications into IMS clients on mobile devices since there is no universal standard. Web 2.0 applications are thus either Web and AJAX-based, are delivered as Java applets or are developed for major mobile device operating systems such as Windows Mobile or Symbian OS. Chapter 6 will discuss this topic in more detail.

4.4.1 IMS Basic Architecture

One of the major goals of the IMS was to create a flexible session establishment platform that can be scaled for networks with tens of thousands to tens of millions of subscribers. The IMS standards define logical components, the messaging between them and how external applications can use the IMS to offer services to users. In practice it is then up to infrastructure vendors and network operators to decide which logical components they want to combine into one physical device depending on the size of the network. It is likely that first implementations will co-locate many logical functions in a single device as there will only be a few IMS users at the beginning, which in turn does not require a large distributed system. As the network grows some functions are then over time migrated into standalone physical devices and a single entity might even be distributed over several devices for load sharing and redundancy purposes.

The IMS has been defined by two standards bodies in close co-operation. The main architecture, the logical components and the interworking between them are standardized by the 3GPP. Most of the protocols used between the components, such as SIP, Diameter, Megaco, Cops and so on, which will be discussed in this section, are standardized by the Internet Society's Internet Engineering Task Force (IETF). This split has been done on purpose to use as many open and freely available Internet protocols for the IMS as possible rather than to use proprietary and closed standards. This allows interworking with other session-based systems using open standards in the future. The split is also beneficial since the IETF has a strong expertise in IP protocols while 3GPP is focused on mobility aspects and network architecture. References to the specifications will be made throughout this section to allow the reader to go into the details if required. All documents of both standards bodies are available online at no cost. To start exploring the standards, 3GPP TS 22.228 [9] and 3GPP TS 23.228 [10] are recommended, which introduce the requirements for IMS and the general system architecture in detail.

Figure 4.11 shows the main components required on the application layer for a basic IMS solution. The IMS standards define several additional functional entities which are only introduced later in this section and are therefore not shown. The figure also docs not show the underlying transport functions of fixed and wireless core and access networks. Again, this has been done for clarity reasons and due to the fact that large parts of the IMS are completely access and transport network-independent. Only a few parts of the IMS communicate with the transport network to ensure quality of service and to prevent service misuse. These interfaces will also be discussed separately below.

Home Network of an IMS Subscriber

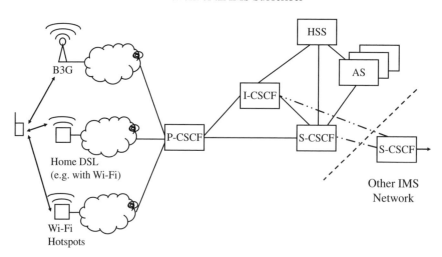

Figure 4.11 The basic components of the IMS framework.

4.4.2 The P-CSCF

From the user's point of view the Proxy Call Session Control Function (P-CSCF) is the entry point into the IMS network. The P-CSCF is a SIP proxy and additionally handles the following IMS specific tasks:

- User proxy – the P-CSCF received its name not because it is a SIP proxy but because it represents the user (it acts as a proxy) in the network. All signaling messages to and from the user always traverse the P-CSCF assigned to a user during the registration process. In wireless networks such a function is required since a user can suddenly drop out of the network if they roam out of coverage. The P-CSCF is notified of such an event by the access network while a session is ongoing and can thus terminate the session in a graceful way. How this is done depends on the type of access network. In the case of a 3G UMTS/HSDPA network (cf. Chapter 2) the RNC informs the SGSN that the radio contact with a device was lost. If a streaming or conversational class connection was established the SGSN then informs the gateway node (GGSN) of the radio link failure by setting the maximum bitrate in uplink and downlink to zero [11]. The GGSN will then stop forwarding packets to the user in the case that a remote subscriber involved in a session is still sending its media stream to the subscriber. The GGSN then informs the P-CSCF via the Policy Decision Function (PDF), which is further discussed below, that the subscriber has lost the network connection. The P-CSCF will then act on the IMS application layer on the user's behalf and terminates the ongoing session by sending a SIP 'bye' message to the remote user or users in the case of a conference. In WiMAX and LTE networks the process is a bit more efficient since there are fewer nodes between the base station and the P-CSCF. Here, the radio base station signals the radio bearer loss to the ASN-GW (WiMAX) or the MME in LTE, which in turn directly forwards the information via the PDF to the P-CSCF.
- Confidentiality – SIP transmits all signaling messages in clear text. This is a big security problem as it allows potential attackers who have gained access to the network at any point between the user and the SIP network to read and even modify the content of messages. The IMS thus requires SIP signaling between the IMS terminal and the P-CSCF to be encrypted. This is done by establishing an encrypted IPSec [12] connection between an IMS terminal and the P-CSCF during the registration process [13]. This encrypted connection will then be used while the user remains registered. The network behind the P-CSCF is considered to be secure and many network operators will thus not encrypt SIP messages and other signaling traffic between different components of an IMS network.
- Signalling compression – as has been shown above, SIP signaling messages are quite large. In the IMS the message size grows even more due to additional parameters for functionalities described later on. The larger a message the more time is required for transmitting it. As sessions (e.g. voice calls) should be established as quickly as possible and as there are several SIP messages required before a session is established, it is vital that each message is transferred as quickly as possible. An IMS terminal may therefore request during the registration process to activate Sigcomp [14] compression to reduce the size of the SIP signaling messages.
- Quality of service and policy control – the P-CSCF has an interface to the underlying wireless network infrastructure (GGSN, ASN-GW, MME, etc.) to ensure a certain quality

of service for a media stream (e.g. minimum bandwidth for a voice call to be ensured during a session). Policy control ensures that the link between two subscribers for a session is only used for the types of media negotiated between the subscribers and the network.
- Billing – like all IMS components the P-CSCF function can generate billing records for post-paid subscribers which are sent to a billing server for offline processing.

While traditional SIP User Agents are usually configured by the user or out of the box to be aware of the IP address of the first SIP server they need to contact for registration it was felt in 3GPP that this approach is not flexible enough for evolving networks. Here, individual P-CSCFs will be added over time to be able to handle the increasing number of users. Furthermore, self-configuration on startup of the User Agent is highly desired since users should not be required to configure their terminals for IMS should they decide to buy them from a source other than the network operator. The IMS standard therefore defines a number of additional options for a User Agent to obtain the IP address of the P-CSCF at startup. In 3GPP networks (UMTS, HSDPA, LTE), the P-CSCF IP address can be sent to a device during the establishment of a network connection, which is referred to as the Packet Data Protocol context activation. During this process the device receives amongst other information its own IP address and additionally the IP address of the P-CSCF. Since a PDP context activation is a general process for establishing a packet-switched connection to a 3GPP network, the device stores this information so the IMS User Agent software can request the P-CSCF identity from the device later on. Another possibility to retrieve the P-CSCF's IP address from the network is via a DHCP (Dynamic Host Configuration Protocol) lookup. This protocol is commonly used for another purpose today by Wireless LAN and Ethernet devices to request their IP address and the default gateway IP address from the network when they enter the network. Since DHCP is a flexible protocol it was extended for the use in the IMS to send the P-CSCF IP address to a User Agent upon request as well.

4.4.3 The S-CSCF and Application Servers

The central component of the IMS framework is the Serving Call Session Control Function (S-CSCF). It combines the functionality of a SIP registrar and a SIP proxy as defined in Sections 4.3.1 and 4.3.2. High-capacity networks will require several physical S-CSCFs. One user will be managed by a single S-CSCF while registered to the IMS network and all SIP requests have to be sent to this S-CSCF. Users can be assigned by a load-sharing algorithm to a particular S-CSCF at registration time. It is also possible to assign a particular S-CSCF depending on the services a subscriber is allowed to use since some S-CSCF may only support a subset of services available in the network.

When a user first registers with the IMS network, the S-CSCF requests the profile of the subscriber from a centralized database. The profile contains the user's authentication information and information about which services the user is allowed to invoke. In SIP terminology this database is known as the SIP location server (cf. Figure 4.3) and in the IMS world it is known as the Home Subscriber Server. This is required as the S-CSCF only stores the subscriber context while he is registered. Downloading user data from a centralized database is also required since the S-CSCF can be a highly distributed system in the case of a large network and it is possible and even likely that a subscriber is assigned a different S-CSCF during a subsequent registration.

After registration the P-CSCF's main task is to forward the SIP messages between the IMS terminal and the S-CSCF. It is the S-CSCF that will then decide how to handle a message and where to forward it. In the case of a SIP 'invite' message the S-CSCF has to decide how to proceed with the session setup request. For a simple post-paid voice session establishment for which no additional services are invoked the S-CSCF's main high level tasks are as follows:

- The S-CSCF recognizes from the SDP payload in the invite message that the user requests a voice session and checks the subscriber's service record to confirm they are allowed to originate calls.
- The S-CSCF then analyzes the destination address, which could either be a SIP URI (e.g. sip:martin.sauter@mynetwork.com) or a TEL URL (tel: + 49123444456).
- If the user wants to establish a session with a TEL URL the S-CSCF needs to perform a database lookup to determine if the destination is an IMS subscriber, that is if the TEL URL can be converted into a SIP URI. Otherwise the telephone number belongs to a circuit-switched service subscriber and the session has to be forwarded to a circuit-switched network. For details on this process see Section 4.4.9 on IMS voice telephony interworking with circuit-switched networks.
- In the case of a SIP URI, the S-CSCF checks if the destination is a subscriber of the local or a remote IMS network by looking at the domain part of the SIP URI (e.g. @mynetwork.com). If the destination is outside the local IMS network the S-CSCF will then look up the IP address of the SIP entry proxy of the remote network (the I-CSCF, see below) and forward the invite message to the foreign network.
- The S-CSCF then stays in the loop for all subsequent SIP messages and creates charging records for the offline billing system so the user can later on be invoiced for the call.

The S-CSCF is also responsible for many supplementary services. If the user, for example, requests that the telephone number or SIP URI is hidden from the destination subscriber, it will again check the service record of the subscriber to confirm the operator allows this operation and then modify the SIP header accordingly. Features such as this are already known in circuit-switched networks and are implemented in a similar fashion in the S-CSCF.

While the S-CSCF is the central part of the IMS its actions are limited to SIP message analysis, modification, routing and forking to reach all devices registered to the same identity. This already allows basic services and many supplementary services such as identity hiding as described before without any further equipment. More complicated services and applications, however, are not implemented in the S-CSCF but on external Application Servers (AS). While the application server is responsible for offering the service, the S-CSCF decides which applications are invoked at a session establishment, while a session is ongoing, or when it is terminated. The following example illustrates this approach.

An S-CSCF receives a SIP invite message for a local user. The user however is already engaged in another session and rejects the additional session establishment request. Based on the user's subscription data the S-CSCF can then decide to forward the SIP invite message to a voice mail system. The voice mail system, which is a typical application server, receives the redirected SIP 'invite' message and acts as a User Agent to establish the session. After the caller has left a voice message, the Voice Mail system sends an instant message via the S-CSCF to the original destination User Agent to let the user know that a voice mail message was left. In this example the voice mail system acts as both a passive component when receiving the redirected call and an active component when forwarding an IM message. Application servers can therefore not only receive SIP messaging from S-CSCFs but can also be the originator of messages based on internal or external events.

Another example of an active application server functionality is an automatic wakeup service. After the user has configured the service the AS providing the wakeup service will automatically establish a SIP session to a user at a predefined time to play a wakeup message, the subscriber's favorite music, the latest news and so on.

Other functionalities such as presence, instant messaging, push-to-talk and many other services are also done on an application server rather than in the S-CSCF.

More formally application servers can work in the following modes:

- as a User Agent (as described above);
- as a SIP proxy by changing the content of a SIP message it receives from an S-CSCF and returning the message back to the S-CSCF;
- as a back-to-back User Agent – this means that it terminates a session as a terminating User Agent towards the S-CSCF and establishes a second SIP session as an originating User Agent.

The S-CSCF's decision when to contact an application server is based on 'initial filter criteria' that are part of the user's configuration profile. This topic is discussed in more detail in Section 4.4.6.

4.4.4 The I-CSCF and the HSS

The third logical type of SIP proxy in an IMS network is the Interrogating-CSCF (I-CSCF). SIP messages traverse the I-CSCF in the following cases.

When a User Agent registers with the IMS, for example during startup, the first action is to find the P-CSCF as described above. Afterwards it will send a SIP registration message to network to announce its availability. The P-CSCF then needs to forward the message to an S-CSCF. Since the IMS can be a distributed system and since the user has not yet been assigned an S-CSCF, the P-CSCF forwards the registration message to an Interrogating-CSCF. The I-CSCF of the user is found using a DNS lookup with the user identity in the SIP registration request. The message is then forwarded to the I-CSCF which will then interrogate the Home Subscriber Server for the S-CSCF name or the S-CSCF capabilities required by the subscriber. When the HSS receives the request, it retrieves the user's subscription record and returns the required information. Based on the S-CSCF name or the services the user has signed up to in combination with the knowledge about the capabilities of individual S-CSCFs, the I-CSCF then selects a suitable S-CSCF and forwards the registration.

The second scenario in which an I-CSCF is required is for terminating a session to a subscriber that is served by a different S-CSCF than the originator of the session. In this scenario the S-CSCF of the originator detects that it does not serve the terminating subscriber and therefore forwards the SIP 'invite' message to an I-CSCF. The I-CSCF of the user is found using a DNS lookup with the user identity in the SIP invite request. The I-CSCF in turn interrogates the HSS by forwarding the SIP URI of the destination subscriber and requesting the HSS to return the contact details (e.g. IP address, port) of the responsible S-CSCF. In case the subscriber is currently not registered a suitable terminating S-CSCF is selected since sessions to currently unavailable subscribers might still be terminated, for example by a voice mail system. The I-CSCF then forwards the SIP invite message to the responsible S-CSCF. As the identity of the S-CSCF is now known further messages need not be sent via the I-CSCF. The I-CSCF therefore does not add its identity to the routing information in the SIP header of further messages.

The Home Subscriber Server is an evolution of the Home Location Register, which was specified back in the early 1990s as the central user database in the first release of GSM standards. In the meantime the database structure has been expanded several times, as shown in Figure 4.12. With the inception of the IMS the HLR was renamed to HSS to account for the additional services it offers:

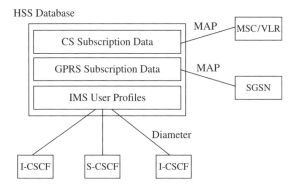

Figure 4.12   The home subscriber server and clients.

For circuit-switched services such as voice calls and SMS the original database entries keep being used. Data stored for a user for the circuit-switched network includes:

- authentication information – a replica of this information is stored on the user's SIM card;
- information about which services the subscriber is allowed to use or is barred from using (e.g. incoming voice calls, outgoing voice calls, incoming/outgoing calls while roaming, presentation or hiding of telephone numbers of incoming calls, SMS send/receive capabilities, Intelligent Network service invocations, e.g. for prepaid services, and many other things).

Towards the circuit-switched mobile switching center/Visitor Location register, the traditional Mobile Application Part protocol continues to be used.

When packet-switched services were added to the GSM network with the General Packet Radio Service in the early 2000s, the database was enhanced to store subscriber information for packet-switched services. For UMTS, HSPA and LTE the database specification was enhanced several times in a backwards-compatible fashion so the information can be used for 2G, 3G and 3.5G access networks. Packet-switched subscriber information in the HLR includes:

- the Access Point Names and their properties – APNs are used by the GPRS network to define the quality of service settings of a connection; implicitly APNs are also used to define services (e.g. full Internet access, only Web access, etc.) and billing methods in the network;
- which APNs a subscriber is allowed to use;
- if packet-switched service roaming is allowed.

For the IMS a third block was added to the HSS/HLR where user profiles of IMS subscribers are stored. An IMS user profile includes information such as the user identities connected to a subscription, which services the user is allowed to invoke and information on how incoming requests should be handled when the user is not registered (e.g. forwarding to voicemail). IMS components such as the I-CSCF and the S-CSCF are connected to the HSS via an IP connection. The protocol used for retrieving and updating records in the HSS is Diameter [15]. Diameter is not an abbreviation but a pun on its predecessor the RADIUS (Remote Access Dial In User Server) protocol.

In practice a network usually stores information for a user in all three parts of the HSS. Circuit-switched subscription information is required in case the user wants to use circuit-switched services such as SMS or voice (e.g. when roaming out of the coverage area in which IMS voice calls are supported). The user also needs to be provisioned in the GPRS part of the HSS since using IMS services is only possible from GSM/UMTS/HSPA/LTE networks once a packet-switched connection to the network (via an APN) has been established. Finally, a user also has to have a user profile for IMS services as otherwise no IMS services can be used.

Large networks can have several HSS entities for capacity reasons. In such a case the CSCFs have to query a database which contains the information on which subscriber is administered by which HSS. This task is managed by the Subscription Locator Function (SLF). The SLF does not sit in the signaling path between the CSCFs and the HSSs but acts as a standalone database that can be queried using the Diameter protocol to retrieve the IP address of the HSS responsible for a subscriber for a subsequent Diameter dialogue with the HSS to retrieve subscription information. The SLF is an optional component and thus not shown in Figure 4.12. It is not required if there is only one HSS in the network.

4.4.5 Media Resource Functions

A supplementary service often used in business communication is conferencing. A conference is established, for example, if a third person is to be included in an ongoing conversation. One of the two users will then put the established call on hold to call the third party. If the third party accepts the call, the caller can then conference all three parties together. In circuit-switched networks a conference bridge function is required in the network to mix the speech path of all participants. In VoIP networks two methods can be used to create a conference call. Some networks do not have conference bridge resources in the network and therefore leave it to one of the subscribers to mix the speech signal and distribute the resulting audio stream to the other participants of the conference. While being simple, the downside of this approach is the additional bandwidth required for the device that is mixing the signal. Instead of sending a media stream to a single device, two (or more where there are more than three subscribers in a conference) streams have to be sent, which doubles the required data rate in the uplink. In the IMS, conferencing is therefore done with a conferencing server in the network. The conferencing facility in the network is implemented on a Media Ressource Function (MRF). The MRF in turn is split into two logical components as shown in Figure 4.13. Towards the S-CSCF the Media Resource Function Controller (MRFC) acts as an application server and terminates the SIP signaling. The media stream is mixed by the Media Resource Function Processor (MRFP), which is controlled by the MRFC. The protocol between the two entities is not defined. The Media Gateway Control Protocol (Megaco, H.248) which is also used for controlling other media gateways as described further below in Section 4.4.9, could be used for the purpose.

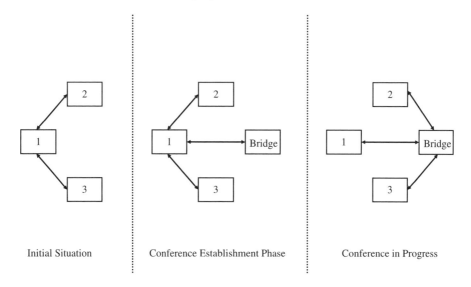

Figure 4.13   IMS conference session establishment between three parties.

A conference session is established during an ongoing voice session as follows: One of the participants puts the current session on hold and calls the third party. When the third party accepts the call, the first party prepares the conference bridge by establishing a session to the conferencing MRFC and MRFP via the S-CSCF. Once the conferencing facility is ready, the first party sends a SIP message to both other parties to transfer the end points of their voice media streams to the conferencing bridge. For this purpose the MRFC has sent the initiator a unique SIP URI which allows it to identify the incoming SIP requests. Once both remote parties agree to re-route their voice media stream to the conference bridge, the MRFP can start mixing the audio signals of all three participants and the conference is established. Figure 4.13 shows in which steps the voice media streams are transferred to the conference bridge. In the standards, this process is described in Chapter 5.11.6.2.3 of [10] and details are given in [16].

Optionally IMS clients and the conference bridge can support conference events. When IMS clients register for conference events they are notified by the conference bridge about the identity of the other subscribers in the conference bridge and receive a message when one of the subscribers terminates its session to the bridge.

If only two parties remain on the conference bridge the MRFC can then choose to either leave the two parties on the conference bridge or re-transfer the media streams back to a point-to-point connection and remove itself from the session.

An interesting feature which is not possible today with circuit-switched conference bridges is the ability to directly invite a subscriber to join an ongoing conference. This way the originator does not have to put the conference leg on hold to call somebody to join the conference.

4.4.6 User Identities, Subscription Profiles and Filter Criteria

As it becomes more and more common that subscribers use more than a single device for communication, the IMS specifications have devised a scheme to spread a single sub-scription over many devices. One private user identity is used per physical device, as shown in Figure 4.14. The private user identity is usually not visible to the subscriber since it is only used during the registration procedure. The private user identity is there-fore used for the same purpose on the application layer as the International Mobile Subscriber Identity on the 2G and 3G network layer.

The identification known to a subscriber and under which he can be contacted is the public user identity which can be compared with the telephone number (MSISDN) in fixed-line and wireless circuit-switched networks. The public user identity is a SIP URI as

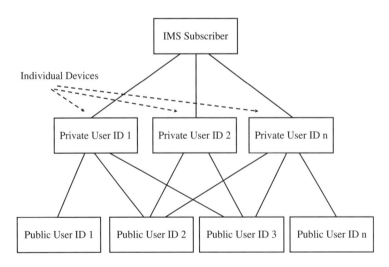

Figure 4.14 IMS subscription information structure.

described above (e.g. sip:martin.sauter@mynetwork.com) or a TEL URL. In the IMS several public user identities can be linked to a single private user identity. This is unlike the 2G and 3G world where a single IMSI is usually tied to a single MSISDN. Due to this 1:n relationship, the subscriber can be reached via several public user identities on the same device. This allows, for example, a public user identity for business purposes and a separate public user identify for private contacts. Since the user can register to single or multiple public user identities simultaneously, they have the choice of when they can be reached under which identity.

An IMS subscriber can also have several private user identities to which a combination of all public user identities can be mapped to. Annex B of [17] describes this concept in detail. In practice this approach gives network operators and subscribers the flexibility to be reachable via several devices. Additionally, subscribers can choose under which public user identities they can be reached at a certain time on a certain device if the IMS operator chooses to offer such a service.

Network services such as managing when a user will be available via which device where it is registered are performed by application servers. These are put into the communication path by S-CSCFs with the help of what is known as initial filter criteria. These are stored on the HSS alongside the public user identities of a subscriber. Initial filter criteria describe the values that parameters have to have in SIP and SDP messages in order for the message to be forwarded to an application server. The application server may then have additional subscriber information to modify the message content or to act as a User Agent on the subscriber's behalf. An application server may, for example, change the destination public user identification to automatically forward a session initiation request to a voice mail system or another public user identity (e.g. that of a colleague) when it detects that a user is online but does not want to be contacted via this user identity at the time. How the application server obtains the information on which to base this decision is not specified in the IMS standards.

Initial filter criteria can be stacked upon each other so a single SIP message can be sent to several application servers and be modified before the S-CSCF forwards it to a subscriber.

On the mobile device side a subset of the subscriber information is stored on the Universal Integrated Circuit Card (UICC) which is usually referred to as the SIM card. A SIM card has three functionalities. First, it serves as a secure data storage card in which both open and secret information can be stored. Data is stored in files in a directory structure. Unlike memory cards and hard drives, which use file systems with file names, the SIM card only uses standardized numbers to identify files. This dates back to the days when SIM cards could only store a small amount of data.

Secrets such as passwords and identification material can be used by internal SIM card applications to generate responses for authentication requests and to generate ciphering keys. These internal applications are invoked from the mobile device via a standardized SIM card interface [18] by requesting authentication and ciphering information, which is then computed by an authentication algorithm based on a random number given to the application and the secret key stored on the SIM card.

Only internal applications are permitted read access to the secret data on the SIM card. This is an important security feature as it prevents cloning of SIM cards. Finally, SIM cards can also host JavaCard applications that can be downloaded to the SIM card by the user or via an Over the Air (OTA) download by the operator. JavaCard applets are used in practice, for example, to create device-independent applications that can be called via an extended menu structure.

For the IMS the following information and internal applications are required:

- To be able to access a 2G or 3G network in the first place the SIM card contains the IMSI and secret key for 2G and 3G networks. This information is contained in the USIM (UMTS SIM) part of the SIM card. Standards refer to this as the USIM application, which is a little misleading. In general terms the USIM application is the combination of files and internal applications required for UMTS.
- A SIM card can also store authentication information and applications for non-3GPP network access, for example for public Wi-Fi authentication via EAP-SIM (cf. Chapter 2).
- For IMS the SIM card should also contain an ISIM (IMS SIM) application (i.e. a directory for IMS and files for a private user identity, one or more public user identities, the home network domain URI and IMS authentication information. Except for the public user identities, it is possible to derive this information from the UMTS parameters for backwards compatibility. This way it is not necessary to replace SIM card(s) to activate users for IMS services.

3GPP2 standards store IMS information on the R-UIM (Removable User Identity Module), which is similar to a SIM card (UICC), but is based on different standards. It is also possible to store IMS subscriber information directly in the mobile device. This, however, is undoubtedly much less secure and it is much more difficult for subscribers to choose for themselves which devices they want to use with an IMS network and to configure them with the necessary account information.

4.4.7 IMS Registration Process

Now that the basic IMS components and the basic concept of user identities have been introduced, it is time to look at how the IMS uses SIP signaling to offer services to subscribers. Before a user is allowed to establish sessions or, in more general terms, to send SIP messages to network components and other subscribers, it is required that the IMS User Agent registers with the network. The IMS registration process follows the lines of the registration as described in Section 4.3.1 and Figure 4.4. In order to provide security and additional features such as multiple identities, several additional steps are required during this process. This section gives an overview of the different processes during the registration process, the involved components and the SIP messages used during the registration process.

Before being able to register, a device first needs to connect to the network. In 3GPP wireless networks, connecting to the packet-switched network is done via a PDP context activation as described in Section 4.2. Once the process is completed the device has an IP address and can start communicating with the IMS.

The next step of the process consists of determining the IP address of the P-CSCF because it is the entry point to the IMS network. The device is either informed of the P-CSCF address during the PDP context activation (in case it accesses the network via a 3GPP radio network) or with a DHCP request. Both methods are described in Section 4.4.2.

Once the P-CSCF IP address is known, the IMS User Agent on the subscriber's device assembles a SIP register message. Registrations are performed with the subscriber's private user identity. If the SIM card does not contain an ISIM application, a temporary private user identity is built with the subscribers IMSI. The registration request also contains one of the public user identities stored on the SIM card by which the user wants to be reachable. Since the device does not yet know the S-CSCF's address, it is not included in the registration message. The P-CSCF therefore forwards the register message, as shown in Figure 4.15, to the I-CSCF. The I-CSCF in turn queries the Home Subscriber Server for the user's subscription profile and selects a suitable S-CSCF. This process is described in Section 4.4.4. Afterwards the message is forwarded to the S-CSCF.

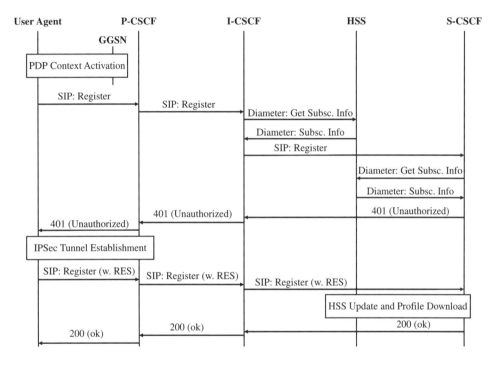

Figure 4.15   IMS registration process – Part 1.

When receiving the registration message the S-CSCF detects that the user is not yet known and also requests the user's subscription information from the HSS. In the process the S-CSCF also requests an authentication vector from the HSS, which consists of the following values:

- RAND – a random number.
- XRES – the random number is used by the HSS and by the SIM card together with a secret key and an authentication algorithm to generate a response. During the authentication process the S-CSCF compares the response received from the User Agent with the XRES value sent by the HSS. The values can only match if both entities have used

the same secret key and authentication algorithm. As the secret key is never sent through the network and the XRES is kept by the S-CSCF, it is possible to securely authenticate a user if a new random number is used for each request.

- AUTN – to prevent malicious attackers from acting as the subscriber's IMS network the HSS generates an authentication token (AUTN) with the random number (RAND) and another authentication algorithm. This token is later on used by the device to verify the authenticity of the network.
- IK – messages between the User Agent and the P-CSCF are integrity checked by adding a checksum that is calculated with an Integrity Key (IK). The IK is calculated by the HSS with an integrity key generation algorithm and the random number as an input value.
- CK – SIP messages are sent through an IPsec encrypted tunnel. The CK is used by the P-CSCF and the User Agent to encrypt and decrypt the messages.

In order to force the user to authenticate, the S-CSCF rejects the User Agent's registration request with a SIP 401 (Unauthorized) message. The body of the message contains, among other values, the five security values listed above. When the message arrives at the P-CSCF the IK and CK values are removed from the message since they are secret and must not fall into the hands of eavesdroppers that have gained access to the potentially unsecured communication network that could be between the User Agent and the P-CSCF. Sending these values to the User Agent is not necessary since the SIM card or the User Agent (in case there is no SIM card) calculates the values itself in the same way as the HSS, that is, with the random number, the secret key and the security algorithms.

When the SIP 401 unauthorized message arrives at the IMS User Agent, it will still contain the RAND and the AUTN. Internally the User Agent will then send a request to the SIM card to calculate the AUTN with the RAND. If the result matches the AUTN sent by the S-CSCF, the network is authenticated. In the next step the User Agent then asks the SIM card to calculate the XRES, which the S-CSCF will use later on to authenticate the User Agent. The SIM card also calculates the Ciphering Key (CK) and Integrity Key (IK), which will be used to establish a secure tunnel to the P-CSCF. At this point the User Agent can also activate SIP Signaling Compression (SIGCOMP) to speed up signaling procedures.

Once the secure tunnel to the P-CSCF is in place the User Agent sends another SIP registration request, this time including the XRES. The message is encrypted and an integrity check code ensures that the message is not tampered with in the very unlikely case the ciphering is broken. Additionally, the message also contains location information in the P-Access-Network-Info parameter. In case of a 3GPP network the parameter contains the Cell Global Identity (CGI) of the cell the device is currently using. The CGI is internationally unique. The P-Access-Network-Info parameter will be included in all future originating messages from the User Agent except for SIP ACK and CANCEL messages. This enables the IMS to offer location-based services via trusted application servers. The S-CSCF also ensures that this parameter is deleted from all messages before they are forwarded to other destinations (e.g. a terminating device).

When the second SIP register message arrives at the S-CSCF the XRES value is compared with the value included in the registration message. If the values match, the subscriber is authenticated and the S-CSCF sets the state of the public user identity to 'registered'. The user's profile can also indicate to the S-CSCF to implicitly register some (or all) of the other public user identities at this point. These are then also set to 'registered'. Afterwards, the S-CSCF marks the user as registered in the HSS, downloads the user profile and returns a SIP '200 ok' message to the User Agent to finalize the first part of the registration process. The message also contains all public user identities that are linked to the private user identity but not their current registration state. In a parallel action the S-CSCF also informs the presence server that the user has registered (i.e. that they are now online). This is done by sending an independent register message to the presence server. The presence service is discussed in Section 4.4.10.

The User Agent analyzes the incoming registration response message from the S-CSCF and thus learns which public user identities are attached to the private user identity it has used for the registration process. This is necessary since the SIM card may not contain the complete set of public user identities. As the message does not contain the registration state of these public user identities, the User Agent now has to query the S-CSCF for this information. This is done by subscribing to the registration state information of all public user identities contained in the registration response message. Subscribing to subscription state information is done by sending a SIP subscribe message to the S-CSCF, as shown in Figure 4.16. The S-CSCF responds with a SIP '200 ok' message. Shortly afterwards the S-CSCF will then send a SIP 'notify' message with the requested registration state. Further notify messages will be sent to the User Agent if the user registers with other devices (i.e. with a different private user identity) to the same

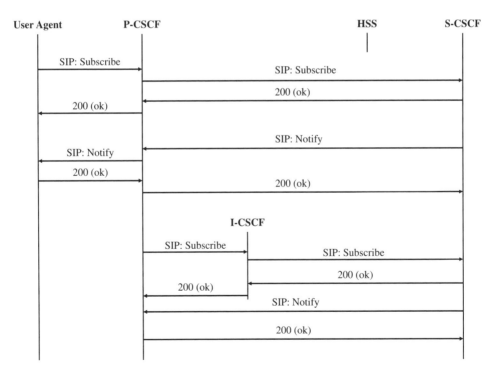

Figure 4.16   IMS registration process – Part 2.

public user identify. Each device is therefore aware which other devices have registered to a public user identity. It should be noted at this point that subscribing to registration state information has nothing to do with subscribing to presence information of other subscribers, which is described below.

The P-CSCF also needs to learn about the registration state of all public user identities that may later on be used by a User Agent as it verifies this parameter before forwarding messages to the S-CSCF. The P-CSCF therefore also registers to the state information.

A device stays registered until it deregisters from the IMS service. A deregistration is performed with a SIP register message in which the expiry (timer) parameter is set to zero. A User Agent can also be implicitly deregistered by the network if it fails to periodically update its registration.

4.4.8 IMS Session Establishment

At the beginning of this chapter, Figure 4.6 showed the basic SIP session establishment messaging. This session establishment takes it for granted that sufficient transmission resources are available and that the media streams in both directions can be sent as soon as required. While this is usually the case in fixed-line IP networks and in Wi-Fi networks, the capacity of cellular wireless base stations in relation to the number of subscribers is much lower (cf. Chapter 3). In addition, packets of real-time media streams should have a higher priority than other packets, like those of a Web browsing session, to reduce jitter. How a certain quality of service for the media stream can be ensured depends on the type of network used:

- Fixed-line IP networks – SIP clients can set the DSCP (Differentiated Services Code Point) value to 'expedited forwarding'. Routers in the path can then forward these packets with a higher priority in case of congestion.
- 802.11/ Wi-Fi – as shown in Chapter 2, the Wi-Fi Multimedia extension of the standard takes the DSCP value to decide which transmission queue a packet should be put into. VoIP packets are classified as media streaming packets and are thus always preferred to other packets.
- UMTS, HSDPA and LTE – due to their architecture, these systems have no visibility of the IP header of a packet and therefore cannot use priority information from this layer to decide which transmission priority the packet should have. As these networks are wide area networks, it would be unwise to do that since a single cell is potentially shared by hundreds of subscribers simultaneously. With some knowledge of the IP protocol, some subscribers could give all of their packets a higher priority independent of the application.

If the IMS is used via a cellular network it is advisable (but not mandatory) to reserve resources for the media stream. For resource reservation, a User Agent needs to know the required bandwidth for a media stream. This information, however, is only available once both ends of the session have negotiated a common codec for the session. As a consequence, media resource reservation cannot be done before the session establishment signaling but only while in progress. This presents the problem that the destination device should not start alerting the user of an incoming session straight away but only

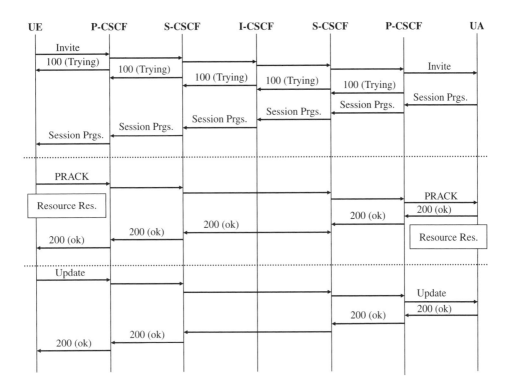

Figure 4.17   IMS session establishment – Part 1.

once it is certain that both parties have successfully acquired the required bandwidth and quality of service for the media stream. This means that the SIP session establishment signaling has to be extended.

Figure 4.17 shows how a SIP session is established, which includes resource reservation at both ends. As in the simple SIP example shown earlier the SIP session is initiated with a SIP 'invite' message. One of the parameters of this SIP invitation is the precondition that the originator needs to reserve resources for the media stream. This tells the called device that it should not start alerting the user until the originator has confirmed that the resources have been assigned. The SIP invitation also contains the supported media codecs of the originator for the types of media streams to be established. The destination then answers with a SIP '183 (Session Progress)' message which contains the list of media codecs supported at this end. As the destination also needs to reserve resources the session progress message conveys this information to the originator as well. At this point the destination is not yet able to reserve resources since it is still not clear which media codec(s) will be used for the session.

When the session progress message arrives at the other end the originating User Agent is then able to select a suitable codec for each media stream. The User Agent then returns a provisional acknowledgement (PRACK) with the information on which codecs were selected back to the terminator and starts the resource reservation process on its side.

Once the PRACK message is received at the destination side the resource reservation process can start there as well, which is confirmed to the originator with a SIP 200 (ok) message.

When the SIP 200 (ok) message arrives at the originator side, a SIP Update message is sent as soon as the resource reservation has been performed successfully. At the terminator side the device starts alerting the user once the SIP 'update' message has been received and the required resources for the media stream have been allocated. If resource reservation fails on either side, the session establishment is aborted and the user on the destination side will not be notified.

If resource reservation on the destination side was successful, the device starts alerting the user and a SIP 180 (ringing) message will be sent back to the originator. The reception of the ringing message is acknowledged by the originator and the media stream is fully put into place once the destination user accepts the session in which case the device sends a SIP 200 (ok) message to the originator. The originator returns as a final step a SIP ACK message to confirm the previous message.

In UMTS, HSPA and LTE networks resources for media streams are requested from the network as follows: as discussed above, the aim of the initial PDP context activation is to obtain access to the network and to receive an IP address. This logical connection is then used for SIP signaling. If no resource reservation is required by the IMS, the media stream will simply use this connection as well. Where resource reservation is required, the mobile will request the required bandwidth and quality of service from the network via a secondary PDP context activation, as described in more detail in [19] and [20]. Secondary PDP contexts are then used to logically separate the real-time data traffic from background or signaling traffic on the air interface while keeping a single IP address on the mobile device. A secondary PDP context is activated by the User Agent, providing the mobile device and the GGSN with a Traffic Flow Template that contains the IP address and UDP port of the destination device that will be used for the media streaming. The mobile device and the gateway router (GGSN) in the network will then screen all packets and treat those with the specified IP addresses and port number differently, such as giving them preference over other packets of the same and other users. These mechanisms are applied below the IP layer and thus no changes are required to the IP protocol stack or the way the User Agent handles the media streaming over any kind of bearer. The only mechanism required in the User Agent is the command to activate a secondary PDP context. Afterwards the quality of service handling is completely transparent to the IP stack and the application.

If several different media streams are used for a session (e.g. voice and video), the P-CSCF decides if all streams can use the same secondary PDP context or if individual contexts have to be established. This is done by the P-CSCF by modifying the SIP session establishment messages according to the network's policy.

To prevent misuse of the quality of service feature, the P-CSCF communicates with the Policy Decision Function to get a media authorization token for the requested media streams. This token is then included in the SIP messages to the mobile device. The mobile device then includes the media authorization token in the secondary PDP activation message to the GGSN. When the GGSN receives this request it will query the PDF to see if the media authorization token is valid and if the amount of requested resources has not been modified by the terminal. In this way the IMS application is not able to request more resources than is required for the media stream. This also prevents other applications from misusing the quality of service features of the network for their own purposes.

The P-CSCFs and S-CSCFs in the network of the calling party and also in the network of the called party can reject a session establishment in the case where a User Agent requests a media codec which is not allowed in the network. This could be the case, for example, if an IMS client suggests using the outdated 64 kbit/s G.711 codec for voice transmission. If the network does not allow this codec to be used, the SIP Invite message is rejected with a SIP '488 (not acceptable here)' message. The User Agent then has to modify the codec list and start another session establishment procedure.

If one of the session participants uses the IMS from a network in which no media authorization is required, some of the SIP messages shown in Figures 4.17 and 4.18 are not required. If the originator, for example, uses a Wi-Fi network for which no media authorization is required, the SIP update message is not required. In this scenario the User Agent on the other side does not expect this message since the originator has indicated in the session invitation that resource reservation is not required on that side. The terminating User Agent can therefore start alerting the user as soon as their own resources have been granted without waiting for the confirmation from the other end.

When a user is registered they can have several simultaneously registered public user identities. When establishing a session the user can decide which of those public user identities shall be shown to the called party. Since the P-CSCF is aware which public user

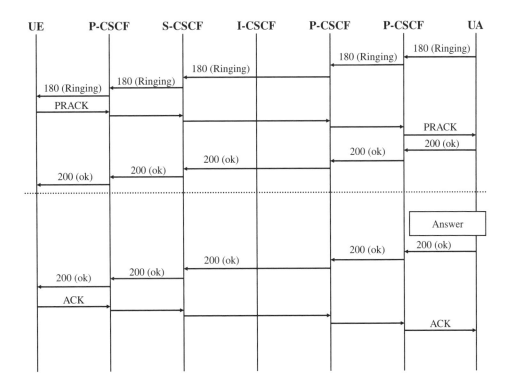

Figure 4.18 IMS session establishment – Part 2.

identities are registered, the network is able to verify this choice to ensure a subscriber only uses an identity that is registered and that belongs to them. When the P-CSCF approves the use of a public identity for the session establishment, it modifies the SIP invite message and thus all other SIP routers in the network are aware that the identity has been verified by the network. The originator can also request the network to remove their identity from the SIP invitation before it is sent to the final destination. This way the network ensures anonymity and does not leave this functionality to the terminating User Agent.

Since Release 6 of the 3GPP IMS standards, session establishment messaging flows have been defined to allow the establishment of a session between an IMS terminal and a standard SIP User Agent connected to an external non-IMS SIP network. As was shown at the beginning of this chapter, the signaling flow for a session establishment in a non-IMS network is quite different since no resource reservation is required. As an IMS User Agent cannot know in advance that a called party is a member of a standard SIP network, it attempts to establish a session with preconditions for resource reservation. As the protocol extensions in the SIP invite message required for this are not understood by the terminating User Agent, the invitation is rejected by the SIP client with a SIP '420 (bad extension)' message. The IMS client then assumes that the terminator is a standard SIP client and will construct a SIP invite message without requiring resource reservation preconditions. In most cases not using preconditions for resource reservation should not have an impact since the reservation process is usually very quick and can be finished before the terminating user has a chance to accept the session.

Charging the user for a session is a complicated process since there are many ways to do this. The price for a session could depend for example on the type of session (voice, voice + video, etc.), the duration of the session, the destination or the kind of codecs (high bandwidth, low bandwidth, high resolution, etc.) used. For post-paid customers billing information is first collected and then sent to a billing server for analysis. To give the billing system the highest flexibility, each SIP router in the network collects billing information for a session and forwards this information to the billing system. Furthermore, the core network routers also collect billing information (e.g. amount of data transferred) and also forward this information to the billing system. In order to be able to correlate the different messages, they all have to contain the same charging identification. A charging ID for a session is generated by the P-CSCF when it receives the first SIP invite message. The ID is then distributed to all other IMS network elements via the SIP messages. The charging ID is never directly delivered to the originating or terminating User Agent as the P-CSCFs and S-CSCFs at both ends of the connection remove the information from all SIP messages before they leave the IMS network.

To be able to correlate the charging information of the core network, the GGSN notifies the Policy Decision Function of the GPRS charging ID for the session. The PDF can then correlate the GPRS charging ID to the IMS charging ID and report the correlation to the billing system.

It is also possible to perform online charging in the IMS system, for example for prepaid users. In this case billing records are not sent to an offline charging system but to a prepaid charging system which decides in real time if the user is allowed to establish a session or not. There are a number of interfaces from different IMS and transport layer components to the online charging system. For a voice or video session establishment it is likely that the charging request to the prepaid billing system will be initiated by a CSCF.

The billing system can then allow the session establishment and instructs the CSCF to report back after a certain time or once the session is finished. If the timer expires and the session is still in progress the prepaid system then has the possibility to check if the user still has enough credit to continue the session or if it should be interrupted.

Not shown in this section was the forwarding of SIP messages to application servers in case initial filter criteria are met by one of the SIP messages of the session establishment dialogue.

4.4.9 Voice Telephony Interworking with Circuit-switched Networks

Interworking with legacy circuit-switched fixed and mobile telephony networks is an important feature for IMS subscribers since, for the foreseeable future, most wireless subscribers will continue to communicate over circuit-switched networks. A number of logical entities have been defined in the IMS standards for the interworking between the SIP-based IMS networks and external circuit-switched networks. Figure 4.19 shows how these logical entities are connected with each other and how they interact with the S-CSCF when an IMS subscriber wants to establish a call to a circuit-switched network subscriber. For small-scale IMS systems all entities could be implemented in the same physical device while for large-scale systems each logical entity could reside in a separate physical device. For scalability and redundancy each component can also be present several times in a single network.

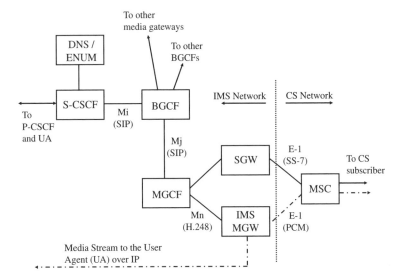

Figure 4.19   IMS circuit-switched interworking with a classical circuit-switched network.

To reach a circuit-switched user, an IMS user sends a SIP 'invite' message with the telephone number of the circuit-switched user to the S-CSCF. The telephone number is formatted as a TEL URL which could either represent a circuit-switched user or the telephone number of another IMS user. IMS users have telephone numbers as well in order to be reachable by circuit-switched users and also in order to make calling IMS users easier from devices that only have a numeric keypad. In a first step the S-CSCF therefore has to discover if the TEL URL sent in the SIP 'invite' message belongs to a local IMS user or to a user of an external circuit-switched network. This is done with an ENUM (Electronic Numbering) Domain Name System (DNS) lookup [21].

Note: The DNS is also used for applications such as Web browsing to convert a domain name (e.g. www.wiley.co.uk) a user has typed into the browser's address line into the IP address of the Web server. The reuse of DNS to convert a telephone number into a SIP URI fits into the overall concept of SIP and IMS to re-use and extend existing protocols whenever possible rather than to define new ones.

For an ENUM DNS lookup the telephone number (e.g. + 443337790) is converted to a domain name which is extended by '.e164.arpa' (e.g. 0.9.7.7.3.3.3.4.4.e164.arpa). The extension signals to the DNS server that the domain name given in a DNS request is not an Internet domain name but a telephone number for which a SIP URI is requested. 'E.164' is the numbering plan defined by the ITU for international public telephone numbers. The DNS server then executes a database lookup and if successful returns the corresponding SIP URI to the S-CSCF. In the case of an IMS to circuit-switched call the ENUM lookup is not successful since circuit-switched users do not have a SIP URI. As a consequence the S-CSCF forwards the SIP 'invite' request to the IMS circuit-switched gateway functionality.

The IMS circuit-switched gateway functionality consists of four logical entities. The first one is the Breakout Gateway Control Function (BGCF), which decides which Media Gateway Control Function (MGCF) to forward the SIP 'invite' message to. The BGCF can also decide to forward the 'invite' to a BGCF in another network. In this example the BGCF forwards the invite message to a local Media Gateway Control Function. The MGCF acts like a SIP User Agent and terminates the SIP signaling in the IMS network. The components behind the MGCF and the processes invoked there are thus transparent to the S-CSCF.

The MGCF is responsible for controlling both the user plane (the voice stream) and the signaling plane for a connection in a similar way to a User Agent on a mobile device. To be flexible in terms of redundancy and network size, the MGCF does not do these tasks itself. Instead, two logical entities have been defined in the IMS standards which are controlled by the MGCF.

The first logical entity is the Signalling Gateway (SGW). The SGW is, as the name suggests, responsible for converting the SIP messaging used in the IMS network into the signaling protocols used in circuit-switched networks. Two different protocols are currently used in circuit-switched networks, depending on what kind of switching center is used:

- Classic fixed-line mobile switching centers use the SS-7. SS-7 messages are usually carried over one or more 64 kbit/s timeslots in E-1 links (2 Mbit/s). Voice data streams are equally carried in timeslots of E-1 links over short distances or over optical STM-1 connections (155 Mbit/s) over longer distances. This setup is shown in Figure 4.13.
- Newer circuit-switched architectures have separated the circuit-switched mobile switching center functionality into an MSC Call Server and a Media Gateway as described in Section 1.6. For IMS interworking the MSC call server communicates with the IMS signaling gateway with the Bearer Independent Call Control (BICC) protocol, which is usually carried over IP instead of a circuit-switched connection. The IMS media gateway and the circuit-switched media are then

connected via an IP connection as the media stream has already been converted from circuit-switched to packet-switched at the circuit-switched media gateway. This setup is shown in Figure 4.20.

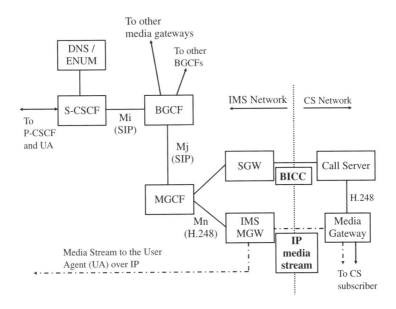

Figure 4.20   IMS circuit-switched networking with a bearer independent core network.

The SGW and the MGCF are usually implemented in the same physical device as the protocol between the MGCF and the SGW has not been standardized.

The transcoding of the circuit-switched voice data stream into an IP/RTP data stream is done by the IMS Media Gateway, which is also controlled by the MGCF. The MGCF and the IMS-MGW communicate with the H.248 Media Gateway Control Protocol (MEGACO) [22]. H.248 is used for the following purposes:

- To instruct the media gateway to start/stop converting a voice data stream between a certain timeslot of an E-1 circuit and an IP RTP data stream. The instruction also contains the information on which IP address and UDP port of a User Agent the IP data stream is to be sent to and received from.
- To define which voice codec transcoder is required for a connection. While circuit-switched networks mostly use the 64 kbit/s Pulse Code Modulated (PCM) speech codec, the IMS prefers to use Adaptive Multi Rate with bitrates of 12 kbit/s or less to increase the number of simultaneous voice calls per radio base station.
- To insert tones into the voice stream. For incoming calls this functionality can be used to signal to the originating circuit-switched user that the destination user is currently alerted (alerting tone).
- To insert announcements (e.g. 'The number you have dialed is not assigned', etc.) into the voice stream.

- To convert a Dual Tone Multiple Frequency (DTMF) tone message into audible signals. DTMF tones are used, for example, when a user types a code to get access to a voice mail system. A conversion is necessary as some systems carry DTFM tones inside the voice data stream, while other systems such as GSM use DTFM messages and only generate audible signals once the network has been left.
- To insert echo cancellation functionality into the speech path if not already done in the circuit-switched network.

Note that the MSC call server also uses H.248 to communicate with its media gateway. In fact, the IMS has re-used this protocol for its own purposes and has added a number of new functionalities.

Figure 4.21 shows the components involved when a circuit-switched subscriber calls an IMS subscriber by using the IMS subscriber's TEL URL. During the establishment of the call the fixed-line or mobile switching center selects a free timeslot on an E-1 line and sends an SS-7 Initial Address Message (IAM) to the IMS Signalling Gateway. The message contains the telephone number and the information on which timeslot on which E-1 line the switching center has selected for transmitting the audio stream. The SGW then forwards the information to the MGCF, which starts preparing the IMS-Media Gateway and starts the SIP 'invite' dialogue. As the MGCF is not aware which S-CSCF is responsible for the subscriber, the 'invite' message is first sent to the I-CSCF. The I-CSCF then queries the HSS for this information and forwards the message to the responsible S-CSCF. The S-CSCF then checks the session invitation and if the session is allowed to proceed, it forwards the invitation via the P-CSCF to the IMS device. If the IMS device accepts the session, the MGCF instructs the IMS-MGW to reserve the required resources and to

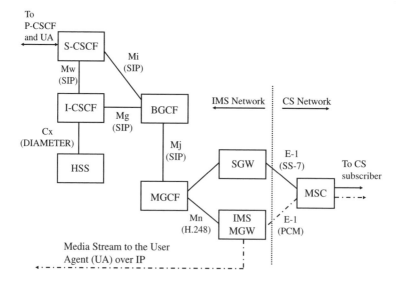

Figure 4.21   IMS components involved in a call setup from a circuit-switched to an IMS subscriber.

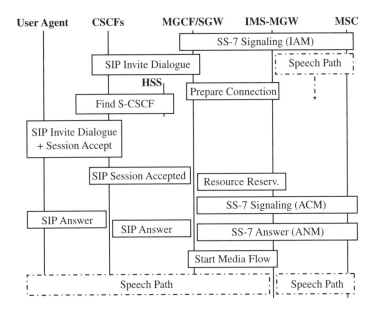

Figure 4.22   Simplified messaging sequence of a voice call establishment from a circuit-switched user to an IMS user.

prepare the codec transcoder that the MGCF and the IMS client want to use for the session. When the IMS user answers the call, the IMS terminals sends a SIP '200 ok' message to the MGCF and the MGCF in turn instructs the media gateway to start the media flow. Figure 4.22, which is based on Figure 5.16 in [9], shows a high-level overview of this procedure by combining several messages into a single box.

4.4.10 Push-to-talk, Presence and Instant Messaging

In the previous sections the IMS has been presented from a voice session point of view. The IMS, however, is a general session setup platform and can offer, with the help of application servers, a wide variety of session-based communication. The most important session applications that have been specified for the IMS are presence, push-to-talk and instant messaging.

While the IMS has been standardized by 3GPP, the Push-to-talk over Cellular (PoC) service has been standardized by the Open Mobile Alliance (OMA). The first version of this standard was published in June 2006 and the basic PoC architecture, functionalities and call flows are described in [23]. IMS interworking aspects, basic PoC session establishment and timing considerations can be found in [24].

As the PoC service is implemented as an application sever in the IMS environment, the SIP protocol is used for session establishment, talker control, leaving, joining sessions and so on. The Session Description Protocol (SDP) is used for negotiating media parameters such as the voice codec. For transport of the audio stream, the Real-time

Transport Protocol is used in combination with the Real-time Transport Control Protocol. It is interesting to note that the PoC service uses the same protocols and standards as other IMS services such as voice and video calls. In PoC sessions, however, speech is exchanged in talk bursts which are originated by the current talker and duplicated and forwarded to one or more listening subscribers by the PoC application server.

The PoC Service has been defined in a very flexible manner and offers the following types of sessions:

- One-to-one PoC session – these sessions, which are also referred to as 'Personal Instant Alerts' are always established between two subscribers. To establish a 1-1 PoC session, the initiator selects the identity of another PoC subscriber. The identities of other users are usually stored on the terminal and shown to the user in a list. If the requested subscriber is online and attached to the PoC server, the session is established and the two subscribers can start exchanging talk bursts in half duplex mode. In half duplex mode, only one person can talk at a time.
- One-to-many PoC session – in this type of PoC session many subscribers can exchange talk bursts with each other in half duplex mode. There can only be one talker at a time. A subscriber can request permission to talk by sending a notification to the PoC server. Once the current talker relinquishes the uplink, the PoC server assigns the uplink to the next person. Details such as priorities, talker pre-emption and so on are further described below. A voice channel, that is a data flow, only exists to listeners in the PoC session while there is a talker.

  A one-to-many PoC session can be established in one of the following modes:

  - Ad-hoc group session – for this session mode, the initiator sends a list of subscribers to the PoC server which will form the group. The PoC server then informs all subscribers on the list that they have been invited. The session is running once at least one of the requested participants has joined the session.
  - Pre-arranged group session – this session mode is initiated by sending a group ID to the PoC server instead of a list of participants. The group ID identifies the responsible PoC server and the group itself. This way it is possible to establish group calls over several PoC servers, which is required in practice as participants might have different national or international home networks.
  - Chat group session – such sessions are established without notifying possible participants. Subscribers can join and leave chat group sessions at any time and access to a chat session can be restricted to authorized members. More information on group and subscriber management is given below.
- One-to-many-to-one PoC session – such talk groups are always pre-arranged. While in other types of sessions all users are equal and all users can hear each other, talk bursts in one-to-many-to-one PoC sessions are only sent to a distinguished participant. Therefore, ordinary subscribers can only hear the distinguished participant while talk bursts sent by other participants are not forwarded to them. Furthermore, only group members that can act as distinguished members are allowed to establish a group session. In practice, such sessions can serve as unidirectional broadcasts from the distinguished participant with an optional uplink.

The PoC standard specifies session establishment as the process and the time it takes from initiation of a new group session until the time the initiator is granted the right to speak. The right to speak is granted once at least one participant has joined the call. Subscribers can join a call by manually accepting it or by setting the device into auto-answer mode.

The right to use the uplink to talk, that is to send a talk burst, is managed by the PoC server. Talk burst control signaling is done via the Talk Burst Control Protocol (TBCP), which defines the messages exchanged between the talker, the PoC server and listeners. In order to receive the right to talk, a subscriber has to send a talk burst request message. The PoC sever acknowledges the request to the subscriber and simultaneously sends out a receiving talk burst indication to all other subscribers to let them know that the uplink is taken and that they shall prepare for the incoming talk burst. The receiving talk burst indication message also contains the identity of the talker unless the talker has chosen to remain anonymous. Once the talker relinquishes the right to speak, a stop talk burst indication is sent to all participants of the session. Furthermore, a no talk burst indication message is sent to inform session participants that the uplink is free.

A subscriber can participate in more than one PoC session at a time. One of the simultaneous PoC sessions is then declared as the primary session. All others are of secondary priority. When a user has joined several PoC sessions at once, the PoC server only forwards media information for one ongoing PoC session at a time. The media stream of the primary session has the highest priority. Other media streams are only forwarded to the user if there is currently no media sent in the primary session.

A user can leave and re-join a group call at any time during an ongoing session. This can be done in two ways. If the user wants to only temporarily leave the call a message is sent to the PoC server to deactivate incoming talk bursts for a specific session until the user reactivates talk burst forwarding again. This can be compared with putting a point-to-point call on hold. The user can also leave an ongoing PoC call. If the user later on wants to participate again, a re-join message is sent to the PoC server. If the session is not ongoing any more at the time, the re-join request will be rejected and the call has to be established again.

A release policy controls the behavior of the PoC server once the originator of the session leaves. Depending on the policy, the PoC server either closes the session or keeps it up until all users have left.

A centralized PoC database referred to as the XDMS (Shared XML Document Management Server) is used to store information about groups, individual PoC parameters per user and other PoC-related parameters. Users can query and modify the database via a standardized XML-based interface over IP. A query or modification request can be sent directly from the handset or by other means, for example via a Web browser and an AJAX or Java application embedded in the Web page. A set of rules govern which entries of the database can be queried or modified by which users. A subscriber can be the owner of a group in the database and has the right to decide which subscribers are added to the group in the database and what rights they have. Rights management includes assigning the right to initiate and join a group call and if users are allowed to hide their identity on the call (anonymity status).

The XDMS database is also queried by the PoC server, for example for predefined group session setups. For such a setup a user only sends the group ID, which is then used by the PoC server to retrieve information about all participants from the XDMS database. In addition to the configuration stored in the XDMS database, a number of parameters also have to be configured in the user terminals. While this can be done partly by users, it is also possible to configure the PoC service remotely. This is described in the OMA Device Management [25] and OMA Client Provisioning [26] specifications.

The presence service is another important element in the overall IMS service architecture. In essence it allows subscribers to be informed about the online status of others. Similar services exist in the Internet as part of the Yahoo and Microsoft messengers. In combination with PoC, the presence service is quite valuable for one-to-one and ad-hoc group sessions. Before establishing such sessions it is beneficial to know whether the other parties for the session are online or not. From a technical point of view the presence functionality is independent from the PoC service. Thus, the service requires its own IMS presence application server. Subscribers using the presence service are referred to as presentities. A presentity can, for example, be online or offline. The component on a subscriber terminal that monitors which presentities are online or offline is referred to as the 'watcher'.

The presence functionality is not only useful in combination with the PoC service but also with Instant Messaging (IM). In the IMS there are two ways to send messages containing text, speech and other multimedia information between subscribers. The simple approach is known as immediate messaging which is done by sending a SIP 'message' instruction via the IMS to another subscriber. If the subscriber is online, the instruction is delivered right away and the receiving terminal answers with a SIP '200 (ok)' message. If the subscriber is not online, the message could be rerouted by the S-CSCF to an application server which stores the message and forwards it as soon as the subscriber comes back online.

The session-based messaging on the other hand establishes a session between two parties by using the SIP 'invite' process. Since no resource reservation is required, the messaging is not as complex as shown above for establishing a voice session. Messages containing different kinds of media information are then exchanged with SIP 'send' messages. Such a messaging session could, for example, run alongside an established voice session to exchange text messages or pictures while communicating over a voice or video channel. When no more messages are to be sent, the session is terminated with a SIP 'bye' message, which is answered by the other party with a SIP '200 (ok)'.

4.4.11 Voice Call Continuity

As the IMS supports different types of access networks, it would be quite beneficial in many situations to keep a session ongoing when the user moves between different networks. This is especially the case for voice calls when the user leaves or enters their personal Wi-Fi Internet bubble during a conversation. Instead of dropping the session, an intelligent transfer function could quickly move the session to UMTS/HSPA/LTE or even to a circuit-switched 2G network. For this purpose an IMS extension referred to as Voice Call Continuity (VCC) has been standardized in 3GPP [27].

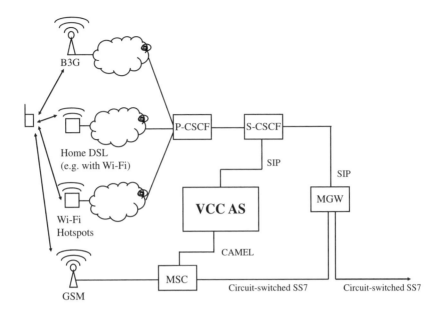

Figure 4.23 VCC architecture.

In the network, VCC is implemented as an application server as shown in Figure 4.23, which sits in the signaling path and receives all SIP signaling for voice sessions. Furthermore, VCC-capable mobile devices are required, as the request to switch to a different access network must come from the mobile device. This is necessary as the IMS itself has no information concerning signal levels and cannot therefore trigger a handover of the session to a different access network.

The actions taken by the VCC server in the IMS network for a transfer of a session to another access network depend on the types of networks involved. Roaming into a Wi-Fi bubble while having a voice session established via a HSPA or LTE network works as follows: a basic requirement for VCC devices is the ability to be connected (attached) to several radio networks simultaneously. This is usually the case since most 3G/Wi-Fi enabled devices can use both types of networks simultaneously. When the VCC-capable device detects the Wi-Fi network, it establishes an IP connection, registers to the IMS a second time and then sends a SIP Invite request with a special SIP URI. This request is routed to the VCC application server that detects, by analyzing the SIP URI and the identity of the user, that this is a request to hand over the ongoing voice session to, from its point of view, a different IP address. It then instructs both parties to initiate a VCC domain transfer by sending a SIP Update or SIP Re-Invite message with the new IP address and UDP port numbers. It is important to note that this message is originated from the VCC application server and not from the VCC device as the VCC application server acts as an anchor for the SIP communication. This means that the VCC application server terminates the SIP signaling from both parties to allow the VCC subscriber to change IP addresses when transferring from one access network to another. The process in the opposite direction, that is of handing over an ongoing IMS voice session from Wi-Fi to a packet-switched cellular network, is identical.

There are also many cases in which the user roams out of a Wi-Fi area and no high speed packet-switched cellular network is available. In this case the session has to be transferred to a circuit-switched 2G network. For this purpose the VCC application is not only connected to the IMS network but also to the circuit-switched MSC, as shown in Figure 4.23. This signaling link is based on the CAMEL protocol, which is already used by the MSC for other purposes such as the prepaid service for which the MSC has to query the prepaid service to verify if the user has enough credit left to make a call. When the VCC mobile reaches the limit of the Wi-Fi network while a voice session is ongoing, it establishes a circuit-switched voice call to a special VCC roaming number. The MSC then informs the VCC application via the CAMEL interface of the incoming call and the VCC application returns an instruction to the MSC to connect the incoming voice call to the IMS media gateway. The IMS media gateway will then send a SIP invite message to the IMS core, which in turn forwards the message to the VCC application. When the VCC application receives the SIP invite message, it already knows which data stream it should redirect as it has already been informed via the CAMEL interface. It will then immediately send a SIP Update or Re-Invite message that instructs the destination to redirect the media stream away from the current IP address of the transferring subscriber to the IP address and assigned port number of the IMS media gateway.

Transferring a voice call that has initially been established in the circuit-switched cellular network to a Wi-Fi connection has also been specified. For this purpose it is required that outgoing circuit-switched calls of a VCC subscriber are not delivered directly to the destination but instead are first connected to the IMS media gateway. The IMS media gateway in turn communicates with the VCC application via the IMS system since it cannot establish the call to the destination on its own. This way the VCC application is part of the circuit-switched call and can thus later on instruct the media gateway to change the routing of the media stream as required. A 2G to Wi-Fi transfer is initiated by a VCC-capable device that detects during an ongoing circuit-switched voice call that a suitable Wi-Fi network is available. It will then connect to the Wi-Fi network, register to the IMS system and send a SIP Invite message to the VCC application via the S-CSCF, which contains the information on which ongoing call on the media gateway should be redirected. The VCC application will then send the necessary commands to the media gateway to redirect the call away from the circuit-switched bearer to an IP connection and to drop the connection to the MSC. This process is completely transparent for the circuit-switched terminator of the call since their circuit-switched connection to the IMS media gateway remains in place.

VCC handovers between 2G and 3G networks are currently not specified since UMTS/ HSPA/GSM devices cannot be simultaneously connected to a 2G and a 3G network. This, however, is a precondition for VCC since the exchange of signaling messages with the new network have to occur while the voice call is still established in the other network. Consequently, VCC devices will preferably transfer a VCC call from a Wi-Fi network to a 2G network instead of to a 3G network.

It can be envisaged for the future that the VCC functionality will be extended and combined with intelligent session transfer functions between devices. A call could then, for example, be started at home as a voice + video call on a high-resolution device that is connected via Ethernet to the home network and then transferred by the user to a mobile device that uses the personal Wi-Fi network. In the process the resolution of the media stream could be reduced due to the smaller display size. Once the user roams out of their home the call could then automatically be transferred to the 2G network by first removing the video component and then using the VCC functionality to transfer the call to the 2G network. The IMS offers all the tools required for such complicated scenarios which should, however, be as seamless for the user as possible. Before such scenarios become reality, however, more work is required to standardize such inter-device transfers in combination with VCC and to develop network and device software.

4.4.12 IMS with Wireless LAN Hotspots and Private Wi-Fi Networks

Many mobile network operators are also deploying Wi-Fi hotspots in hotels, airports and other places where there is a high concentration of people who communicate outside their home and office. These Wi-Fi hotspots complement B3G cellular networks and therefore increase the capacity available at these locations. Current hotspot deployments, however, suffer from a number of disadvantages:

- Wi-Fi hotspots operated by mobile network operators are not connected to their cellular network subscriber database (HLR, HSS). Consequently, a separate authentication and billing system is required. From the user's point of view, this is also less than ideal as they are usually required to pay for access by entering their credit card details.
- Public Wi-Fi hotspots are usually not encrypted on the network layer as the Wi-Fi standards do not include methods to establish an encrypted connection to users who have not previously installed an authentication certificate or password. This makes it very easy for eavesdroppers to intercept data the traffic of others. While the login procedure is usually protected by an encrypted HTTPS (HTTP secure) Web session, many other Web services later on use unencrypted connections. IPSec software can secure the use of a nonencrypted network by establishing a tunnel to an endpoint, usually at a company over which all data traffic is encrypted. Again, this is usually something the user has to start manually as well. Even if such a product is used, most people are not aware of whether all data packets are sent through the encrypted tunnel or if only data packets to IP addresses of the company are protected while traffic to and from the Internet bypasses the tunnel.
- Many Web services such as Web-based e-mail interfaces do not use an encrypted connection. This makes it very easy for attackers to collect usernames and passwords.
- Active attackers have direct access to the device that has attached to the network and can therefore try to exploit operating system weaknesses.
- Current wireless network operator hotspot deployments are also not suitable for connecting to the IMS, which is usually deployed in a secure network environment and cannot be accessed via an unsecured public network.

As shown in Chapter 3, it is likely that Wi-Fi could play an important role in the future to satisfy overall wireless capacity demands and thus a solution is required for secure, easy and automated use of public Wi-Fi hotspots. While Wi-Fi networks at home and in the office can be secured using authentication and WPA or WPA2 encryption on the radio interface, a different method has to be found for public Wi-Fi hotspots. One of the solutions designed with access to the IMS in mind is 3GPP's Interworking-Wireless LAN (I-WLAN) specification [28].

I-WLAN removes the disadvantages discussed above by standardizing authentication and encryption over public Wi-Fi hotspots as follows. When an I-WLAN-capable device such as a notebook or a B3G/Wi-Fi mobile device detects an I-WLAN capable public Wi-Fi hotspot, it first of all contacts the network as before to establish a nonencrypted connection. In a second step the mobile device then authenticates itself using the EAP-SIM or EAP-AKA protocol [29].

EAP-SIM uses the same authentication framework as described for WPA personal and enterprise authentication as shown in Chapter 2. Figure 4.24 shows the messages exchanged between the mobile station and the authentication server via an EAP-SIM capable access point during authentication. After the Wi-Fi open system authentication and association, the access point starts the EAP procedure by sending an EAP Identity Request, to which the mobile device has to respond to with an EAP Identity Response message. The identity returned to the network in this message is composed of the IMSI (International Mobile Subscriber Identity), which is taken from the SIM card, and an operator-specific postfix. Alternatively, the mobile device can also send a temporary identity (pseudonym) which has been agreed with the network during a pervious authentication procedure. The pseudonym is similar to the TMSI (Temporary Mobile Subscriber Identity) used in GSM networks to hide the subscriber's real identity from eavesdroppers but has a different format.

In the next step, the network sends an EAP SIM start request which contains a list of different versions of supported EAP SIM authentication algorithms. The client device selects one of the algorithms it supports and sends an EAP SIM start response message

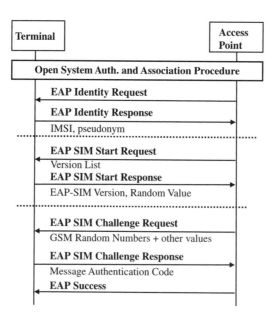

Figure 4.24   EAP dialogue between a mobile device and an I-WLAN access point.

back to the network. This message also contains a random number which is used for a number of subsequent calculations on the network side in combination with a secret (the Kc), which is shared between the mobile device and the network. In this way the network is also able to authenticate itself to the client.

At this point the authentication server in the network uses the subscriber's IMSI to request authentication triplets from the GSM/UMTS Home Location Register/ Authentication Center (AuC). Two or three GSM random values and GSM cipher-ing keys returned by the HLR are then used to generate EAP SIM authentication keys, EAP SIM encryption keys and other values required for the EAP-SIM authentication process. These are sent in encrypted form together with the two or three GSM random values in plain text to the client device in an EAP SIM challenge request to the mobile device.

The mobile device then uses the GSM random values received in the message and forwards them to the SIM card. The SIM card then generates the GSM signed response and GSM ciphering keys which are used afterwards to decipher the received EAP SIM parameters. If those values are identical to the values used by the network, the mobile device is able to send a correct response message that is then verified on the network side. If verification was successful, an EAP success message is returned and the client is admitted to the network. The Wi-Fi network then also delivers charging related informa-tion to the authentication server, such as consumed traffic and online time for online (prepaid) or offline (post-paid) charging.

Figure 4.25 shows the different devices and protocols used during authentication. On the left side the mobile client sends its EAP messages via the EAPOL protocol. For the messaging between the access point and the authentication server, RADIUS or DIAMETER is used. The authentication sever communicates with the HLR/AuC via the SS-7 circuit-switched signaling network and the Mobile Application Part protocol.

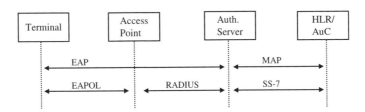

Figure 4.25 Protocol used for I-WLAN authentication.

The authentication and network admission procedure above only allows the device to access the network behind the access point and a direct gateway to the Internet. Access to the IMS network, which is usually located in the secured network of the subscriber's home network operator is not granted. This is because the home network operator has no control or visibility concerning the security of the Wi-Fi network and the interconnection to the home network. For access to the IMS and other services provided by the home network operator it is therefore necessary to establish a secure and encrypted tunnel between the subscriber's device and a gateway located at the border of home network operator. This scenario is shown in Figure 4.26. The secure tunnel can either be established directly after

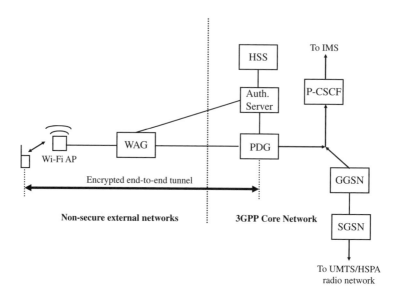

Figure 4.26 IPsec tunnel architecture to the B3G network.

association to the public hotspot or after the EAP-SIM authentication procedure shown above. The method used depends on the policy of the home network operator.

There are several possibilities to establish a tunnel with the gateway router, which is referred to as the Packet Data Gateway (PDG). If the mobile device has roamed to a Wi-Fi access point of the home network operator, a DNS (Domain Name Server) query is made to retrieve the IP address of the home network PDG. If the mobile has roamed to a Wi-Fi Access Point of another national or international network operator, the mobile will first try to get access to a PDG in the visited network. The home network PDG is only contacted if this fails. The domain name of the PDG can be built by the mobile device from information stored on the SIM card or in the mobile terminal.

Once the IP address of the PDG is known, the mobile then starts the tunnel establishment procedure with the Internet Key Exchange (IKE) protocol. The PDG in turn retrieves the authentication information required for the user from the authentication server, which acts as a gateway to the HLR/HSS as described above. During the key exchange messaging, the PDG and the mobile device authenticate each other. Once the keys for the tunnel are exchanged, IPSec ciphering is activated and the mobile device gains access to the home operator's IP network. As can be seen in Figure 4.26, the P-CSCF is now reachable by a Wi-Fi IMS device. As a final step an IMS Wi-Fi device will now start an IMS registration procedure, which starts with the P-CSCF discovery as described earlier.

Since the establishment of the IPSec tunnel is transparent for the Wi-Fi access point, it can be envisioned that this or a similar process could also be used in the future for IMS access via personal Wi-Fi access points connected to the public Internet via IP. For this purpose the B3G network operator must therefore connect a PDG to the public Internet.

The Wireless Access Gateway (WAG), which is also shown in Figure 4.26, has not been mentioned so far. Its responsibility is routing enforcement; that is, it ensures that the mobile device can only exchange IP packets with the PDG it has an encrypted tunnel with. The WAG is able to enforce such a rule since it can be informed by the PDG during the tunnel establishment process of how to filter packets to and from a certain user device.

Tunnel establishment and the use of IPsec to create an encrypted tunnel between a device and a gateway as shown in this section for B3G core network access is very similar to proprietary security products available for notebooks and mobile devices today.

4.4.13 IMS and TISPAN

In the previous section it has been shown how the 3GPP standard defines how mobile devices can use public and potentially also private Wi-Fi hotspots to connect to a 3GPP IMS network. For private Wi-Fi hotspots, however, the solution offers no quality of service control, since the 3GPP IMS system does not have interfaces to control network elements outside its domain. As such it does not fully meet the requirements of fixed-line network operators who would also like to use IMS as their future service platform. The European Standards Institute (ETSI) has therefore decided to define a broader service architecture in their TISPAN (Telecommunications and Internet converged Services and Protocols for Advanced Networking) standards project. In the meantime 3GPP and TISPAN are working in close co-operation and it is expected that Release 8 of the 3GPP standard will contain a common architecture. In its core, a TISPAN Next-Generation Network (NGN) consists of the following entities:

- a subset of the IP Multimedia Subsystem as defined by 3GPP;
- a PSTN/ISDN Emulation Subsystem (PES);
- other non-SIP subsystems for IPTV, Video on Demand and other services.

As can be seen in the list, the IMS is only one of several core subsystems of TISPAN. The reason for this is the fact that many services today are not based on SIP and sessions such as IPTV or Video-on-Demand. Many different approaches exist on the market to deliver such services to the user. TISPAN aims to standardize the way such services are delivered, controlled and billed and how such applications can interact with the transport network to request a certain quality of service level for their data packets.

It should be noted at this point that there is a fierce discussion ongoing about the advantages and disadvantages of prioritizing some packets over others. While from a technical point of view this has advantages, many fear that network operator-controlled preference of some packets over others has a negative effect on the evolution of the Internet. The proponents of the 'all packets are equal' approach do not allow any kind of prioritization in the network. This, they argue, would potentially allow network operators to dictate terms to Internet companies such as Google, Yahoo and many new startup companies who are offering innovative services over the Internet with high bandwidth requirements. Network operators, on the other hand, argue that quality of service for applications such as VoIP, IPTV and video on demand can only be achieved during times of network congestion by giving higher priority to data packets of such applications. The debate over this topic is referred to as 'network neutrality' and includes a number of other topics, such as whether network operators should be allowed to block packets of certain applications at their borders. Further information can be found in [30].

While in theory the IMS has been defined to be access network agnostic, the 3GPP standards still make a number of assumptions on the kind of access network and subscriber databases to be used. To make IMS usable for DSL and cable operators, it is necessary to fully generalize interaction with the access network and to have a generalized user database in the network. Figure 4.27 shows a simplified model of how the IMS is enhanced by TISPAN for this purpose as defined in [31] and [32].

The first difference between fixed-line and wireless networks is how devices connect to the network. In case of fixed-line access networks, a DSL or cable modem at the customer's premises is usually the gateway device that establishes the connection to the network. One or more devices behind this gateway device will then use the established connection to register with the IMS service. This is quite different to 3GPP, where each device connects both to the transport network (PDP context activation) and to the IMS service (SIP register).

A number of different ways exist today for a DSL or cable modem to attach to the network. TISPAN has made the step to standardize the functionality required in the network for user management at the network layer with the Network Attachment Subsystem (NASS). When a DSL or cable modem is powered on, it first communicates with the NASS to authenticate and to obtain an IP address. Protocols used for this purpose are, for example, PPPoE (Point-to-Point Tunneling Protocol over Ethernet) and PPP over ATM.

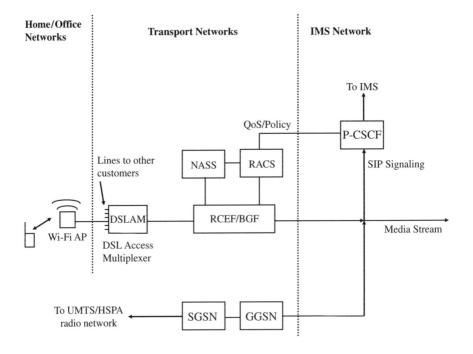

Figure 4.27   TISPAN IMS architecture for xDSL access.

The NASS is reached during the attachment process via the Resource Control Enforcement Function (RCEF)/Border Gateway Function (BGF), which sits between the access network and the core network of the operator. Their tasks are, among others, to route IP traffic between an external network and subscribers and to ensure that quality of service requests coming from the IMS or other core systems listed above are enforced on the transport layer.

The TISPAN Resource and Admission Control Subsystem (RACS) performs a similar task to the IMS Policy Decision Function. When an IMS session is established by a TISPAN device, the P-CSCF contacts the RACS subsystem to reserve the required resources for the session and to allow IP packets to flow between the participants of the session. The RACS then communicates with the RCEF/BGF to see if enough resources are available for the session and configures them accordingly. While in the 3GPP IMS model resources are reserved by the mobile devices and the P-CSCF, once codecs have been agreed on, the TISPAN P-CSCF contacts the RACS initially when receiving the invite message from the originator. If bandwidth requirements change during the session setup because a different codec has been selected by the devices, the P-CSCF contacts the RACS again to modify the policy. This means that, unlike a 3GPP IMS mobile device, which requests a certain bandwidth and QoS with a secondary PDP context activation, TISPAN IMS devices are not involved in QoS processes at all since the P-CSCF takes care of the whole process. This is necessary as TISPAN devices, unlike 3GPP mobile devices, are pure IP devices and therefore cannot influence the quality of service settings of the network themselves.

Besides the IMS subsystem, the PSTN/ISDN (Public Switched Telephone Network/ Integrated Services Digital Network) Emulation Sub-System (PES) is another important element of TISPAN. Its aim is to enable legacy analog and ISDN telephones to be connected to an IP-based next generation network via a media gateway which is either part of the access modem or a standalone device. An IMS independent implementation for PES is described in [33] while [34] defines how a PES can be implemented with IMS. Figure 4.28 shows how legacy analog (PSTN) or digital (ISDN) devices can be included in the IMS. As legacy devices cannot be modified a gateway has to be deployed on the user's location. On the one side of the gateway the analog or digital telephone is connected to a legacy connector. In case of a standalone gateway an Ethernet connector is used for the connection to the DSL or cable modem. The PES specification in [35] knows two types of devices. Voice Gateways (VGWs) emulate a SIP User Agent on the behalf of the legacy device and communicate with SIP commands with the P-CSCF of the IMS. The second approach is to deploy a media gateway that communicates via the H.248 (Media Gateway Protocol, MEGACO) to the Access Gateway Control Function (AGCF). In this approach the SIP User Agent functionality is included in the AGCF, that is, not on the customer device but in the network itself. Additionally, the AGCF includes the P-CSCF functionality. During call establishment, the P-CSCF or the AGCF then communicated with the RACS to reserve the required transport resources to ensure the quality of service for a call. In addition, PES requires an IMS application server to emulate the PSTN or ISDN service logic when the PES User Agent sends SIP messages with embedded ISUP messages.

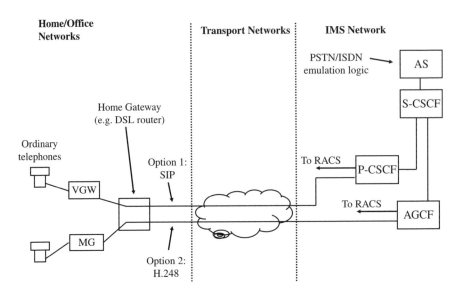

Figure 4.28   High level architecture of the PSTN emulation subsystem with IMS.

Finally, TISPAN also aims to standardize non-IMS subsystems. While out of the scope of the current version of the specification, it is likely that TISPAN will specify a standardized IPTV and Video-on-Demand (VoD) system in the next version.

4.4.14 IMS on the Mobile Device

Up to now this chapter has focused on the network part of the IMS. The IMS implementation on the mobile device, however, is just as important. Deploying IMS applications could be done in practice in several ways. One approach could be to create IMS applications which include the service itself and the necessary IMS software stack to communicate with the network. This approach, however, is unlikely to be seen in practice since the IMS is a highly complex system and the size of the protocol stack would therefore require more memory than the application logic itself. It would therefore be unlikely that mobile devices could execute several IMS applications simultaneously as each application would require a large amount of nonerasable memory and a high amount of Random Access Memory (RAM) to execute applications.

A better alternative is to split IMS applications into the IMS protocol stack, which offers abstract access to the IMS with a simple to use Application Programming Interface (API) for applications on top. This approach has the following advantages:

- Development of an application that uses the IMS as a service messaging framework does not require detailed IMS knowledge by application programmers. This speeds up the development of the application.
- Several IMS applications can run concurrently on a mobile device. All applications then use single instance of the IMS environment; that is, only a single copy of the IMS framework is in the memory of the mobile device.

- By giving some degree of control to the IMS environment, it is possible that applications are only started when a SIP message arrives from the network for a specific IMS application.
- A single IMS environment with a specific API can be available for several hardware platforms. All adaptations required for the hardware are encapsulated in the IMS environment. IMS applications can therefore be written in a device-independent way. This speeds up development and reduces interoperability issues between the same services on different types of devices.

An open question at this point in time is how IMS environments will be deployed. Today, IMS environments and APIs are developed by companies such as Ericsson [34], Comneon, Ecrio and Movial. Most IMS environments are offered for a wide variety of different hardware and software platforms. These range from powerful high-end operating systems such as Windows XP and Linux for notebooks to mobile device operating systems which are streamlined for much more restricted hardware environments such Symbian UIQ, Symbian S60, Windows Mobile, different flavors of mobile Linux (e.g. Google's Android), VxWorks and so on.

While application developers could select one of these IMS environments for their development and later on distribute their application together with the stack to their customers, it is more likely that mobile device manufacturers will at some point decide to deploy one of those products as part of their initial software distribution over their entire range of IMS-capable devices. Some mobile device manufacturers might even go further and acquire one of these companies to be able to control future IMS-specific developments for their devices and integrate the IMS stack into their operating systems.

The IMS specification knows two kinds of applications. Those that are already defined in the 3GPP IMS specifications are likely to be included as part of the IMS environment to a large extent. Applications on the mobile device can then take advantage of these services easily with only a little interaction with the IMS environment. An example of such a service is presence information, which could be used by the mobile device's native address book application to enhance the entry of a user with presence information (online, offline, in meeting, etc.). No IMS knowledge is required as the API function call would reveal presence information to an application while not requiring the application to know where this information came from and how it was obtained.

New proprietary applications require much more interaction with the IMS environment. This includes basic procedures such as registering their service with the IMS framework on the mobile device and interacting with application servers in the network or other service subscribers. All actions are abstracted via the IMS environment's API. This means that application programmers do not need to have in depth IMS knowledge such as, for example, how and which SIP messages are used to interact with external elements, how SIP headers of SDP content inside a message are formatted, how quality of service for a media stream is requested (e.g. via secondary PDP context activation), and so on.

For applications written for a specific operating system, IMS frameworks usually offer a C++ IMS API, as shown in Figure 4.29. Native applications then directly interact with the mobile device's operating system for tasks such as file access and the graphical user interface and use the C++ library from the IMS framework. Many mobile devices also offer a Java virtual machine for platform-independent programming. Such programs,

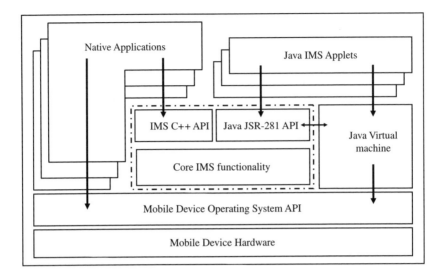

Figure 4.29   IMS framework in the software stack of a mobile device.

also referred to as Java applets, do not directly interact with the mobile device's operating system, which would make them platform-specific, but do so only via the Java virtual machine. IMS frameworks supporting Java applets have a link into the Java Virtual machine, as also shown in Figure 4.29, and offer a Java IMS API for Java applets. This Java API is specified in JSR-281 [36], which ensures that once a Java applet is developed it will work on any mobile device which includes a Java virtual machine and supports Java JSR-281, again independently from the developer of the IMS framework.

Most IMS frameworks are able to interact with many IMS applications running on a mobile device simultaneously where the device supports multitasking of user applications. The IMS framework therefore needs to be able to route an incoming message to the correct application. This is simple for responses to messages origi-nating from a specific application. For unsolicited incoming messages a different method has to be used. Release 7 of [32] introduces the IMS Communication Service ID (ICSI) and the IMS Application Reference ID (IARI) for this purpose. When applications are executed for the first time on an IMS device they register themselves with the IMS framework by specifying their ICSI and IARI. When an unsolicited message arrives at the mobile device that contains these IDs, the IMS framework then routes the message to the corresponding application. If the appli-cation is not executed at the time a message arrives, the IMS framework can also launch the application and then hand over the message.

The IMS framework running on a mobile device also takes care of basic IMS proce-dures on behalf of all applications. For example, the registration process when the mobile device is switched on is such a basic procedure. Depending on the operating system, hardware capabilities, types of currently available networks (HSPA, Wi-Fi, etc.), and user preferences, advanced IMS frameworks can decide if and over which network it should try to register with the IMS network. This process is completely transparent to

IMS applications as they just use the API to query the framework for the current registration state or be informed if the registration state changes.

4.4.15 Challenges for IMS Rollouts

While work on the IMS specification has already started in the year 2000, there are only few IMS systems deployed today. Likewise, there are only a few handsets with an IMS stack and IMS applications. This slow uptake is due to a number of reasons that make it very difficult for operators to introduce IMS and services to their customers.

4.4.15.1 Circuit-switched Voice and SMS

The main applications for IMS, namely voice calls and instant messaging, are difficult to introduce due to two in-house competitors, circuit-switched voice telephony and SMS messaging. These applications work well in wireless networks today and networks have been optimized for these services over many years. Voice and instant messaging over IMS would have to perform at least as well to be accepted by customers. Mobile operators are therefore in no hurry to replace or complement them. In addition, to be fully embraced by customers, IMS voice and messaging would have to be enriched to a point where users clearly see an advantage over using circuit-switched voice or SMS. This is certainly possible for both services. With IMS, rich media can be added to a voice call while the conversation is ongoing by adding a video stream or by allowing the sending of pictures while a voice session is ongoing. Web integration for instant messaging, presence information and Web 2.0 community-style extensions for instant messaging (for details see Chapter 6) could similarly make IMS instant messaging preferable to SMS for customers. These enhancements, however, make the service even more complex to develop, test and deploy to a large customer base.

4.4.15.2 Network Capabilities

Until recently, cellular networks did not have the capability or the capacity to support large-scale migration of voice and video streaming services to IP. With the introduction of B3G networks such has HSPA and WiMAX, this has certainly improved. The coverage area and in-building penetration of most B3G networks today, however, is far inferior to that of 2G networks. This is due to fewer base stations covering an area and the higher frequency range used for B3G than for 2G networks in many parts of the world (cf. Chatper 3). This means that IMS voice sessions in mobility scenarios will not work as well as today's circuit-switched wireless calls. As shown above, the Voice Call Continuity (VCC) function could reduce this issue to a certain extent, as it allows an IMS voice session to fall back to a 2G circuit-switched channel.

4.4.15.3 Solution Complexity

The IMS has been designed to be a basic and secure messaging framework for a wide variety of services. As a consequence, development of IMS network components and software stacks for handsets is very complex and time-consuming. While companies are working on IMS developments, standards continue to enhance existing features and specify new features for the framework even before existing features have been tested in networks on a large scale. This makes it difficult for implementers to choose which of the features to implement and from which version of the standard.

4.4.15.4 Network Interoperability

It is likely that each network operator will want to get its own IMS network to be in control of services and revenue. This means that several IMS networks will be deployed in most countries. Consequently these networks do not only have to allow subscribers to establish sessions to subscribers of other IMS networks but also have to allow them to exchange other information such as presence updates and instant messaging. In today's global environment, many people also want to communicate with people living in other countries and there should be no difference if another person uses the same IMS network, an IMS network of a different national network operator or an IMS network of a network operator on the other side of the world. While the IMS standard specifies how IMS networks and even non-IMS SIP-based networks can exchange SIP messages between each other, an IMS interconnection infrastructure has to be established in practice, since it is impossible to have a dedicated interconnection from each IMS network to all other IMS networks. From a transport layer point of view, this is easy to achieve because the Internet is already in place. From a service point of view, however, this is a complex task because interconnections between IMS networks must be secure. Furthermore, network operators need to reach agreements on how to charge each other for services that are established between subscribers of different IMS networks. If inter-IMS charging models in the future are based on charging models of circuit-switched voice calls, full international IMS interconnection will be a difficult to achieve because operators have to reach agreement on how much to charge for each particular service.

4.4.15.5 Mobile Device Capabilities

Mobile devices for circuit-switched wireless telephony have become very cheap in recent years as the technology has matured, processes are understood and many of the functions required for telephony are embedded in hardware. For IMS multimedia communication, sophisticated mobile device hardware is required with the following capabilities:

- multiband radio capabilities to support several network types (e.g. 2G, HSPA, Wi-Fi);
- a multitasking operating system, ideally, and enough memory to execute several multimedia applications concurrently;
- a sophisticated graphical user interface that supports multimedia applications (video, music, high quality picture capture, etc.);
- high-resolution screens as IMS applications for picture sharing and video streaming only make sense when the multimedia information can be displayed accordingly.

Devices with such potential have only been available since 2005, that is 5 years after the IMS standardization started. It could therefore be said that IMS standardization was well ahead of its time. With devices available on the market today which fulfill the requirements above, it is likely that the development of IMS handset capabilities will accelerate.

4.4.15.6 IMS Frameworks on Handsets

As discussed in the previous section, mobile device vendors have not yet chosen IMS frameworks for their mobile multimedia devices. While devices are not being shipped with an IMS framework as part of the initial software stack and applications, it is difficult to develop and deploy services as a third-party vendor.

4.4.15.7 Mobile Device and Application Interoperability

Today, there is a rigorous process in place to test a huge number of 2G/3G communication scenarios between a new mobile device and network elements of several network vendors before the mobile device is released to the market. Particular emphasis is put on voice telephony as it is an integral part of the mobile device and is in fact in most cases the main function. IMS services, however, are much more complex than pure voice telephony so interoperability testing between applications on mobile devices and IMS network elements from different vendors is likely to be more difficult. Furthermore, the network will no longer act as a buffer between two mobile devices by rewriting all messages exchanged between devices on the application layer. Therefore, an IMS service does not only have to be tested in combination with network elements but also with IMS devices of other manufacturers.

4.4.15.8 Business Model

There is a sound business model in place today for voice and SMS communication based on national networks. As a consequence, mobile operators are likely to require IMS services to generate revenue at the point they are introduced. This is difficult to achieve in practice, which is why most Internet companies first introduce services, see how they develop and evolve and work on a business plan for a particular service only once it has become popular and it has become clear how people use it. Even big mobile network operators only have a small customer base compared with the total number of people using Internet services today and have a strong national focus. As a consequence, development costs for a new service and maintenance once in service have a far greater impact than if a service is available to a global audience. As is shown in more detail in Chapter 6, many communication services only become useful if a critical mass of subscribers is reached. This is much more difficult to achieve in national markets served by mobile operators compared with international markets served by Internet companies. Even if only deployed on a national basis, interoperability agreements for the service need to be put in place between the IMS networks of a particular country to allow subscribers of different networks to use a service to communicate with each other. It is therefore questionable if the current circuit-switched business model will work for IMS or if network operators will at some point have to change their approach.

From a development point of view there are at least two kinds of business models. Some operators may decide to buy services from third-party companies and run them as their own services. For the third-party developer this means that the network operator becomes the customer rather than the end user. This is a difficult constellation for service development as any changes to the service to what was initially agreed require another round of contract negotiations. This makes fast turnaround times very difficult. Therefore this approach is mainly suited to services which, once put into place, are not expected to evolve very much. Another model under which services could be developed is by allowing third parties to develop services for their infrastructure and run them independently with a revenue-sharing contract in place. This gives developers much more freedom on service creation and evolution.

4.4.15.9 Service Development and Processes

As network operators are unlikely to develop services themselves they depend on third-party developers to create services for their IMS. This includes the logic on application servers and the plug-in programs to an IMS framework on the handset, as discussed in Section 4.4.14. Access to the IMS network infrastructure for development and later service deployment is difficult, however, as it first requires a business relationship with a network operator. Furthermore, access to users during the development phase is even more difficult since, as discussed above, network operators only have a limited customer base compared with the number of worldwide Internet users. As a consequence most small startup companies or research groups will rather develop their ideas in an Internet environment by either offering their mobile service via a mobile Web browser, via Java applets, or by developing native applications for mobile handsets and release them to a global audience. It is more likely, therefore, that IMS vendors will partner with third-party companies to develop applications for their IMS framework and sell them to network operators. These third-party companies, however, are unlikely to be startup companies but rather established companies and thus strongly revenue-oriented. The types of applications that can be developed in such an environment are very different from those developed in an environment where new ideas can be tried without financial pressure. As a result it is unlikely that IMS services will be developed in an evolutionary way, that is by launching a first version of the service to a global audience, get customers by recommendation and then rapidly evolve the service step by step from feedback and customer behavior.

4.4.16 Opportunities for IMS Rollouts

As shown above, the IMS faces many challenges on its way to become an established platform for services. There are also, however, a number of reasons why it is likely that IMS networks will establish themselves in the future despite the difficulties.

4.4.16.1 B3G Network Design

Except for HSPA +, B3G networks no longer contain a circuit-switched subsystem. Consequently, network operators will have no other choice than to introduce IMS once launching WiMAX and LTE networks if they want to be the provider of voice telephony service instead of leaving this area to a third party. Some network operators might prefer to use a standard SIP network architecture. Startup costs might therefore be lower, but it is questionable whether such an offer could be competitive over time.

4.4.16.2 Fixed Line Network Evolution as Role Model and Complement

Another angle to look at the evolution of voice telephony in wireless networks is to analyze the current evolution process in fixed-line networks and to see how this change will in the future also apply to wireless networks. In many countries, fixed-line analog telephony is quickly replaced by DSL lines and voice telephony over IP. The incentive for customers to replace their analog telephone line, which they previously had to have as a precondition to getting DSL service over the same line, is usually a lower price when both services are delivered over IP. Part of such an offer is usually a DSL access device (modem/router/Wi-Fi, etc.) with a built-in SIP User Agent or media gateway, as discussed in Section 4.4.13 and Figure 4.28. The DSL access device then offers one or more sockets for analog telephones. Today most fixed-line operators will not use a TISPAN solution, as discussed before, but this might change once such solutions mature.

A similar approach could also be taken by wireless network operators, especially if they also have fixed-line network assets. By offering a unified fixed and mobile voice service, they could use their system to let their subscribers register to the IMS system either via a B3G wireless connection or at home via the DSL line, which is also provided by them. The benefit for the user would be a single telephone number through which they can be reached and Internet access wherever they are. At the same time this would also reduce the overall bandwidth required in cellular wireless networks as most people mainly use services requiring a high data rate for a longer amount of time while they are at home or at the office (cf. Section 3.18). By offering such bundles at an acceptable price level users might be less inclined to select a third-party service provider for their voice and multimedia telephony needs that are not as well integrated to the fixed-line and cellular wireless networks they use. With the IMS network, operators also have the ability to offer advanced voice and multimedia services such as call handover from one device to another, which might be more difficult to do for service providers using a non-IMS platform. How well users take up IMS services other than voice and multimedia communication remains to be seen due to interconnection issues between IMS networks and pricing strategies, as discussed in the previous section.

4.4.16.3 Preconfigured Services

As mobile network operators usually offer their services together with a mobile device, they are in the unique position to preconfigure IMS on those devices. This is a huge advantage as third-party services require users to configure their mobile devices or even download and install software. As this is very difficult for nontechnical people, this is a huge competitive advantage.

4.4.16.4 National Telecom Infrastructure

From a financial point of view, having a centralized international infrastructure for a service is certainly beneficial. The consequences of a service failure, however, are far more serious as users in many countries will be impacted at once. Furthermore, the risk of a service failure in a single country increases due to the higher number of network components between the user and the service in the network. For critical services such as voice communication, it is therefore advantageous from a security and safety point of view to have several national operators with their own independent IMS networks. These are not impacted when national or international Internet connections are interrupted for whatever reason.

4.4.16.5 Conclusion

When taking the pros and cons of the technology, service development, network deployment and business cases into account, the IMS shows its strengths for voice, multimedia and messaging centered services. While the list seems to be short, there is much potential around these services, as discussed in previous sections. Beyond those services, Internet companies are much more flexible in developing new services and in reaching a mass market audience quickly. No contracts between developers and network operators need to be in place to launch a service and feedback and word of mouth from early adopters helps to develop a service. Some of these services might make it into the service portfolio of network operators over time where they do not require an international customer base and a business model can be found that includes at least national interworking between IMS networks of different national operators. Some services might not even require national interworking, such as, for example, an IMS presence and messaging extension for private Web pages, Blogs of users or Web community activities. It will also be interesting to observe if the IMS standard has enough appeal to serve as the basis for independent voice service providers. Quite a number of these companies already exist today and their infrastructure is based on basic SIP. A network operator-independent IMS network would not have the ability to request resources from the transport layer, but could help them expand their service structure beyond today's basic telephony service.

4.5 Voice over DSL and Cable with Femtocells

Many mobile operators are interested in offering voice and data services to users not only while they are on the move but also when they are at home or in the office, without waiting for IMS or advanced terminals supporting IMS and Wi-Fi in addition to B3G cellular networks. The easiest way to do this is to offer special pricing while users are in their home zone, which is defined by operators as the cells around the location of the user's home. The problem with this approach is scalability, as even when the user is at home the cellular network continues to be used. To address this disadvantage, several companies are developing femtocells, also referred to as femto base stations. From the mobile device's point of view, femto base stations look and behave like ordinary B3G UMTS/HSPA or CDMA/EvDO base stations. In practice, however, they have very limited transmission power and their size is similar to a DSL or cable modem. Instead of being connected via E-1, ATM or fiber Ethernet, femtocells are connected to the cellular network infrastructure via the Internet and a DSL or cable connection. Femtocells are either integrated into DSL or cable modems or are connected via an Ethernet cable. Once connected to the Internet, a femtocell automatically establishes an

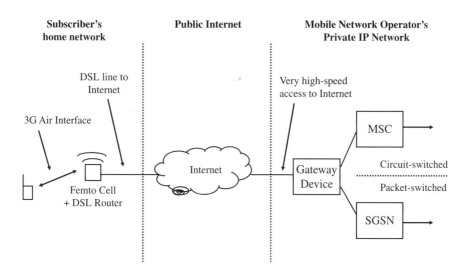

Figure 4.30   Femtocell network scenario.

encrypted tunnel to a gateway node of the cellular home network and connects to a specialized RNC. This specialized RNC terminates the encrypted tunnel and the femto-cells are treated like any other cell of the network. Figure 4.30 shows how such a setup would look in practice for a UMTS/HSPA femtocell. Since until recently there was no common femtocell backhaul standard, the network component that terminates the IP tunnels is usually from the same manufacturer as the femtocells or from a partner company. As femtocells behave like standard B3G cells for a mobile telephone, circuit and packet-switched voice calls as well as any other kinds of data applications offered by the mobile network operators can be tunnelled over a DSL or cable Internet connection.

In practice it is extremely important to integrate femtocells with DSL or cable modems for several reasons. First, femtocells are installed by the user and such an approach therefore ensures that the installation is easy and is done properly. Additionally, an integrated device is the only way to ensure quality of service for the femtocell since data traffic generated by B3G voice calls must be prioritized on the fixed-line link over any other traffic. If a femtocell was attached to an existing DSL or cable router which already served other users, the uplink data traffic from these users could severely impact B3G voice calls since ordinary DSL or cable routers do not have quality of service features to ensure that traffic from the femtocell is prioritized. This behavior can already be observed in practice today in other situations. If an ordinary DSL or cable router is used for a VoIP call in addition to a simultaneous file upload, voice quality is usually severely degraded due to the packet delay and insufficient bandwidth availability.

As a consequence, a mobile operator deploying femtocells ideally owns DSL or cable access as well or is at least partnering with a company owning such assets. In this way a single fixed-line gateway could be deployed with Wi-Fi for PCs and other devices and a femto radio module for 3G mobile devices. The single telephone per user idea also benefits from such an approach since owning or partnering for DSL or cable access removes the competition between fixed and wireless voice. This also ensures that a femtocell is only used in locations where the mobile operator has licenses to operate its wireless network since femtocells use licensed frequency bands.

In practice it can be observed today that a number of mobile operators are taking this route already by either buying DSL access provider companies or are at least partnering with them (e.g. Vodafone/Arcor or O2/Telefonica in Germany). It is unlikely that this is done specifically to roll out 3G femtocells at some point, but it seems that such companies have understood that it is vital for the future of a telecommunication company to have both wireless and fixed assets in order to be more than a mere bit-pipe for services running over the network. This completely reverses an earlier trend of splitting up fixed and mobile access of a company into separate entities.

Another technical aspect concerning femtocells is interference. In B3G networks, all cells usually transmit on the same frequency and interference is managed by having enough space between them and by adjusting output power and antenna angles. Most 3G operators have at least two frequencies they can use so femtocells could, for example, use the least used second frequency. However, there is still an issue with interference between femtocells of users living in the same apartment building and who have therefore installed their equipment in close proximity. This will result in lower capacity of each cell and might impact quality of service.

The following two sections take a look at femtocells from the operator's point of view and from the user's point of view to analyze in which scenarios femtocells could be successful in the future and how they could fit into an overall B3G network architecture.

4.5.1 Femtocells from the Network Operator's Point of View

In Europe and Asia 3G networks are operated on the 2100 MHz frequency band and in the USA on the 1900 MHz band, which is not ideal for in-house coverage. Even in cities it can be observed in practice that dual-mode 2G/3G mobiles frequently fall back to the 2G network. This is because many GSM operators use the 900 MHz band in Europe, which is much better suited for in-house coverage as lower frequencies penetrate walls much better. Some proponents of femtocells claim that in-house coverage for voice calls is greatly improved by femtocells. In cities, however, this benefit is rather small since GSM in-house coverage is usually not an issue. As most users are mainly interested in voice service, most do not care if the mobile device falls back into 2G mode.

An improvement could be seen in cases in which the mobile device cannot decide to stick with either the 2G or the 3G network due to changing 3G signal levels. This creates small availability outages while the mobile selects the other network type. During these times, incoming voice calls are either rejected or forwarded to voicemail.

It can also often be observed in practice that a mobile device with weak 3G in-house coverage changes to the 2G network once a connection to the Internet is established (e.g. to retrieve e-mails or to browse the Web on the mobile telephone) and sometimes changes back to the 3G network during the connection. The reason for these ping pong network selections is the changing reception levels due to the mobility of the user and changing environmental conditions. Such network changes result in outage times which the user

noti ... because it takes a long time for a
Wel

A ... ...he use of the 900 MHz frequency
ban ... ...850 MHz band in the USA. It is
likel ... ...many regulators have opened or
are i ... ... G technologies in Europe. It will
take ... ...network operators will have deployed their B3G
networks in those lower frequency ranges and until devices for these bands are available.
It is also likely that 900 MHz B3G cells would first be used to cover rural areas instead of
enhancing coverage in areas already covered by B3G networks in the 2100 MHz band. In
the meantime, femtocells could be an interesting alternative.

As the above weaknesses of B3G technologies in higher frequency bands show, it is
likely that femtocells can improve customer satisfaction. Putting a femtocell in the user's
home would have the additional advantage for network operators of reducing churn,
that is customers changing contracts and changing the network operator in the process.
Customer retention is all the more reinforced if the femtocell comes in a bundle with DSL
access, as further described below since changing wireless contracts also has consequences for the fixed-line Internet access at home.

Another advantage of femtocells is to reduce the gradual load increase on the B3G
macro networks as more people start using B3G terminals for voice and data applications. This could result in a cost benefit since, should the right balance of macro and
femtocells be reached, fewer expensive macro cells would be necessary to handle overall
network traffic.

The question is how much these advantages are worth to a network operator since
femtocells do not come for free. The options for network operators therefore range from
selling femtocells to their customers or over subsidizing them, to giving them away for
free as they benefit from a decreased churn or higher monthly usage and revenue.

4.5.2 Femtocells from the User's Point of View

While from the network operator's point of view femtocells have quite a number of
advantages, it is far from certain if users will perceive femtocells as equally beneficial.
While the user shares all operator advantages discussed in the previous section, increasing customer retention and thus churn is not necessarily in the interests of users since it
could reduce their choice. Also, it is unlikely that all family members use the same mobile
operator and thus could benefit from a single femtocell.

In addition, mobile multimedia users are usually still early adopters who tend to use
sophisticated devices, many of which include Wi-Fi. With such devices a femtocell for
multimedia content is not required since Wi-Fi offers a similar or better experience for
Internet content. Multimedia services offered by mobile network operators, however, are
usually not available over Wi-Fi which, from the end user perspective, is not a huge loss
since early adopters tend to prefer services available on the Internet. The reason for this is
that operator services are usually more expensive or come with limitations, such as being
limited to national boundaries, which are not acceptable to many early adopters.

An advantage not mentioned before is that better B3G in-house penetration would
increase the call establishment success rate for circuit-switched or IMS video calls, as

mobiles reselecting to the 2G network, because reception quality is better, cannot be used for incoming or outgoing video calls. Thus, femtocells could become an important element in the future to make video calls and IMS service more popular as the service still fights with the famous chicken/egg problem of B3G network availability and number of users with compatible handsets.

Monetary incentives could persuade users to install femtocells. Operators could, for example, offer cheaper rates for voice calls via femtocells. Also, the operator could propose to share revenue with femto 'owners' if other subscribers use the cell for voice and data communication instead of a macro cell.

Often the argument is brought forward that femtocells allow the marketing of single telephone solutions in which the user no longer has a fixed-line telephone and uses his mobile telephone both at home and on the go. However, such solutions which use the macro layer instead of femtocells have already been available for several years in countries such as Germany (O2's famous home zone for example) and are already very popular. Also, it is unlikely that mobile network operators would have competitive prices for all types of calls so many users would still use a SIP telephone or software client on a PC for such calls at home. Calling a mobile number is still more expensive in most parts of the world (excluding the USA) than calling fixed-line telephones, so single telephone offers have to include a fixed-line number. Again, this is already done in practice, for example by O2 in Germany for a number of years. Femtocells, however, might enable mobile network operators to deliver such services more cheaply than with a macro network approach.

It should also be mentioned that using a femtocell would have a configuration and usability advantage over SIP Wi-Fi telephones. However, it is likely that the configuration process for SIP and Wi-Fi on handsets will improve over the next few years, thus decreasing this advantage.

4.5.3 Conclusion

When looking at the arguments presented above, femtocells are not likely to be an immediate and outright success. A number of iterations will probably be needed before the form factor, usability and quality of service are adequate. This is likely to take several years. Furthermore, mobile operators need to continue their path of buying or partnering with companies owning fixed-line DSL or cable access. This is unlikely to happen quickly. However, there is currently still enough capacity available in the macro layer of the network so femtocells are not immediately needed to reduce the load on the network. Therefore, the major immediate benefit of femtocells is improving in-house coverage especially in rural regions. Femtocells are therefore likely to remain a niche market for now, since 2G and 3G coverage and capacity for urban users are usually sufficient even for in-house use.

4.6 Unlicensed Mobile Access and Generic Access Network

Unlicensed Mobile Access (UMA) is another approach to improving in-house coverage and offloading cellular traffic from the macro network to DSL or cable IP connections. The big difference to femtocells, however, is that UMA does not require any special network equipment on the user side. Instead, the mobile device is equipped with a Wi-Fi interface that it uses to communicate with a standard Wi-Fi access point. Furthermore, UMA simulates 2.5G GSM/GPRS connections through the Internet while femtocells tunnel B3G UMTS/HSPA or CDMA/EvDO traffic through the Internet.

4.6.1 Technical Background

UMA is a 3GPP standard and defined in [37] and is referred to in the standards as Generic Access Network (GAN). The principle of UMA/GAN is simple: it replaces the GSM radio technology on the lower protocol layers with Wireless LAN. A call is then tunnelled via a Wi-Fi Access Point connected to a DSL/cable modem via the Internet to a gateway node. The gateway then connects to the mobile switching center for voice calls and SMS and to the Serving GPRS Support Node for packet data. The gateway between the Internet and the network of the mobile operator is called a GAN Network Controller (GANC), as shown in Figure 4.31.

In practice, a GAN capable mobile can attach to GSM networks, 3G UMTS/HSPA networks where it is 3G enabled, and Wi-Fi networks. To take advantage of Wi-Fi, it is usually configured to prefer using Wi-Fi networks over 2G or 3G networks. Where a GAN mobile detects a Wi-Fi network (e.g. the user's home network) over which it can connect to the 2G/3G core network, it will attach to the Access Point and establish an encrypted connection to the GANC. Moving between a cellular network and a Wi-Fi network is referred to as 'roving' in the standards. Once the encrypted IP connection to the GANC is in place, a 2G/3G Location Update Message is sent over the encrypted IP

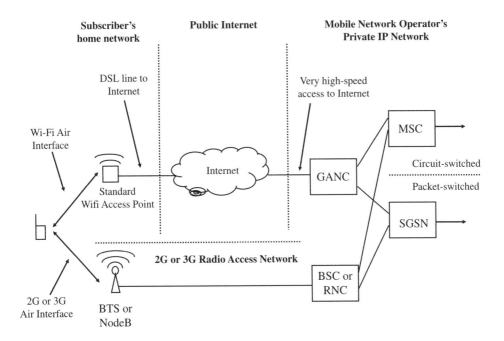

Figure 4.31 UMA/GAN network architecture.

link and the MSC and SGSN will update their subscriber information and the location in the Home Location Register. In practice, the GANC simulates a 2G Base Station Controller for these network elements. Thus, no specific changes are required on GSM core network elements for GAN.

From the network's point of view, signaling for incoming (or outgoing) calls to the subscriber is the same up to the GANC as for a traditional circuit-switched mobile terminated call. The GANC encapsulates the signaling messages into IP packets and forwards them to the mobile device via the encrypted IP connection. The mobile receives the messages over the IP link and presents them to higher protocol layers in the same way as if they had been sent via the GSM/UMTS network. In the other direction, the higher layers of the GSM protocol stack assemble GSM signaling messages as before. Once these messages reach the lower layers of the GSM protocol stack on the mobile device they are not sent via the GSM protocol stack but via the GAN protocol stack over Wi-Fi via the IP connection. Once the connection is established, the audio stream is also sent over the encrypted connection, encapsulated in RTP (Real-time Transfer Protocol), UDP (User Datagram Protocol) and IP. Inside RTP, the audio stream is encoded by using the standard GSM Adaptive Multi Rate codec. It is interesting to note at this point that the same protocols are used for audio transmission as in SIP and IMS.

Cell reselection is an action controlled by the mobile and thus does not require any network assistance. Handovers from one cell to another, however, are under the control of the network. In order not to change the software of the existing network infrastructure, a handover between the cellular 2G or 3G network and a Wi-Fi connection works as follows: each cell in a 2G or 3G network has a set of neighboring cells. These are identified with a Location Area Code (LAC), the Cell ID, and on the lower protocol layers with a Base Station Colour Code (BCC) and a Network Colour Code (NCC). Additionally, each cell operates on a distinct frequency, the ARFCN (Absolute Radio Frequency Number). To enable handovers to Wi-Fi, each cell of the GSM network has to have a neighbor cell in its neighbor list for this purpose. The parameters above are set to predefined dummy values. The same set of dummy values is used for all cells of the 2G/3G network, that is all cells have the same dummy entry in their neighboring cell list. If the mobile decides it wants to handover to a Wi-Fi cell, it first establishes a connection to the GANC over Wi-Fi and creates an encrypted tunnel. Once this is in place it will start sending measurement results to the cellular network that indicate that the dummy cell is received with a much stronger signal then the detected 2G or 3G cells. Based on these measurement reports the 2G or 3G radio network then decides to initiate a handover. Since the dummy cell is under the control of a different BSC (the GANC), the BSC that controls the current 2G cell or the RNC that controls the current 3G cell sends a request to the MSC to initiate an inter-BSC handover, or, in the case of 3G, an inter radio access technology (3G to 2G) handover. The MSC then sends a handover request message to the GANC. The GANC is not yet aware of to which subscriber the handover is to be performed, but nevertheless acknowledges the request. In the next step the current BSC sends a handover command to the GAN mobile. The GAN mobile in turn sends the information contained in the handover command to the GANC over the encrypted connection. This allows the GANC to correlate the handover request from the mobile with the handover request from the network. The handover sequence is finalized with the mobile switching the voice path to the Wi-Fi connection and the GANC connecting the speech path of the GAN connection to the MSC. On the cellular side the previous BSC will then free the timeslot used for the speech path and also finishes the handover procedure. As the GANC acts as a standard BSC, only a simple datafill change per cell is required in the existing network.

The handover from a Wi-Fi cell to the 2G network is a little simpler since the mobile can supply the correct cell parameters of the macro cell to the GANC once it detects that the Wi-Fi signal is getting weaker. The GANC then requests a handover from the MSC which in turn communicates with the BSC/RNC responsible for the 2G or 3G cell the mobile requests to be handed over to.

A mobile is also able to move between the cellular network and a Wi-Fi cell while a packet connection is established. This is done while no data is transferred between the mobile and the network as during those times the mobile device is allowed to select the most suitable cell on its own without the support of the network. Once the access network type has changed, the mobile then sends a location update to the MSC and a routing area update to the SGSN to inform the network of its new location.

4.6.2 Advantages, Disadvantages and Pricing Strategies

While SIP and IMS are real end-to-end VoIP technologies, UMA can only be considered a semi-VoIP service, as a call is only transported over IP on the link between the mobile telephone and the UMA Network Controller. On the mobile telephone and after the gateway, a traditional circuit-switched connection and a Mobile Switching Centre are used to connect the call to the destination.

Apart from reducing the traffic in the macro network, an additional benefit of UMA for network operators is the fact that a voice call always traverses the core network and an MSC of the mobile operator. This strengthens the relationship between network operator and subscriber and is an interesting way to replace fixed-line telephony by offering a DSL or cable connection and only using the circuit-switched mobile infrastructure. A fixed-line circuit switching center is no longer required. This approach is especially interesting for network operators that can offer DSL and cable access together with a cellular subscription.

As described above, UMA replaces one radio technology with another and otherwise leaves the rest of the system unaltered. This makes it difficult to price incoming calls differently for a caller while the called party is at home and using their (cheaper) Wi-Fi/DSL/cable connection compared with calls the called party receives while roaming in the cellular network. This is due to the fact that mobile operators in Europe use special national destination codes in order to be able to charge a caller a different tariff for calls to a mobile telephone user. Therefore, it might make sense for a network operator to also assign fixed-line numbers to their UMA subscribers for incoming calls while they are in their UMA home cell. When roaming in the macro network, such calls could then either be automatically rejected, forwarded to a voicemail system or forwarded to the subscriber. In the latter case the terminating subscriber could then be charged for forwarding the call.

It should be noted at this point that in the USA this problem does not exist since both fixed and mobile networks use the same national destination codes. There is no additional charge for the caller as the mobile telephone user gets charged for incoming calls.

As the mobile network is aware that the user is currently in their (cheaper) home Wi-Fi cell, incoming calls could then be charged at a different rate to the terminating subscriber.

Outgoing calls made via the Wi-Fi access point and a DSL or cable connection are also under the control of the mobile operator. It is unlikely that mobile operators will offer outgoing calls for free as is usually the case for connections between two VoIP subscribers, as the call will always be routed through a mobile switching center and a circuit-switched connection instead of being transported via IP end to end.

A slight disadvantage of UMA compared with a full VoIP approach such as SIP and IMS is the fact that UMA is not an end-to-end VoIP technology. Consequently, there is no presence information and built-in instant messaging capabilities as in other systems.

On the positive side, UMA offers a seamless experience. From an application point of view UMA is transparent to the user on the mobile as the GSM/UMTS telephony application is used for both cellular and Wi-Fi calls. The standard even offers seamless roaming between the two access technologies for ongoing calls, that is, a call is handed over from Wi-Fi to the cellular network and vice versa when a user leaves the coverage area of a Wi-Fi access point or detects the presence of a suitable access point.

UMA also tunnels GPRS services into the core network of the mobile operator. Data speeds are much higher though, which results in a seamless or even better experience for the user while in a UMA Wi-Fi cell, for example for Web browsing on the telephone, operator portal access or music downloads.

As UMA/GAN has been present on the market for a few years now, quite a number of mobile network operators around the world have started to offer services. While T-Mobile in the USA and Orange in France, for example, see the service as an important and flourishing addition to their portfolio, others like British Telecom and Telecom Italia have since backed away again and have discontinued the service.

References

1. Sauter, M. (2006) *Communication Systems for the Mobile Information Society*, John Wiley and Sons, Ltd, Chichester.
2. Rosenberg, J. and Schulzrinne, H. (2002) SIP: Session Initiation Protocol, RFC 3261. The Internet Society.
3. Berners-Lee, T. (2005) Uniform Resource Identifier (URI): generic syntax, RFC 3986. The Internet Society.
4. Handley, M., Jacobsen, V. and Perkins, C. (2006) SDP: Session Description Protocol, RFC 4566. The Internet Society, July.
5. Schulzrinne, H., Casner, S., Frederic, R. and Jacobsen, V. (2003) RTP: a transport protocol for real-time applications, RFC 3550. The Internet Society, July.
6. Schulzrinne, H. and Casner, S. (2003) RTP profile for audio and video conferences with minimal control. The Internet Society.
7. Rosenberg, J., Weinberger, J., Huitema, C. and Mahy, R. (2003) Simple traversal of User Datagram Protocol (UDP) through network address translators (NATs), RFC 3489. The Internet Society, July.
8. Daidalos (2008) Designing advanced network interfaces for the delivery and administration of location independent, optimised personal services, An EU Framework Programme 6 Integrated Project, http://www.ist-daidalos.org.
9. 3GPP (11 June 2008) Service requirements for the Internet Protocol (IP) multimedia core network subsystem; Stage 1, TS 22. 228.
10. 3GPP IP Multimedia Subsystem (IMS); stage 2, TS 23. 228.
11. 3GPP (4 October 2008) General Packet Radio Service (GPRS); service description; stage 2, Release 7.5.0, TS 23. 060, Chapter 9.2.3.4.

12. Kent, S. and Atkinson, R. (September 1998) Security architecture for the Internet Protocol, RFC 2401. The Internet Society.
13. 3GPP (17 June 2008) Access security for IP-based services, TS 33. 203.
14. Price, R., Borman, C., Christoffersson, J. *et al.* (January 2003) Signaling Compression (SigComp), RFC 3320. The Internet Society.
15. Calhoun, P., Loughney, J. and Guttman, E. (September 2003) Diameter Base Protocol, RFC 3588.
16. 3GPP (25 March 2006) Conferencing using the IP Multimedia (IM) Core Network (CN) subsystem; Stage 3, TS 24. 147.
17. 3GPP (9 June 2008) IP Multimedia (IM) Subsystem Cx and Dx interfaces; signalling flows and message contents, TS 29. 228.
18. 3GPP (12 June 2008) Subscriber Identity Module – Mobile Equipment (SIM-ME) interface, TS 11. 11.
19. 3GPP (9 June 2008) General Packet Radio Service (GPRS); service description; Stage 2, Release 6, TS 23. 060, Chapter 9.2.2.1.1.
20. 3GPP (6 June 2008) Mobile radio interface Layer 3 specification; Core network protocols; Stage 3, Release 6, TS 24. 008, Chapter 10.5.6.12.
21. Faltstrom, P. and Mealling, M. (April 2004) The E.164 to Uniform Resource Identifiers (URI) Dynamic Delegation Discovery System (DDDS) application (ENUM), RFC 3761.
22. Cuervo, F., Greene, N., Rayhan, A. *et al.* (September 2000) Megaco Protocol Version 1.0, RFC 3015.
23. Open Mobile Alliance (2008) OMA-AD_PoC-V1_0- 20060519-C, Push to Talk over Cellular (PoC) – architecture.
24. 3GPP Enablers for OMA PoC services stage 2, TR 23. 979.
25. Open Mobile Alliance (19 June 2007) OMA device management, http://www.openmobilealliance.org/release_program/dm_v112.aspx.
26. Open Mobile Alliance (2008) OMA client provisioning, http://www.openmobilealliance.org/release_program/cp_v1_1.aspx.
27. 3GPP (9 June 2008) Voice Call Continuity (VCC) between Circuit Switched (CS) and IP Multimedia Subsystem (IMS), Release 7, TS 23. 204.
28. 3GPP (9 June 2008) 3GPP system to Wireless Local Area Network (WLAN) interworking; System description, TS 23. 234.
29. 3GPP (20 March 2008) 3G Security; Wireless Local Area Network (WLAN) interworking security, TS 33. 234.
30. Wikipedia (2008) Network neutrality, http://en.wikipedia.org/wiki/Network_neutrality.
31. ETSI TISPAN (August 2005) TISPAN NGN functional architecture release 1, ETSI ES 282 001 V1.1.1.
32. ETSI TISPAN (March 2006) IP Multimedia Subsystem (IMS); Stage 2 description, [3GPP TS 23.228 v7.2.0, modified], ETSI ES TS 182 006 V1.1.1.
33. ETSI TISPAN Protocols for Advanced Networking (TISPAN); PSTN/ISDN Emulation Sub-system (PES); functional architecture, ETSI ES 282 002 V1.1.1.
34. Kessler, P. (2007) Ericsson IMS client platform. *Ericsson Review* **2**, 50–59.
35. ETSI TISPAN (April 2006) Protocols for Advanced Networking (TISPAN); IMS-based PSTN/ISDN emulation subsystem; functional architecture, ETSI TS 182 012 V1.1.1.
36. The JSR 281 Expert Group (2008) JSR 281 IMS Services API for Java™ Micro Edition. Public Draft Version 0.9.
37. 3GPP (16 June 2008) Generic access to the A/Gb interface; Stage 2, Release 7, TS 43. 318.

5

Evolution of Mobile Devices and Operating Systems

5.1 Introduction

Mobile devices with wireless network interfaces have gone through a tremendous evolution in recent years. From around 1992 to 2002 the main development goal was to make these devices smaller. While during that time the form factor of phones shrank considerably, voice telephony and SMS texting remained the only two real applications and overall functionality changed very little. By around 2002, technology had developed to a point where it became impractical to shrink phones any further from a usability point of view. The Panasonic GD55 is one of the smallest mobile phones ever produced, with a weight of just 65 g, and is smaller than a credit card [1]. Since then, development has concentrated on adding additional multimedia functionality to mobile devices. At first, black and white displays were replaced by color displays and display resolutions quickly rose from 100×64 pixels to 240×320 pixels, 480×320 pixels and even higher. High-resolution color displays are a prerequisite for all other functionalities that have been added to mobile phones since. These functionalities include cameras, multimedia messaging, e-mail and mobile Web browsing, just to name a few.

High-resolution color displays, high processing power with low power consumption and an increase in available resident memory and storage space have given rise to a number of other wireless mobile device categories:

- Smartphones – a smartphone is a combination of a mobile phone and a PDA. Smartphones often include a high-resolution camera for taking pictures and videos, and various network interfaces such as high-speed cellular interfaces, Bluetooth and Wi-Fi. These devices are usually shaped like mobile phones, but are slightly bigger to accommodate a larger screen and additional hardware.
- Connected PDAs – such devices are an evolution of PDAs, with operating systems such as Microsoft's Windows Mobile. Historically, these devices had no network

---

Beyond 3G – Bringing Networks, Terminals and the Web Together: LTE, WiMAX, IMS, 4G Devices and the Mobile Web 2.0   Martin Sauter   © 2009 John Wiley & Sons, Ltd

interface and were optimized for personal information management (calendar, address book, etc.). Over time, a Wi-Fi interface was added, and not long thereafter the step into the cellular world followed. Today, the difference between smartphones and connected PDAs is narrowing but many PDAs retain the physical shape of the original PDA concept.

- Internet tablets and mobile Internet devices – the main applications for these devices are not voice calls or personal information management. Instead, these devices are small computers optimized for accessing the Internet via a Wi-Fi or B3G interface. The user interacts with the device via a touch-sensitive display, a small retractable keyboard or a foldable keyboard connected via Bluetooth. With full-screen Web browsers ported from the desktop world, the user can use most Web-based services including video applications such as YouTube. Internet tablets are also an ideal tool to stay connected with friends via instant messaging applications and social networks, and they can be used as RSS feed readers, e-mail clients and Internet radios. Examples of such devices are, for example, Nokia's Internet tablets and devices built by third-party companies based on Intel chipsets for what the company calls Mobile Internet Devices.

- Ultra mobile PCs – slightly bigger than Internet tablets are what Intel calls Ultra Mobile PCs (UMPCs). The first models were not very successful as storage and processor capacities were too small for the requirements of Microsoft's Windows operating system. Since then, performance has improved and these devices have become more popular. Other manufacturers such as Asus with the eeePC, for example [2], have also developed devices for this category. Instead of using Windows, Asus is using a Linux-based operating system which is less resource hungry. Even though these devices are smaller than notebooks, they are optimized to run desktop Windows and Linux software with no or minimal adaptation. Input concepts for such devices are either touch-sensitive displays as in the case of UMPCs or a miniaturized full keyboard that can still be used with all fingers in the case of the eeePC. For the moment, most of these devices are equipped with a Wi-Fi interface, but it is expected that in the future B3G network interfaces for WiMAX, HSPA and LTE will be included as well.

- Connected music players – Apple's iPhone was the first such device on the market. Its primary application, at least from the vendor's point of view, is to be a portable music player and mobile phone. With the integrated Web browser it is also possible to access the Web. While usability of the iPhone as a connected Internet device was limited at the beginning due to the slow 2.5G network interface, a 3.5G network interface has made later versions of the device much more usable. In the future, it is likely that companies will invent devices that fall between smartphone, connected PDA, music player and Internet tablets. New product categories are thus likely to appear.

- Connected photo and video cameras – cameras built into mobile phones and other wireless devices have reached an amazing quality level, but are still behind in terms of picture quality compared with dedicated photo and video cameras. It is likely to remain this way for the time being as mobile devices are usually optimized to be general purpose devices that fit into a pocket. As will be discussed in more detail in Chapter 6, having connected recording devices enables the user to share pictures and videos created at the point of inspiration, in real time. This is much easier than if pictures or videos have to be downloaded to a PC or notebook first, and then shared via the Internet. Some companies are therefore experimenting with Wi-Fi and other network interfaces in dedicated photo and video devices. GPS chips are another interesting addition to record the location where a picture or video was taken.

- Wireless computing equipment – a recent trend for home and office networks is to untether computer equipment from attached devices such as printers and hard drives (or Network Attached Storage) using Wi-Fi interfaces. With Wi-Fi chips becoming a commodity, the additional price for consumers has dropped significantly and such devices are becoming more and more popular. This device category is different from those previously mentioned because the aim of equipping them with wireless interfaces is not mobility but to reduce the amount of cables in home and office environments. Nevertheless, they should be mentioned at this point since connecting them wirelessly to home and office networks makes them usable from the mobile devices mentioned above.

Most of today's connected devices are based on a chip with a processor design from ARM [3]. Although many companies such as Texas Instruments, Marvell, STM and VLSI design and manufacture chips for small devices, most are based on a CPU core licensed from ARM. On the desktop, Intel's x86 design dominates in a similar way. With both architectures now targeting sophisticated mobile devices, these two worlds are about to collide.

5.1.1 The ARM Architecture

The ARM design was initially targeted at ultra low power embedded devices. As technology evolved so did ARM's processor design and it is estimated that an ARM processor core is used in 95% of mid- to high-end mobile phones today [4]. The ARM-11 platform for example is used in devices such as Nokia's N-series phones like the N95 and in Internet tables like the Nokia N800 and N810. The ARM-11 platform is the result of a bottom-up approach, as it has evolved from earlier platforms for simpler devices. According to ARM, all phones of mobile giants such as Sony Ericsson, Nokia, LG and Samsung are ARM powered [5]. This shows the flexibility of the ARM architecture since requirements range from voice telephony with very low power requirements to multimedia devices that trade in a higher power consumption for higher processing capabilities.

Today, a lot of operating systems support the ARM architecture. Examples are fully embedded operating systems of low-end to mid-range mobile devices to operating systems for smartphones like Symbian, Windows Mobile and also Linux. Linux is a relatively new operating system for mobile devices but is increasing its market share quickly, for example with Nokia's Internet tablets, and in the future with devices built around Google's Android OS. The advantage of using Linux on mobile devices is the wide variety of available software from the Linux desktop world, which often only has to be slightly adapted and recompiled for the ARM processor architecture.

5.1.2 The x86 Architecture for Mobile Devices

Intel is at the other end of the spectrum and seems to be keen to enter the mobile space with its x86 processor architecture. A few years ago Intel tried to get a foothold in the mobile space by licensing ARM technology and building a product line around that architecture. In the meantime, however, Intel has abandoned this approach and is now refining their x86 architecture for low power consumption and size. Intel's development is directly the opposite of ARMs approach as they have to streamline a powerful desktop processor architecture for smaller devices.

Using an x86 platform for mobile devices has the advantage that only a few adaptations are required to run applications on mobile devices written for the desktop. Adaptation is usually only required for smaller screen sizes, mobile device-specific desktop environments and less disk and memory capacity. In theory, Microsoft Windows can also run on x86-based mobile devices but in practice it is too resource hungry. On the downside, Intel's platform for mobile Internet devices and ultra mobile PCs does not have a native cellular interface like ARM. Thus, device manufacturers have use additional chips for B3G connectivity in their devices. Moreover, Intel is likely to use their mobile platform to combine it with their own WiMAX chips.

At the time of publication, Intel and ARM have come quite close in terms of performance and power consumption and the two architectures are now competing for use in high-end mobile devices.

5.1.3 From Hardware to Software

The following sections now take a look at how mobile device hardware has evolved over recent years and give an introduction to both hardware architectures mentioned above. Different parts of the world use different frequency ranges for wireless communication. This chapter therefore takes a look at the global situation and describes the impact on mobile hardware design and global usability of devices. Adding a Wi-Fi interface to mobile devices has been another important step in the evolution of wireless communication and this chapter will discuss the profound impacts of this step on networks and applications. Finally, this chapter takes a look at the Symbian OS and Linux for mobile devices. These two operating systems have been chosen as they represent two different approaches in the evolution of mobile operating systems. Linux as a desktop operating system is being adapted in a top-down approach for mobile devices, while the Symbian OS was originally designed for mobile devices and is now evolving to take advantage of the ever more powerful hardware platforms available for mobile devices.

5.2 The ARM Architecture for Voice-optimized Devices

In the entry level segment, mobile phones are sold today both in developed markets and emerging economies that are optimized for voice communication. While the functionality of such phones has not changed much in the past decade, prices have been on a steady decline due to much higher production volumes and reducing the number of required chips and electronic components. This is referred to in the industry as reducing the Bill of

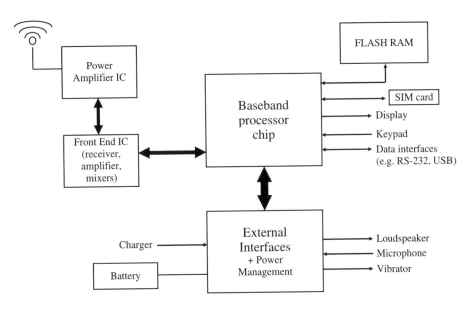

Figure 5.1 Block diagram of a voice-optimized mobile phone hardware platform. (Reproduced from *Communication Systems for the Mobile Information Society*, Martin Sauter, 2006, John Wiley and Sons.)

Materials (BOM). Figure 5.1 shows a block diagram of a typical voice-optimized mobile phone computing platform which is offered by many companies. The example in this book is based on Freescale Semiconductor's GSM i.200-22 hardware platform [6], which is optimized for voice communication and even excludes functionalities such as basic GPRS.

The core of this chipset is the baseband processor chip. It contains a 32 bit ARM7TDMI-S RISC (Reduced Instruction Set Computer) microprocessor but can be used with a 16- and 32-bit instruction set. While operations that can be performed with the 16-bit instruction set are not as versatile, only half the memory space is required for code compared with 32-bit instructions. Especially in memory-limited devices such as basic mobile phones, this is a big advantage. It is also possible to mix 16 and 32 bit instructions which enable software developers to compile their code into 16 bit instructions and profile specific portions of the software by hand to use 32 bit instructions where more performance is required. The maximum clock speed of the ARM processor used in this chipset is 52 MHz. This is very low compared with processor speeds of 2 GHz and beyond used in desktop systems today, but sufficient for this application. For more sophisticated devices more processing power is required. As will be discussed below, ARM thus offers several processor families and multimedia devices use ARM processor types that offer far better performance at the expense of higher production costs and power consumption. According to [7], power consumption at 52 MHz is between 1.5 and 3 mW. This is at least three orders of magnitude less than the power requirements for notebook processors.

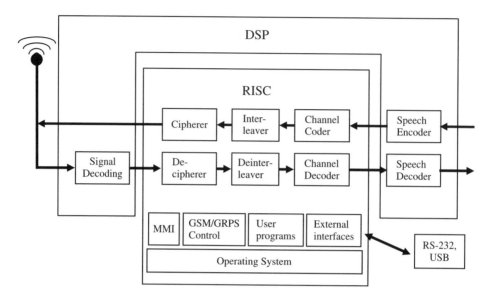

Figure 5.2 Work split for voice telephony in a mobile phone. (Reproduced from *Communication Systems for the Mobile Information Society*, Martin Sauter, 2006, John Wiley and Sons.)

In addition, the baseband chip contains a Digital Signal Processor (DSP) of Motorola's 56x family, which is clocked at 130 MHz. DSP microprocessors are optimized for mathematical operations and run software which is usually designed for specific tasks. Figure 5.2 shows how the RISC CPU and the DSP are used in combination in a mobile phone. The DSP chip is responsible for decoding the received signal from the network and for encoding and decoding the voice signal. There are two main advantages of performing these tasks on the DSP and not on the main processor:

- A digital signal processor has an optimized instruction set for mathematical operations required for dealing with codecs and decoding analog radio signals that have been digitized by an analog-to-digital converter.
- Encoding and decoding external signals is a continuous process and must not be interrupted by other activities such as reacting to user input or updating the display.

A typical voice call is treated by the baseband chip as follows:

- The analog input signal from the microphone is digitized and sent to the DSP chip.
- The DSP applies speech coding and forwards the data to the ARM RISC CPU.
- The ARM processor then packetizes the data stream, adds redundancy to the data (channel coding), changes the order of the bits so block errors can be more easily corrected on the other end (interleaving), encrypts the result and then sends the packet over the air interface.

In the reverse direction, the same actions are performed in the reverse order. In addition, the DSP performs signal decoding. This is a complicated task since the signal sent by the base station is usually distorted by interference. To counter these effects, packets contain training bits (in the case of GSM) that are set to predefined values [8]. These are used by the DSP to build a mathematical model of how the signal was distorted. The mathematical model is then applied to the user data around the training bits to decrease the transmission error rate.

In addition to the tasks above, the ARM CPU is responsible for interaction with the user (keyboard, display), to execute user programs such as Java applications, and to communicate with external devices (e.g. a computer) via interfaces such as USB and RS-232. As all of these tasks have to run in parallel; a multitasking operating system is required that is able to give precedence to repetitive actions concerning communication with the network and assign the remaining time to less time critical tasks.

For executing programs, about 250 kb of RAM is typically available on the baseband processor chip. In addition, about 1.7 Mb of nonvolatile memory (ROM, Read Only Memory) is available. If more memory is required, the chipset offers an external memory interface that can be used to connect additional RAM and ROM (e.g. flash memory). A 225-pin multiarray ball grid array connects the baseband chip via a 13×13 mm connection field to the other components of the device (cf. Figure 5.1). Other important components of the baseband chip are the module to access the SIM card and a display module for a monochrome or color display.

In addition to the digital processing functionality of the baseband chip, other analog components such as power amplifiers, signal modulators and functionalities to convert and control power for the device are required. These are implemented in separate chips as analog functionalities require a different manufacturing technology from the purely digital functions of the baseband chip.

5.3 The ARM Architecture for Multimedia Devices

The design intent for a voice centric mobile device chip set is to strip down the functionality to the bare minimum to reduce the price as much as possible. For high-end wireless mobile multimedia devices, however, the aim is to include as many functions as possible in the chipset. At the same time the device must consume as little power as possible in idle mode in order to achieve acceptable standby times. The chipset has to find a balance between power efficiency and performance while the user interacts with the device. Figure 5.3 shows a simplified block diagram of such a mobile multimedia chipset. The diagram is based on the design of the Texas Instruments OMAP 24xx and 34xx chipsets [9]. This chipset family has been selected for discussion as it is used in many high-end mobile devices today, such as the Nokia Nseries mobile phones and Internet tablets, which appear in the screenshots in Chapter 6. These devices also use the mobile operating systems that are discussed at the end of this chapter.

The core of the baseband processor chip shown in Figure 5.3 is the ARM CPU. While most voice-optimized mobile devices are based on the ARM7 architecture, multimedia devices require considerably more processing power. OMAP chipsets thus use ARM11 or ARM Cortex CPUs with clock rates between 220 and 550 MHz. Like in the PC world,

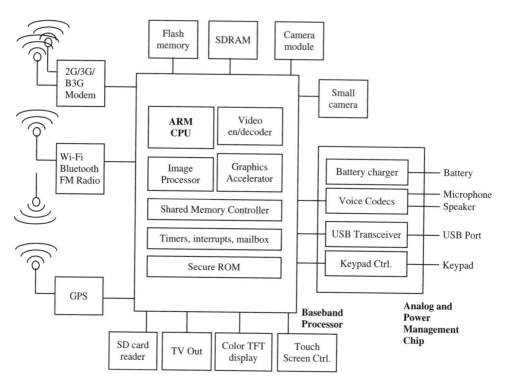

Figure 5.3   Block diagram of a multimedia chipset for a mobile device.

the clock rate is not the only means by which performance can be increased. Especially with the Cortex CPU, ARM has increased performance by introducing a super scalar design that increases the number of execution units a machine instruction passes during its execution. This way, several machine instructions can be processed simultaneously as each can be in a different stage of execution. According to ARM this increases performance by a factor of 2–3 compared with the previous ARM processor generation at the same clock frequency.

Java programs have become quite popular on mobile devices in recent years, especially for games and applications requiring Internet access such as Google maps or picture upload applications such as Shozu (see Chapter 6). This is due to the fact that Java ME (micro edition) programs are not compiled for a specific hardware platform but are instead translated into a machine-independent 'byte code'. This enables Java applications to run on a wide variety of Java enabled devices with no or only small modifications. To execute machine-independent byte code, a Java Virtual Machine (JVM) is required. On the PC, for example, JVMs are implemented in software. On mobile devices, however, this would be very inefficient due the lower overall processing power and memory availability compared with desktop CPUs. ARM processors thus offer a CPU operating mode which the company refers to as Jazelle that executes Java byte code using hardware. This way, Java applications can be executed much faster than if a software-based JVM translates the byte code into native machine instructions. Together with the graphics acceleration unit described below, sophisticated 2D and 3G games and other graphical applications can be written in Java and executed with a similar performance to native operating system software.

In addition, the baseband chip contains a number of other units for specific functions. The image processor unit, for example, is used for processing the input stream delivered from external camera modules. The chipset supports camera sensors with resolutions of up to 12 mega-pixels and converts pictures to compressed jpeg formats on the fly. This significantly reduces the load on the CPU since this computationally intensive task is performed entirely in this dedicated unit. The video encoder/decoder unit supports the CPU by encoding video streams delivered from the camera to MPEG4 formats. The unit is also used to decode and display videos that have been recorded earlier or downloaded to the device and stored on either internal flash memory or a memory card. Most high-end wireless devices have a main camera on the back of the device to take high-quality pictures and videos and a small camera on the front for video calls. Consequently the baseband chip has two camera interfaces.

Gaming is another important application for mobile devices. As games are usually very graphics-intensive, the chipset contains a dedicated graphics accelerator unit to speed up calculations for two- and three-dimensional graphics rendering. The processor is thus free to work on the program logic of the game while the graphics accelerator with its special 2D and 3D graphics capabilities takes care of rendering the graphical output.

All processing units require access to the main memory to store data and to communicate with each other. A shared memory controller synchronizes the memory requests of the different units. The chipset also contains dedicated hardware for mailboxes, which are used for communicating between different tasks, and interrupts, which are used to inform special program handlers of the operating system of external events (e.g. the user has pressed a button).

Finally, the baseband chip also contains a secure ROM to store confidential information and software which cannot be accessed from outside the chip. This can be used, for example, to protect the unique equipment identity of the device (IMEI, International Mobile Equipment Identity) and to enforce a SIM lock limitation to bind the device to a specific network operator. Furthermore, the software loaded from the secure ROM when the device is reset can also be used to ensure that only a certified version of the operating system is loaded into memory [10].

In addition to the battery control and a USB transceiver, the analog and power management support chips, shown on the right in Figure 5.3, handle microphone, speaker and keypad input and output. Displays on the other hand are directly connected to the baseband chip, which in addition also features a TV out signal to connect the device to a television set. Since the success of Apple's iPhone, touch screen input has become more important and the baseband processors chip also contains a direct input for touch panel sensors.

While the baseband chip includes only a small amount of both RAM and flash memory for program execution and long-term storage, most mobile multimedia devices require external memory chips due to high memory requirements. Typical configurations are 128 Mb of RAM for the execution of several data-intensive user applications simultaneously (multitasking) and several gigabytes of flash memory, which can usually be extended via memory cards of up to 32 Gb depending on the device.

The left side of Figure 5.3 shows how the baseband chip is connected to typical network interfaces required for high-end mobile multimedia devices. At the top of the figure a modem chip is shown that includes all functionalities to communicate with 2G, 3G and B3G networks. The baseband chip becomes radio technology-independent and can be used with a variety of different cellular network technologies. More than one antenna is shown as it is usually required to be able to communicate with different types of cellular networks using different frequency ranges. The impact of the multifrequency approach is further discussed below.

A Bluetooth network interface has become indispensable for most mobile devices today and Wi-Fi interfaces are also becoming more popular. In TI's OMAP chipset, both functionalities are included on the same chip and require only a single antenna as both technologies use the 2.4 GHz ISM (Industrial, Scientific and Medical) frequency band [11]. In addition the chip contains an FM radio receiver for which a dedicated antenna is required as FM radio is broadcast between 87 and 108 MHz. As this requires a long antenna due to the longer wavelength of the signals, the headset cable is usually used for the purpose. This means that the FM radio application can only be used when a headset is plugged in.

In recent years, navigation systems have become smaller and there is a clear trend to move from dedicated pocket-sized navigation devices to integrating the functionality as one of the many applications running on multimedia phones, Internet tablets, and so on. This becomes more and more feasible as mobile multimedia devices are now powerful enough for route calculation and display sizes are now suitable for showing maps. In addition, navigation applications require a built in GPS module. As shown in Figure 5.3, multimedia chipsets now offer the possibility to directly connect a GPS module to the baseband processor chip. In this way the use of the GPS module is not restricted to navigation applications. As will be discussed in more detail in Chapter 6, location information is also very useful for geotagging, that is to attach location information to pictures and videos taken with a mobile device. Web services can then use the location information embedded in pictures and videos to show where they were taken and can furthermore search for other pictures and videos that were taken nearby.

5.4 The x86 Architecture for Multimedia Devices

Figure 5.4 shows the chipset architecture Intel is offering to mobile device developers. It should be noted at this point that, unlike the OMAP chipset described above, Intel's solution is new on the market and only a few devices are presently using this chipset. It is likely however, that as the chipset evolves, many third parties will become interested in an Intel x86-based chipset for mobile devices for the reasons discussed at the beginning of this chapter.

Like desktop and notebook chipsets from Intel, the chipset for small mobile devices consists of three chips. The main chip is an Intel x86 processor that has been optimized for use in battery-based small mobile devices. According to [12], these processors are referred to as A100 and A110 and are based on the Intel Pentium M processor architecture. They are clocked at 400 or 600 MHz and have an on-chip 2×32 kb layer 1 cache and a 512 kb layer 2 cache to store program instructions and data which reduces the

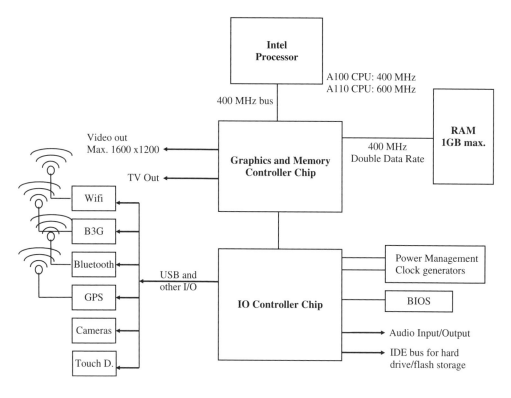

Figure 5.4   The intel chipset architecture for mobile devices.

waiting time when accessing memory. Like in other Pentium processors, a built-in multimedia coprocessor speeds up video streaming and other computationally intensive multimedia processes.

The second chip is the memory controller and graphics chip which is connected to the processor via an external 400 MHz bus [13]. As can be seen in Figure 5.4 the processor is not directly connected to the external RAM, but has to transfer all data via this chip. A memory controller is necessary since the processor is not the only component in the system that needs to access the system's memory. The graphics card, which has been integrated into the memory controller chip, also reserves part of system memory instead of using a dedicated memory bank. This is especially advantageous for small mobile devices as fewer chips are required. The drawback is that the data bus has to be shared between the processor and the memory card. The bus is a 400 MHz double data rate bus and transfers 2 bits per line, per clock impulse to and from memory. Consequently, the effective data rate of the memory bus for each component is lower. The built in graphics card is capable of resolutions of up to 1600 × 1200 pixels (UXGA, 1.92 million pixels). In practice, however, small devices use a much lower display resolution like 320 × 240 or 800 × 480 pixels. In addition the chip features a TV out signal to connect the device to an analog input of a television set. This is an interesting feature for mobile devices as a way to present pictures and videos to a larger audience.

The third chip is the Input/Output (IO) controller chip which connects peripheral components to the processor and memory. Like the processor and the graphics card, peripheral components can also access the system's memory without the help of the processor. This reduces the load of the processor. A Wi-Fi adapter can, for example, transfer a received data packet to the system memory and only afterwards inform the processor that new data has arrived. While data is transferred between the peripheral and memory, the processor usually continues working on a different task with data and instructions stored in the first and second level cache. In addition, the IO chip also includes a number of built-in components to reduce the number of additional chips. The two most important interfaces are for a hard drive and an audio card. Hard drives are available in the form of traditional magnetic media, although somewhat smaller for mobile devices, or solid-state flash memory, and are used to store the operating system, programs, data files, and so on.

All other peripheral components require their own chips and are connected to the chipset via a bus system such as USB. The chipset currently supports USB 2.0 with transmission speeds of 480 Mbit/s. USB 3.0 with even higher data rates is unlikely to play a major role in mobile devices in the coming years as the technology has to be optimized first for mobile use. The following external components are usually required in a mobile device and are shown in Figure 5.4 on the left:

- A cellular network interface such as GSM, GPRS, UMTS, HSPA, EvDO, LTE or WiMAX – how such modules are connected to the IO controller chip is implementation-dependent. A common way of connecting such modules is via USB. The maximum speed of 480 Mbit/s is fast enough for any cellular wireless network technology currently deployed or under development. Software drivers delivered by the manufacturer for such modules simulate either a high-speed serial port or a network card for the operating system. In case the module is accessed via a virtual serial port, the dial-up networking subsystem of the operating system is used for connecting the device to the network. In case the driver implements a network card, the module is used directly via the network subsystem of the operating system. In both cases it is necessary to deliver additional utility programs or APIs to perform tasks such as network selection and getting information from the module such as current signal strength and available networks.
- A Wireless LAN (Wi-Fi) – integrating a Wi-Fi adapter into the chipset via USB is straightforward since USB Wi-Fi adapters are readily available on the market. For small and mobile devices, power-efficient Wi-Fi modules are required beyond what is currently used in off-the-shelf Wi-Fi adapters. Since the potential market size for such a product is significant, it is very likely that many silicon companies are continuously enhancing their products for this category.
- Bluetooth – the design goal for Bluetooth chips has been low power consumption from the beginning. Therefore, no special adaptations are required for small mobile devices. The Bluetooth Host Controller Interface (HCI) uses USB or a UART (Universal Asynchronous Receiver and Transmitter) connection to exchange data with the host system [14] so a specification is already in place for easy connectivity.
- GPS – first introduced in mainstream mobile telephony devices in 2007, it is likely that global positioning system chips will become common place in high-end

mobile devices since a wide variety of applications benefit from accurate location information. As has been discussed in the previous sections, several of the described components can be integrated into a single chip in practice. While Texas Instruments has developed a combined Wi-Fi/Bluetooth/FM chip for the OMAP ARM chipset, other companies such as SiRF have developed combined GPS/Bluetooth chips [15].

• Cameras and touch panel input – in addition to external hardware that is similar or identical to that used in desktop and notebook computers, mobile device-specific components such as high-resolution camera modules and touch pads must also be connected to the chipset.

5.5 Hardware Evolution

5.5.1 Chipset

In the case of ARM-based chipsets such as the OMAP platform from Texas Instruments described in Section 5.3, the next few years will see further performance enhancement of all subcomponents. In 2008, most high-end devices using ARM and OMAP products are based on the ARM 11 processor family and the OMAP 2 platform with processor speeds between 330 and 400 MHz. The OMAP 3 architecture will exceed this performance by using an ARM Cortex processor which is expected to be up to three times faster at the same clock speed as current ARM 11 processors. Together with an increased processor speed beyond 500 MHz, a five times increase in performance over the next few years is likely. As RAM and flash modules continue to become cheaper, it is also likely that in the next few years, built-in RAM will go far beyond the state of the art of 128 Mb (e.g. Nokia N810) and eventually reach 512 Mb to 1 Gb. With prices for flash memory cut in half roughly every 12 months, the current 2–8 Gb used in mobile devices will soon be replaced by 16 and 32 Gb.

On the Intel side, the next step is to combine the memory controller and graphics chip with the IO controller chip into a single support chip for the processor. Even though there are already sophisticated mechanisms in place to reduce power consumption, research is continuing in this direction to counter the otherwise rising power consumption of integrating more functionality, of supporting several wireless technologies on a single device at once and of using faster and faster interfaces. While in idle mode, chipset components including the processor have several sleep states with varying power consumption and wakeup times. Less power being used in a sleep state means longer times for the component to be returned to an active state. Intel's next generation of mobile processors, codenamed Silverthorne, are expected to consume between 0.6 and 2W under full processing load depending on the clock speed, which can be up to 2 GHz. These values are reached by reducing the number of transistors on the chip from over 400 million (currently used for notebook processors) to just about 40 million by removing performance enhancing features such as out-of-order execution of commands [16]. In addition, enhanced power saving functionalities ensure that the processor automatically switches off unused functional blocks to reduce energy consumption as much as possible. In the deepest power saving mode, registers are saved to a special memory bank which only requires a voltage of 0.3 V to store the information.

While in an active state, processors and memory have to become faster in the future in order to be able to handle more data and to further enhance the user experience. This is especially important for Web page rendering, which is a very computing-intensive task. Even high-end mobile processors require a noticeable amount of time before a Web page is fully rendered. Another critical delay that impacts the user experience is the time between clicking on a program's icon and when the main window of the program is displayed and available for input. On PCs and notebooks, there is usually audible and/or visible feedback to the user in the form of noise or an LED indication until the program is displayed on the screen. Such feedback, however, is missing on mobile devices so users get impatient much more quickly. Faster processors will help to some degree, but more available memory that allows more programs to execute at the same time will be even more beneficial. Once a program is loaded into memory, the delay when switching from one program to another is almost zero, independent of the processor's clock speed. The capability to use flash memory in order to extend main memory also helps to improve the user experience when moving between different applications executing simultaneously. While the use of the hard drive or flash memory to increase the amount of main memory has been part of Linux almost since the beginning, such features have only recently been added to mobile operating systems such as Symbian. This is because flash memory, until recently, was too expensive to be used as main memory extension.

5.5.2 Process Shrinking

Decreasing power consumption and increasing the clock speed and the number of components and subsystems on a chip requires transistors on the chip to be smaller. Most microprocessors, static RAM (memory) and image sensors are based on CMOS (Complementary Metal–Oxide Semiconductor) technology today. Chips using ARM11 cores or the Intel A100 and A110 processors are manufactured with a 90 nm CMOS process. The 90 nm length refers to the average half-pitch size of a memory cell. Traditionally this value has considerably shrunk over time. In 1972, the Intel 4004 CPU was manufactured in a 10 μm CMOS process, which is 10 000 nm [17]. The CPU speed at the time was 108 kHz or around 2000–3000 times slower than the current clock rates between 2 and 3 GHz for high-end desktop and notebook processors. Since then transistor sizes have been shrinking around 70% every 2–3 years. Chips based on the ARM Cortex design, the successor to the ARM11 processor, are manufactured in 65 nm technology and a further step down to 45 nm is foreseen. Successors to the Intel A100 and A110 processors are equally expected to be manufactured with the 45 nm process [18]. Further manufacturing process enhancements are foreseen and it is predicted that memory cell sizes will shrink to 32, 22 and afterwards to 16 nm within the next 10 years.

Early CMOS transistors only required power to switch from one state to another while otherwise drawing almost no current. As a consequence the power requirement of a processor could be reduced during idle times by lowering the clock speed. Owing to continuing size reductions, however, the layers between the different parts of transistors have reached a thickness of less than 1.5 nm; that is, only a few atoms separate the different parts of the transistor. This leads to increasing leakage currents while a transistor is idle. This effect is undesired since the leakage is independent of the clock speed. This means that the leakage power, also referred to as static power, remains the same even if the system clock speed is reduced to 1 Hz. In theory, reducing the voltage also reduces the leakage power. In practice, however, chip voltage has already reached a very low level and a further decrease would result in unwanted interference from external components such as conducting paths between different chips. For the moment leakage power can still be controlled by new manufacturing methods which, however, slow down the possible switching speeds of transistors and hence reduce the clock speed improvements that would otherwise result from reducing the size of transistors. At some point below a 45 nm production process it is expected that leakage prevention techniques will eat up any performance gains in terms of increased clock speeds that a reduction in transistor size would normally bring about. So while for the moment reducing the size of components still results in faster processors, lower power consumption and smaller chips, this trend is unlikely to continue with current technologies. The only benefit from reducing transistor sizes in the future is thus a reduction of the size of the chip unless new methods are found to prevent leakage that do not interfere with the speed a transistor can switch from one state to another.

5.5.3 Displays and Batteries

The evolution of conventional displays is limited by the size of the mobile device. While on Internet tables there is room for displays beyond VGA resolution, traditional screen sizes of pocket-sized devices cannot exceed 2.5–3 inches. Most devices have a resolution of 320×240 pixels at this size which might be increased to full VGA (640×480) over time. Beyond this resolution the human eye will not be able to see a difference. For some applications, however, larger screens would be preferable. Foldable or rollable displays offer an interesting evolution path. Polymer Vision [19], a spin-off of Philips Electronics, develops such displays. Their display approach is based on organic semiconductors that are applied on a thin plastic foil. Currently, this method is suitable for black and white displays that do not require a background light and are thus very power-efficient. Unlike other displays that are difficult to read under sunlight conditions, this display technology delivers excellent results outdoors. It is thus especially suitable for applications such as eBook reading. As the thin plastic foil display can be folded or rolled up, devices that can be carried in the pocket are likely to be developed in the future with extendable display sizes exceeding 5 inches.

As power requirements of high-end mobile multimedia devices are unlikely to decline in the future, another research focus is better batteries or a different kind of energy storage for mobile devices. Current lithium ion technology is unlikely to get significantly more efficient in the future as physical limits of the technology are almost reached. A possible solution for the future could be fuel cells producing electrical energy from methanol, water and air [20]. Research has been ongoing for many years now and it is expected that fuel cells can store 10 times the amount of energy compared with current battery technology. However, no major breakthroughs have been reported so far and it seems unlikely that fuel cells will be used in mobile devices during the next few years. It should be noted at this point that new battery or power cell technology would mostly be used to extend the operating time of a mobile device and not to supply more energy to the device itself. Unlike notebooks and PCs, which tend to get warm and require active cooling, fans are not acceptable for mobile devices.

5.5.4 Other Additional Functionalities

In the previous sections, a number of features from cameras to GPS chips have been mentioned which are embedded in connected mobile devices today. In the future, it is likely that additional functionalities will also become small enough to be included in mobile devices or act as external add-ons.

One of the main issues of mobile devices due to their size is the missing or miniaturized keyboard, which makes text input difficult and slow. A solution is external keyboards that are connected to devices via wireless technologies such as Bluetooth. Figure 5.5 shows a foldable keyboard as it exists today, which connects via Bluetooth to Internet tablets (left) and high-end multimedia mobile phones (right). When folded together the keyboard is the size of a chocolate bar. When unfolded the user has an almost full QUERTY keyboard available. Even though the keys are slightly smaller than on a full notebook or desktop keyboard, typing with 10 fingers is nevertheless possible. This makes text input for e-mail, mobile blogging and many other activities very simple and convenient while preserving mobility. Alternatives to foldable keyboards are rollable keyboards and laser-projected keyboards. Compared with foldable keyboards, which have been available for a number of years, such keyboards have been introduced relatively recently and still require a number of improvements before being universally usable.

Figure 5.5 Mobile Internet devices and add-ons. (Reproduced by Permission of Nokia, Keilalahdentie 2-4, FI-02150 Espoo, Finland.)

With increasing multimedia capabilities and storage capacities, mobile devices are commonly used today for storing pictures and videos and for presenting them to other people. The biggest disadvantage compared with prints or to using a notebook for this purpose is the size of the screen, which is limited by the portability and mobility of the device. One current solution is to include a TV out port so the device can be connected to a standard television set (cf. Figure 5.3). Currently, several companies are working on miniaturized projectors which are small enough to fit into ultra mobile devices such as smartphones and Internet tablets. First working samples have been shown on occasion [21], but based on their current size it is likely that it will take a number of years before first models will appear on the market.

Another area of ongoing development is embedding mobile payment functionality in mobile devices, sometimes also referred to as the mobile wallet. In Japan, Sony's FeliCa RFID system [22] is already widely used as a payment system for public transportation, convenience stores and vending machines. Embedding the chip in a mobile phone and making it accessible to applications running on the mobile phone extends the use of the payment system for online ticketing for flights and sporting events via the mobile phone. Tickets can then be printed out at the airport, at the stadium, or the embedded FeliCa chip is used directly to gain entry. In addition, FeliCa equipped mobile phones can be used as door keys. If the door lock is connected to the Internet, users can even check remotely with the mobile phone if the door is locked. Since the FeliCa system is designed as a micro-payment system, linking the system to the Internet via a mobile device extends the payment system beyond direct interaction with an RFID chip card reader and it is likely that many online services beyond the two examples given above will make use of such a system in the future. In other parts of the world it will be more difficult to introduce a standardized RFID-based payment system due to the number of different market players and the uncountable number of public transport companies each using their own ticketing system. Thus, it will take much longer for a single system to reach a critical mass if this is possible at all. Having a universally adopted standard, however, is critical for mobile device manufacturers to consider including an RFID technology in their mobile phones, which are produced for a global market without national hardware variants. In the past, Japan had proprietary wireless network standards and consequently proprietary mobile phones which made embedding a national RFID solution in phones easier. Recently, DoCoMo's FOMA network has been upgraded to full 3GPP UMTS and HSPA compliancy and other Japanese networks have also followed this trend in order to profit from lower network equipment and handset prices produced for a global market as well as global roaming capabilities. As a consequence it might be much more difficult in the future to include national Japanese applications that require specific hardware into mobile phones.

Motion sensors are also beginning to be used in mobile devices. Currently, they are used to change the orientation of the display between landscape and portrait mode when the user turns the device and to advance from one music track to the next as a response to the user shaking the device. In the future, more than a single motion sensor might be used for three-dimensional interaction with the device and to react to more complex gestures. Early demonstrations have also shown the ability of proximity or pressure sensors embedded in the casing of the device itself to recognize the way users hold the device to automatically activate specific applications. For taking a photo for example a device is held in a different way and in different places then for Web browsing or for making a phone call.

A multimedia application that has not been mentioned so far is mobile TV. Several standards are currently developed or are under deployment. In Europe, DVB-H has been selected by most countries and network operators. In Korea the mobile-optimized version of DMB has become very successful. Both technologies are direct descendants of digital television standards. Other mobile TV technologies such as MediaFlo and MBMS are also still in the race. As it is likely that several different standards gain the upper hand in different parts of the world, it will not be possible to develop mobile devices with TV functionalities for a global market. In practice this will greatly inhibit the adoption of this service as only few device manufacturers are likely to be willing to develop mobile devices that can only be sold in local markets.

Mobile TV broadcasting is done either via a mobile cellular network by using a part of the overall capacity of a cell or independent of the wireless network via a dedicated broadcasting system similar to traditional broadcast networks. DVB-H and DMB are examples of traditional broadcasting systems adapted for mobile use, that is, for smaller screen sizes, limited processing power and battery capacity. As a consequence mobile devices require dedicated receiver hardware. Standards such as MBMS that use the cellular network, on the other hand, only require additional software in a 3G handset. Whether mobile broadcasting will be successful in the future remains to be seen as it faces strong competition from individual audio and video streaming applications and downloading. These services are in many cases better suited for mobile use since content can be consumed at exactly the time the user has time to spend (e.g. while waiting for the bus) rather than only at predefined times. In addition, content that has previously been retrieved can be consumed at places where no or only sporadic network coverage does not allow receiving a continuous data stream. The main advantage of mobile TV is thus broadcasting of live events such as football games, Formula 1 races, and so on.

5.6 Multimode, Multifrequency Terminals

While in the past only a few frequency bands were used for cellular wireless systems, their number is now rising rapidly. Table 5.1 shows the bands currently assigned in different parts of the world for 3GPP B3G networks for GSM/UMTS/HSPA and LTE terminals based on [23]. Table 5.2 shows current and future frequency ranges for WiMAX based on [24].

In practice, the growing number of bands has a number of undesired implications for both users and network operators. Typically, even sophisticated mobile devices today support only one or two 3G frequency bands like, for example, the 2100 MHz band for European models or 850/1900 MHz for US models. In addition, such high-end phones usually also support four bands for 2.5G GSM/GPRS, that is, 900 and 1800 MHz for Europe and 850 and 1900 MHz for the US. Data cards and B3G USB adapters are slightly ahead and usually support the 2100, 1900 and 850 MHz bands. Mobile phones are also moving in this direction, with Sony Ericsson having been the first company to offer such a Tri-Band B3G + Quadband 2.5G band model [26]. While connectivity with

Table 5.1 Frequency bands for 3GPP LTE, HSPA.

Band	Uplink bands (MHz)	Associated downlink bands (MHz)	Locations used (not exhaustive)
I	1920–1980	2110–2170	Europe, Asia, Australia
II	1850–1910	1930–1990	North America
III	1710–1785	1805–1880	Europe (GSM refarming)
IV	1710–1755	2110–2155	USA, Japan
V	824–849	869–894	USA, Australia
VI	830–840	875–885	Japan
VII	2500–2570	2620–2690	For future use in Europe
VIII	880–915	925–960	Most of the world, except North America (GSM refarming)
IX	1749.9–1784.9	1844.9–1879.9	USA, Japan
X	1710–1770	2110–2170	—
XI	1427.9–1452.9	1475.9–1500.9	Japan

Table 5.2 Frequency bands for WiMAX.

Band	Location used (not exhaustive)
700	USA
2300	Asia
2500	USA, Europe
3500	Europe
3700	—
5800	ISM band, worldwide, nonlicensed, low-power
1500, 3100, 4900, 5100, 5400, 5900	Bands with potentially local licenses [25]

such devices is currently guaranteed in most places in the world, the support of three B3G bands is far from the 11 bands currently defined in 3GPP. In the future, mobile phones thus have to support additional bands in order to ensure that travelers can use their wireless equipment globally.

For device manufacturers and network operators, the increase in the number of bands is equally disadvantageous. T-Mobile, for example, who have bought spectrum in the USA in band IV (1710–1755 MHz and 2110–2155) [27], will struggle to get a wide variety of devices for its B3G services in this band since it is specific to the USA and Japan. This means that the volume of devices using this band is very limited compared with global sales of billions of devices for more popular bands such as 900 and 1800 MHz. Furthermore, incentives are small to include this local band, only used by a few operators, in mainstream devices since the support of each new frequency band adds to the production cost per device. Since the market grows only insignificantly for a device, if this band is included this reduces sales margins and will thus inhibit production of such mobile devices for global sales. In addition, operators using local bands cannot profit from users roaming into the country from abroad who have at best a tri-band 2.5G or quad-band B3G device. Already today this is felt by network operators not using GSM or UMTS, since they cannot profit from visiting GSM and UMTS roamers [28].

Adding support for an additional frequency band in a mobile device has no impact on the digital components of the phone since processing of signals once they are digitized are independent of the band. The software for the digital signal processor and the radio protocol stack software of the baseband chip, however, have to be adapted to the additional frequency bands. Examples of such changes are the support of the additional channel numbers of a new frequency band, scanning of all available frequency bands for networks at power-up or while in idle mode and changing messages, and parameters to inform the network of the additional capabilities of the device.

The analog part of the mobile consists of antennas, front end filters and RF chips. Here, each additional frequency requires additional hardware components. To limit the number of additional components, hardware designs are usually multiplexing certain components for use with more than one frequency band. As multiplexing and switching components reduce sensitivity, each additionally supported frequency band further decreases the reception sensitivity of the device. As technology improves, this is usually compensated for by improved hardware designs.

It is estimated in [29] that the analog part of a mobile device is responsible for 7–10% of the cost of a mobile device, independent of whether it is a low-, mid- or high-end handset. This is due to the fact that, while low-end devices are cheaper, they do not support as many radio interfaces and frequency bands as high-end devices. The report also estimates that the hardware cost per supported frequency band is around US$2. For GSM devices this includes the frond end switching and routing functionality and an additional duplex filter. For an additional 3.5G CDMA frequency, additional duplex filtering is required. In addition, a separate antenna is required if the band is too far away from other bands. In addition, high-end phones require a special chip and board design to prevent unwanted inter-modulation effects between the cellular, Wi-Fi, Bluetooth and GPS radio units. The white paper further estimates that, in addition to the hardware costs, the engineering cost for a new band is in the order of US$6 million. This includes development costs, type approval and testing. With annual global GSM 900 MHz phone sales of several hundred million devices, the development costs per device are only a few cents. For national frequencies for which only a few million devices are sold per year, the development costs for supporting the frequency band per device can easily exceed the price of the hardware.

From a development point of view, many manufacturers thus prefer to focus their design resources to decrease the price of their next-generation designs and to improve sensitivity rather than to add exotic frequency bands. In other words mainstream bands attract much more engineering effort resulting in less expensive hardware with higher sensitivity.

In the future, analog hardware in mobile devices is likely to get more expensive since advanced B3G technologies such as HSPA+, LTE and WiMAX require at least two antennas and receiver chains for the MIMO transmission (cf. Chapter 2). Standards even include data transmission modes for 4×4 MIMO, which requires four separate antennas and receiver chains in mobile devices. As the hardware components cannot be shared between receiver chains, this increases the number of required components and thus increases costs. If in addition to Wi-Fi, Bluetooth and GPS, several frequency bands for cellular B3G networks are supported, physical limits will limit either the number of MIMO antennas per band or the number of supported bands per device.

5.7 Wireless Notebook Connectivity

One of the most important factors of the success of Wi-Fi has been Intel's push for an embedded Wi-Fi chip in all notebooks with its mobile Centrino chipset. For some time, Intel was considering repeating this approach with a combined Wi-Fi/HSPA wireless chipset, but they gave up on these plans when their alternative WiMAX strategy emerged. Since then, several other possibilities have emerged to connect notebooks to cellular networks. These have gained significant popularity among travelers and users who are using a B3G Internet connections as an alternative to Internet access via DSL or cable:

- A built-in 3G network card via a mini PCI expansion slot. Such notebooks are sold, for example, by mobile network operators in combination with a 12 or 24 month contract and a monthly fee. While an interesting concept, this is likely to remain a niche market as the majority of people will continue buying their notebooks from other sources.
- Removable PC cards via an expansion slot. The advantage compared with the built in card is that the user can remove the card when not needed and thus increase the battery life of the notebook. A slight disadvantage is the external antenna.
- A USB add-on. This sort of B3G network adapter is completely external and has the advantage that it can be placed in a convenient position and orientation, which is especially important when the signal strength to the B3G network is weak (cf. Chapter 3). Sizes of external B3G USB adapters have shrunk considerably in recent years to the size of a memory stick.

For WiMAX similar options exist but Intel is likely to tie WiMAX closer into the chipset to create a widespread availability of WiMAX devices. It is then up to the wireless network operators, both HSPA/LTE and WiMAX, to make attractive offers to customers. Even today, prepaid and low monthly service charges with several gigabytes of data volume included are available in countries such as Austria and Italy. Combined with cheap HSPA adapters for notebooks, this has triggered a significant adoption of cellular network connectivity with notebooks. In Austria, for example, about 30% of broadband subscriptions are wireless (mostly over HSPA networks) instead of DSL and cable [30].

5.8 Impact of Hardware Evolution on Future Data Traffic

In 2003, one of the most sophisticated connected mobile devices available in Europe was the Siemens S55 GSM/GPRS phone. It was one of the first mass market devices with a Bluetooth interface and a stable GPRS protocol stack already in the first version of the device's software. The S55 was also the first device in this product line with a real color display with a resolution of 101×180 pixels and 256 colors. Memory card slots in mobile phones were not yet available, so file storage space in the built-in 1 Mb flash memory was limited to a few hundred kilobytes. Even if the device would have had a multimegabit B3G network interface, it would not have been possible to take full advantage of such capabilities. The simple Web browser was only suitable to load and display small Web pages designed for mobile use with a size of only a few kilobytes.

Five years later, the mobile device landscape has changed completely. Cellular data speeds have evolved from 45 kbit/s to several megabits per second, onboard memory has grown from a few hundred kilobytes to several gigabytes and memory slots allow an expansion of the storage capacity to tens of gigabytes. Device sizes on the other hand have only increased slightly, mostly to accommodate bigger and higher-resolution displays with resolutions exceeding 320×480 pixels and 16 million colors. At the same time, the sales price has remained the same. Web browsers have evolved to handle standard Web pages composed of hundreds of kilobytes of information. This is more than the total available flash memory storage of the S55 from 2003.

Internet applications have enjoyed mainstream success in recent years and require a high amount of data to be transferred over the network. This include podcasts, videocasts and music downloads. So far, users have mostly used cable and DSL access networks to first download such files to the computer and from there transfer them to a mobile device. This process is also referred to as sideloading. With devices now incorporating Wi-Fi and fast B3G cellular network interfaces, it is no longer required from a technical standpoint to place a computer between the source in the Web and the mobile device. Widespread direct downloading of music files, podcasts and video via Wi-Fi and cellular networks is just a matter of time.

This trend is further strengthened by players such as Intel entering the market who do not see voice telephony as the primary application for a mobile device and a cellular wireless network but rather Internet connectivity itself. Together with alternative network operators using WiMAX to break into the wireless market, this trend will be further reinforced. IT companies using Intel's mobile chipsets as well as established mobile device manufacturers such as Nokia are shipping or planning to release mobile Internet devices and Internet tablets where the majority of applications require continuous Internet connectivity. Without connectivity, the device becomes almost useless. The ideas behind such devices are very different from those of current mobile phone design and also from the PDA approach with a focus on local applications such as calendars and address books that run well even without network connectivity. On mobile Internet devices, applications such as Web browsing, feed reading, Web radio, on-demand videos, VoIP telephony, messaging and gaming are prevalent. Calendars and address books, if present at all, are only playing a secondary role. Such devices, especially once they become more popular and widespread, will increase data volumes in wireless networks by at least an order of magnitude over today's level. In addition, the widespread use of such devices will increase bandwidth requirements even further. Streaming a radio station 8 h a day with a data rate of 128 kbit/s, for example, results in a data volume of over 460 Mb per day or 13 Gb per month. As users mostly listen to radio at home or in the office, devices with both cellular and Wi-Fi interfaces are ideal to offload this type of traffic from the cellular network to DSL and cable connections.

With the examples above it becomes clear that mobile devices today and in the future, especially those dedicated for being used mainly with Internet-based applications, match or even exceed the amount of data that is consumed with notebooks and desktop PCs.

As cellular networks will have difficulties handling such high loads on their own, network operators offering Internet access via cellular and via Wi-Fi over DSL/cable will have a competitive advantage in the future. Another future challenge will be the management of several devices per user. Requiring a separate contract or prepaid SIM subscription for each device will become impractical once users are connected with their notebook, their private phone, their business phone, their dedicated MP3 player, an Internet tablet, their car, and so on.

5.9 The Impact of Hardware Evolution on Networks and Applications

The integration of Wi-Fi and B3G network interfaces in mobile devices is likely to have a significant effect on future network architectures, applications and services. In Chapter 4 the development of femtocells and the Generic Access Network was discussed from a network point of view. Femtocells connected to home networks with DSL or cable backhaul are addressing scenarios in which users access the Internet with mobile devices with a B3G cellular network interface but without Wi-Fi. As combined Wi-Fi and B3G network interface integration becomes more common, the future for femtocells is rather uncertain. Among other reasons, this is due to the fact that femtocells do not allow a mobile device to directly interact with other devices in the local network (e.g. network attached storage, PCs or an MP3 music library). This limits the usefulness of single network technology devices. For details see Section 4.5. GAN serves a similar purpose but uses Wi-Fi to offload traffic from the cellular network, as discussed in Section 4.6. Current GAN devices exclusively use the Wi-Fi interface for bridging cellular traffic over Wi-Fi. It is unlikely, however, that manufacturers will open the Wi-Fi interface for applications on GAN devices, since this is not in the network operator's interest, even though it is possible from a technical point of view.

Most combined B3G/Wi-Fi devices can access to the Internet via cellular B3G networks and Wi-Fi networks and in addition have access to local networks and applications. This trend is further enforced by Intel entering the mobile space with their x86 platform. The resulting interest from device manufacturers from the desktop world will bring about device architectures with Wi-Fi as a primary interface and a B3G interface as a secondary interface. This is different from devices built by traditional handset manufacturers that are primarily built for cellular network operators who for the most part still consider the integration of Wi-Fi and the resulting integration of cellular and home/office networks as a threat rather than an opportunity.

It is therefore likely that mobile device architectures will shape future network architectures as follows:

- Wi-Fi networks at home and in the office will carry the majority of data traffic.
- B3G networks serve the majority users as overlay networks while they are outside of their personal Wi-Fi bubbles.
- Moving between networks must be transparent for the user in order for services to be used from any location, that is, the device can access the same services, no matter whether they are Web-based or running on devices at home or in the office from any type of network. For this, network operators should adopt an open policy towards

services and software that can open a secure and encrypted tunnel to local home and office networks while the user roams in the cellular network.

• Devices with both Wi-Fi and B3G cellular network interfaces can be used as Internet gateways for Wi-Fi-only devices. From a hardware perspective this is already possible today but the required software has not yet been included in devices. This is probably partly due to the fact that this is again not in the immediate interest of cellular network operators, and hence, not very high on the agenda of device manufactures. With open source operating systems such as Linux becoming more popular on mobile devices, the Web community can develop such services themselves without relying on support from network operators and device manufactures.

Powerful connected mobile devices will also have an impact on the future development of applications that store data in the network or locally on a device. Applications such as calendars, address books and e-mail clients will no longer only be used from a single device but used on many devices depending on the circumstances. When using an application with more than one device, it is required to synchronize the data that has changed between the different devices. E-mail messages, for example, which have already been read and answered from a mobile device, should not appear in the inboxes of the Web-based and notebook-based applications. In addition, the response, written on the mobile device, should be available in the outboxes of the Web-based and notebook-based applications. Network operators having both cellular and fixed-line DSL/cable assets will greatly benefit from these trends.

5.10 Mobile Operating Systems and APIs

The middleware between the mobile device hardware and applications is the operating system. It decides how much or how little applications can access device properties and network resources directly, how the user interface looks and how the user can interact with the device. Operating systems thus have a significant impact on the usefulness, popularity and market success of a mobile device. This section now describes the most popular operating systems and application programming interfaces on mobile devices today and discusses how they will evolve in the future. A particular emphasis is put on two different operating systems for mobile devices, Symbian/S60 and Linux. While Symbian/S60 is one of the most successful operating systems for high-end mobile devices today, Linux-based operating systems such as Android have only recently become popular. With Intel and Google pushing into the mobile device market, it is likely that Linux-based operating systems will gain a significant market share in the future. This section will then discuss the approach of Linux-based operating systems compared with traditional mobile device operating systems and the consequences on hardware design and application development.

5.10.1 Java and BREW

For low-end to mid-tier mobile phones, most device manufacturers are using proprietary, or closed, operating systems today. Most of them have no or only limited multitasking support for user applications. Multitasking capabilities are usually restricted to running a single program in the background such as a music player application. Third-party developers have no access to the operating system and can only extend the functionality of such devices with Java programs that are executed in a JVM of the Java Platform Micro Edition environment, originally developed by Sun Microsystems.

5.10.1.1 Java Platform Micro Edition

The advantage of the Java Platform Micro Edition, also known as Java 2 Micro Edition (J2ME), is that a program does not run on only one device or operating system but across a broad range of different devices and operating systems. Since JVM implementations and available Java packages differ slightly between devices, some adaptations are required for support of a broad range of devices. On GSM and UMTS phones, Java is the main platform for games. The downside of a JVM is that applications can only get a generalized access to the operating systems and have only limited capabilities to access data from other applications such as calendar and address book entries. Also, other applications cannot exchange data with Java applications, which makes it difficult, for example, to open a Web page from a link embedded in an application in a Java-based Web browser. This limitation, imposed by the Java sandbox concept, significantly reduces the usability and interaction between different applications on the device. From a security point of view, however, the sandbox ensures that the application cannot gain access to network functions such as sending an SMS or initiating a phone call without the consent of the user. This is of particular importance on mobile devices since, unlike on PCs, programs can accidentally or intentionally cause costs by accessing the network. The Java environment is open to developers and most developers choose to distribute their applications themselves, directly to the users. Users can then download the application file to the mobile phone via the cellular network, via Bluetooth from a PC or by transferring the application file to the device from a PC via a cable.

5.10.2 BREW

Another cross platform application runtime environment is BREW (Binary Runtime Environment for Wireless), developed by Qualcomm. It is mostly used in CDMA-based mobile devices in the USA and Japan. BREW developers have the choice between several programming languages, C, C++ and Java. Similar to the Java Micro Edition (ME) environment discussed above, BREW offers a cross-platform programming environment but without the restrictions imposed by the Java sandbox approach. To reduce potential security problems and to ensure the quality of applications, BREW applications have to pass rigorous tests in a certification laboratory before they can be distributed. This increases the time to market and reduces the number of developers since certification is not free. Furthermore, most CDMA network operators do not allow customers to install BREW applications themselves. Therefore, developers depend on network operators to distribute their applications. This makes developers dependent on network operators and requires negotiations with many network operators to reach a large customer base. Therefore, the BREW environment and ecosystem, which is controlled by the network operator, only attracts few developers compared with the Java ME environment, which is fully open.

5.10.3 Symbian/S60

On high-end phones, sometimes also referred to as smartphones, more sophisticated operating systems are used as users buy those phones for their richer application suites and extendibility. As such devices have more processing capacity, memory and so on their operating systems are usually fully multitasking capable, that is, several programs such as a Web browser, address book, music player, and so on can run simultaneously and interact with each other. The Symbian operating system, used in many high-end mobile devices from Nokia, Sony-Ericsson, Motorola, Samsung, LG and others, has its roots in the PDA market and was adapted for the mobile phone market when mobile phone hardware became powerful enough for smartphones. By design, it has no roots in the PC or notebook world and is thus very well adapted for use on mobile devices with, compared with PCs, limited processing and memory capabilities. Several graphical user interfaces have been developed to run on top of the Symbian operating system. The most popular user interface is S60, developed and used by Nokia and also sold to other companies such as LG and Samsung. Other graphical user interfaces for Symbian are UIQ, mostly used by Sony-Ericsson, and MOAP, mainly used by a number of companies to develop phones for Japanese mobile network operator NTT-DoCoMo. The remainder of this section will focus on the combination of Symbian and S60.

For users, Symbian offers a full multitasking operating system and a rich graphical user interface. From a software developer's point of view, S60 can be compared with Microsoft's Windows Operating System. Like with Microsoft Windows, the source code of the operating system was not disclosed to third parties. It is planned, however, to open source the operating system and the code of the user interfaces in the near future to give the developer community more insight and control over the evolution of the system. At the moment, developers are offered an API for application development in C++, a universal programming language used for writing efficient code for many platforms such as Windows, Linux, and so on. The procedures offered by the API, however, are S60 proprietary and thus not known to a large developer community. To attract more developers, S60 thus also offers a POSIX (Portable Operating System Interface) compatible library which is well known to the Linux programming community [31]. As native S60 applications are given access to local resources and the network, applications must be signed by an S60 certification lab before being distributed to the user. Unlike the BREW approach discussed above, most S60 applications are directly supplied to the end user by the developer without the involvement of the mobile network operator. Developers can also publish noncertified applications. In this case the user is shown a number of warning screens during the installation process to remind them that the application is not certified and which types of sensitive actions the application would like to perform. The installation process only continues if the user agrees. From a practical point of view this is a good middle ground between requiring certification for all applications and having a fully open deployment scheme without any safeguards in terms of quality and security. In practice, native S60 applications have become quite popular with users and also with the developer community due to the high number of mobile devices using the S60 operating system and the open deployment strategy. Figure 5.6 shows three screen shots of third-party S60 programs, Shozu, Nokia Maps and Handy Weather.

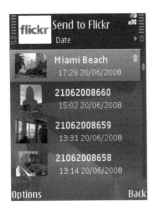  

Figure 5.6 Examples of native S60 applications developed by third-party developers. (Reproduced by Permission of Nokia, Keilalahdentie 2-4, FI-02150 Espoo, Finland.)

S60 also supports the Java ME runtime environment so applications programmed in Java will also work on S60-based devices. As discussed before, however, Java applications do not have the same access to the operating system for local services and the network as is the case for native S60 applications.

In addition, S60 offers a number of other possibilities for developers to create applications. A runtime environment for Python, a script language very well known to Web developers, can be used to develop standalone applications or Web server applications that can be called by the embedded S60 Web Server. Web-based applications using JavaScript and Widget engines are also supported by S60. Both Web servers and Web-based applications are discussed in more detail in Chapter 6. Furthermore, the S60 Web browser includes a flash plugin from Adobe that has become very popular in the desktop world. Flash is used for anything from showing advertisement banners to YouTube video integration.

Today, the combination of the Symbian operating system and the S60 User Interface can also be found in a growing number of mid-tier mobile phones as Nokia and others attempt to increase the number of users that can make use of mobile Internet connectivity and the features of a fully multitasking capable mobile device. As discussed earlier, this is made possible through the continuing decline in hardware prices that allows moving functionality once perceived as high-end into the mid-tier market. For most functionalities the time between introduction on high-end phones to availability in the mid-tier sector is around 18–24 months. Compared with high-end mobile devices, their cheaper price is due to smaller screen sizes, less memory and less sophisticated cameras compared with current high-end devices. Hardware functions that have moved into the mid-tier segment recently are on-board GPS receivers, HSDPA B3G network interfaces and Wi-Fi.

The Symbian/S60 operating system only supports ARM processors. This excludes it from use with the upcoming x86 chipsets from Intel for mobile Internet devices. While in theory it might be possible to port the operating system to this platform, it seems unlikely that this will be done in the future. This is due to the different hardware architecture approach with the network components outside the main chipset, which would require a significant redesign. In addition, development of a number of Linux-based operating systems for x86-based mobile devices is already well underway, as discussed further below.

5.10.4 Windows Mobile

Microsoft's Windows Mobile operating system is similar to Symbian. The origin of Windows Mobile is the PDA market and over the years the system has migrated to the connected mobile device sector. Unlike Symbian, however, Microsoft as a company does not focus on mobile computing. For a long time, the integration of cellular wireless network support in the operating system and across different applications was far behind the seamless integration of cellular networking support on the Symbian platform. In recent years this has improved and Windows Mobile has become popular, especially among business users. Like Symbian, Microsoft does not develop devices itself, but leaves this task to companies such as Hewlett Packard, HTC and Samsung. From a programming point of view, Windows Mobile is a closed source operating system and offers an API for programmers to write native applications. The API offers access to local system resources and the network and programs do not have to pass a certification test from Microsoft before being distributed.

Unlike Symbian/S60, Windows mobile does not support the Java Platform Micro Edition or the BREW environment, which were both described in the section on device-independent applications. In addition, most Windows Mobile devices are PDA-like in both size and functionality and thus mainly address the needs of business users. There have also been attempts to use Windows Mobile in more mobile phone-like devices, but those attempts have not been very successful to date. In addition, Linux-based mobile device operating systems are becoming serious competitors for incumbent mobile operating systems, as discussed below. While Symbian has always been a smartphone operating system and is about to expand into the mid-tier sector for further growth, it seems that Windows Mobile will not be able to do the same. Growth in the mobile Internet device space is also limited since Windows Mobile only supports the ARM processor architecture and the popularity of Linux on the x86 platform is already significant. Windows Mobile is thus entrenched between Symbian for smartphones and the upcoming Linux-based mobile Internet devices. This will make it difficult for Microsoft to grow its market share in mobile devices in the future with this operating system.

5.10.5 Linux: Maemo, Android and Others

Compared with the previously mentioned incumbent operating systems, Linux-based operating systems are new entrants in the connected mobile device market. Over the years, there have been quite a number of attempts by the open source community to create both a hardware platform and a customized Linux distribution for mobile devices. It was only in 2005, however, when Nokia as a major mobile device manufacturer launched an Internet tablet, that Linux emerged as an interesting operating system alternative for a new breed of connected mobile devices. Figure 5.7 shows the graphical

Figure 5.7 Idle screen of Maemo, a Linux-based mobile operating system. (Reproduced by Permission of Nokia, Keilalahdentie 2-4, FI-02150 Espoo, Finland.)

user interface of Maemo, the operating system for Nokia's Internet tablets, which is based on GNU/Debian Linux.

Nokia had only a few alternatives when selecting Linux as the basis for this new device. Symbian and S60, on the one hand, were focused on the smartphone market. Windows Mobile, on the other, might have been another option, but was directly competing with Symbian/S60 and was thus no alternative for Nokia. Another advantage for Nokia of using Linux for a new product category compared with closed source operating systems was that the majority of software components required were freely available and only had to be adapted to the hardware limitations such as slower processors, less available memory and smaller screen sizes. In addition, Nokia mentions the following advantages over closed source proprietary operating systems for this new product category in [32]:

- Reduced development cost – by using freely available open source components, development costs can be significantly reduced.
- Speed – using already existing components and using them as a base for further development reduces the time to market as the development of a new product does not have to start from scratch.
- Flexibility – compared with using a third-party mobile operating system for a device, the manufacturer can react much more quickly to reported software defects. The problem can be fixed by the manufacturer without being dependent on a third-party company.
- Software licensing – no complicated legal negotiation process is necessary between a manufacturer and a third-party company before the product can be shipped. Terms and conditions for the use of open source software is known in advance. In addition, no licensing fees have to be paid, which reduces the cost of the product.

- Developer community – using a popular open source operating system as basis for a mobile device opens the door to a large developer community. This helps in finding talented employees, and devices enjoy widespread support of application developers as there is a broad knowledge of how to program for such operating systems. This helps to foster application development for the device by third-party developers.

The paper also lists a number of disadvantages, mainly in the area of code stabilization and architecture management. Compared with the advantages listed above, however, these seem to be small.

Since the launch of the first Internet tablet, the N-770, Nokia has introduced a number of other Internet tablets based on the same Linux platform which have become very popular in a niche market that Nokia and others hope to extend in the future. As discussed at the beginning of this chapter, Intel has also become interested in the mobile Internet device market. On Intel platforms, Linux seems to be without competition for the moment.

As Linux is an open source operating system, other mainstream companies have followed Nokia's lead and have started their own Linux mobile operating system development. The most prominent of those is Google with its Linux-based Android operating system. Google also counts on the popularity of Linux and open source to attract developers. The application programming interfaces are identical and the source code of the operating system is readily available. Developers therefore have far greater access to the operating system compared with Symbian or Windows Mobile and can even extend or modify parts of the operating system with features that cannot be implemented in the application layer. To help with application development in the mobile domain, Google has built a development framework that uses a derivative of the Java programming language. The Linux-based operating system Android is distributed under the GNU open source license and thus the source code and all changes made by device manufacturers must be made available to the developer community. The Java-based application programming interface, however, is distributed under a BSD open source license. This means that, while Google gives out the toolkit including the source code, third-party companies are not required to open their modifications to the developer community. More on this topic can be found in Section 6.6.5 on the terms and conditions of different open source licenses. While Nokia's Maemo Linux for Internet tablets has been specifically designed for Nokia devices, Google's goal is to get as many manufacturers as possible using their operating system and the Java-like application programming interface toolkit for their products. Maemo is still the most popular Linux-based mobile operating system, but this could shift quickly towards Android if Nokia does not take steps to encourage third-party developers to use their platform.

Like Windows Mobile, current Linux operating systems do not support the BREW or J2ME runtime environments described above. As Linux is mostly used in Internet tablets or similar devices, they would not significantly benefit from such a move. This is because the goals of such devices are different from those of smartphones and mid-range mobile phone. It is also due to the availability of well-known application programming interfaces in Linux.

5.10.6 Fracturization

Compared with the desktop computing world with its three main operating systems, Windows, Mac OS and Linux, the mobile device landscape is much more diverse. In addition to the operating systems and application programming interfaces discussed in this chapter, there are many other proprietary operating systems used in mid-tier and low-end mobile phones. In addition, Apple has chosen to use a proprietary operating system for its iPhones. In practice, this makes it extremely difficult for developers to design mobile applications across a wide range of different devices. It remains to be seen if, in the future, some of those operating systems will dominate and force others out of the market or if the diversity remains. On the positive side, diversity helps to reduce the effect of malware. While today the threat from malware such as viruses on mobile devices is still small, it is likely that this area will get more attention in the future as the number of users and devices increases. The more different operating systems and device combinations are on the market, the more difficult it is for a virus or other harmful program to propagate from one system to another.

5.10.7 Operating System Tasks

Today, operating systems for mobile devices have reached a level of functionality and complexity equal to operating systems for PCs and notebooks. With the introduction of Linux as an operating system for mobile devices, there is no longer even a difference from a practical point of view. The following section now takes a look at the basic building blocks and functionalities of a high-end mobile operating system such as Linux, Symbain/S60 and Windows Mobile. If there are significant differences between operating systems for a function, they will be mentioned as well.

5.10.7.1 Multitasking

Operating systems of all mobile devices, from entry level to high-end devices, must be capable of multitasking as there are a number of tasks that need to be performed quasi simultaneously. The most important task of a connected mobile device, even while not communicating with the network, is to monitor periodic transmissions of broadcast information from the network. This is important to stay synchronized and to receive paging messages for incoming calls and SMS messages. In addition, the processor also needs to react to user input and to execute the code for the required action, like for example updating the display as the user moves between applications or from one menu level to another. In other words, the operating system has to switch between tasks responsible for the communication with the network and tasks responsible for interacting with the user. While, on simple devices, multitasking is limited and the execution of several user applications is not possible, high-end mobile operating systems such as Linux and Symbian/S60 offer full multitasking support, which includes the execution of several user applications in addition to all tasks required for staying connected with the network and dealing with all other external interfaces such as Wi-Fi, Bluetooth, USB, and so on.

While multitasking on the application layer is usually not time-critical, monitoring the network and making decisions about moving to another cell while in idle mode is a time-critical process and the processor has to be available at specific times to analyse incoming information. Depending on the hardware architecture, there are different ways to ensure that the processor is available for executing the radio interface-specific code. In the case of Symbian/S60 and the OMAP chipset, the ARM processor is part of the processing chain for the B3G interface. This means that the operating system has been designed to support real-time constraints for these tasks. Many mobile devices using Linux as an operating systems use a different approach. Here, the mobile interface is completely separated from the main chipset and time-critical tasks are all performed in a dedicated network interface chip. This chip is then connected to the chipset and the main processor via a standardized interface, e.g. via USB. This way, no modifications are required in the Linux kernel to support this form of time-critical real-time multi-tasking. Both Symbian and Linux use pre-emptive multitasking. This means that a task cannot block other tasks from running as each is interrupted by the processor when its allocated time slice has been used up and the application has not yet returned control to the operating system.

As mobile operating systems such as Symbian/S60 are moving down from high-end devices into the mid-tier sector, sophisticated mobile operating systems are likely to become widespread even in small, inexpensive mobile devices in the next few years. For low-end devices, it is likely that the hardware will continue to be optimized for cost rather than functionality, thus preventing more sophisticated operating systems from being used in such devices for some time to come.

5.10.7.2 Memory Management

Programs, also referred to as tasks, running in a multitasking environment do not only share the processor but also the available main memory (RAM), where programs and data are loaded from flash memory, sometimes also referred to as the flash disk, before they can be executed. Management of the main memory is therefore another important function of the operating system.

The operating system must ensure that programs can be executed no matter where they were loaded in memory. This is required since the order in which programs are started and stopped is not known to the operating system in advance. To make an unpredictable place in memory predictable for a program, virtual memory addresses are used in high-end mobile operating systems in the same way that they are used in PC operating systems. When a program is prepared to run by the operating system for its timeslice, the microprocessor's memory management unit is instructed by the operating system how to map the virtual memory addresses known to the program to real memory addresses. While the program is executed the memory management unit of the processor transparently translates the virtual memory addresses used by the program into physical addresses for each command. This mapping has the additional benefit that a program cannot access the memory of another task since the memory management unit would never map a virtual memory address to a physical memory address belonging to another task. For additional security, the memory management unit can only be configured by the operating system, as code running outside the operating system's scheduler does not have the permission to access the unit.

As RAM is expensive and thus a scarce resource, most high-end operating systems have the ability to use a part of the flash disk as a swap space. If the operating system detects it is running out of memory, it starts removing parts of programs and data to the swap space, which cannot directly be accessed by the processor. If a program requires access to data that has previously been swapped out, the operating system interrupts the program and retrieves the data from the swap space. Afterwards, the program is allowed to continue. In practice this works quite well in a multitasking system because only a few applications are actively running, although many applications might be loaded into the main memory. Most applications are usually in a dormant state while the user does not interact with them. A practical example of this is a Web browser that the user starts once and then leaves running while using other applications like the calendar, the notes application or the photo application to take a picture. While the Web browser application is still in memory, it is not scheduled for execution by the operating system as it does not interact with the network or the user (assuming a static Web page without Java Script or flash content, both of which can run in the background). Thus, the application is completely dormant. If in such a situation the main memory is almost fully used, the operating system can start swapping out parts of the memory required by the Web browser to the flash disk and use it for another program. For this purpose the main memory is divided into pages. The operating system is aware when each page was last used and it can decide which pages to swap out to the flash disk once memory gets scarce. As memory is organized in pages, a program does not have to be fully dormant before the operating system can swap out some of its pages. Even if active in the background, there is usually always some program code or data that has not been used recently and can thus be swapped out in the hope that not being used recently also means that this part of memory will also not be used again soon.

While Linux performs swapping of pages to flash memory by default, this capability was only added to Symbian/S60-based devices in 2007. This step was necessary as in the years before it became clear that this feature was urgently needed due to the increasing versatility of the devices and growing program sizes.

5.10.7.3 File Systems and Storage

In the past few years the amount of internal storage space on mobile devices for applications and data such as pictures, videos and music files has skyrocketed. Today, device internal memory for storage has reached 16 Gb in high-end devices and this trend is likely to continue in the future. For a short time, hard disks were used in mobile devices to reach high capacities. With falling flash memory prices, the popularity of hard disks has decreased as flash memory is much smaller, requires less energy and is more robust against vibrations and shocks. Most devices also have an external memory slot for removable flash memory cards. As these cards can be used to exchange data with other devices including PCs and notebooks, the FAT (File Allocation Table) file system is used on such cards, which was originally developed by Microsoft many years ago. While due to its age it is not the most sophisticated file system standard, it is supported by all major operating systems today including Microsoft Windows, Unix/Linux and Apple Mac OS. It is thus the best choice for use in mobile devices.

5.10.7.4 Input and Output

Like any other operating system, mobile operating systems are abstracting devices attached to the system for applications. The display is the first example that comes to mind as it is the main output device to interact with the user. The display is connected to the graphics device which in turn is connected to the processor via a bus system. The operating system then abstracts the function of the graphics card into an application programming interface that can be used by applications. These APIs offer a wide variety of functions to applications ranging from simple primitives of drawing lines and shapes, to printing text at a certain location on the screen to generating graphical menus and buttons. The keyboard is a typical input device, again abstracted for programs by the operating system. A keyboard driver receives keyboard input (a key was pressed) and the operating system is then responsible for passing that information to the currently active program.

Other external input and output devices in mobile devices are, for example, Wi-Fi and Bluetooth interfaces, GPS receivers, FM radios, cameras, TV video output, touch screens, and so on, as shown in Figures 5.3 and 5.4. To connect these external devices with the chipset, a number of different I/O bus systems are used. Slower devices such as the interface to the Bluetooth and the GPS chip use interfaces such as serial UART (Universal Asynchronous Receiver/Transmitter), I2C and SDIO.

Only a few years ago, a UART interface was also used to connect mobile devices via a cable to the serial interface of a PC or notebook for exchanging address book entries and to establish data calls to the Internet, to another computer or to a FAX machine via the mobile phone. On the PC, the UART interface is limited to speeds of around 110 kbit/s. While sufficient for many applications, such a standard serial interface is no longer capable of transporting data exchanged via B3G networks as data rates now exceed several megabits per second. In recent years, serial interfaces have been replaced by USB, which is also a serial bus system but capable of much higher speeds. Until recently, most mobile devices were equipped with a USB 1.1 interface with speeds of up to 11 Mbit/s. While almost increasing transfer rates by a factor of 10 and being sufficient for using the mobile device as a B3G network interface for a PC or notebook, USB 1.1 quickly became a bottleneck for applications such as transferring music files, videos, pictures and maps between a mobile device and a PC. Consequently, high-end mobile devices now use USB 2.0 (also referred to as USB Hi-Speed) as an external interface. With data rates of up to 480 Mbit/s, the bottleneck has now moved from the transfer capabilities of the interface to how fast the processor and operating system can send and receive data and how fast that data can be written to the flash disk.

While USB 2.0 will be sufficient for connecting mobile devices to PCs for the next few years, USB 3.0 is already on the horizon and promises data rates exceeding 4 Gbit/s. First products based on this standard are foreseen in the 2010 timeframe. At this point it will then take some time to miniaturize the technology and to optimize power consumption before the technology is suitable for integration in small, portable, battery-powered devices.

5.10.7.5 Network Support

As one of the main purposes of a mobile Internet device is to connect the user with other devices and people via a wireless network, support of different network types and interfaces and their abstraction for applications is another important task of mobile operating systems. Applications are usually not aware of the type of a network interface and instead request the creation of a TCP or UDP IP connection to another device from the operating system. If no network connection is established at the time of the request, the operating system either decides on its own to connect to a network via one of the wireless interfaces or opens a dialog box to allow the user to select an appropriate network and configuration. Once a connection to a network is established, the operating system processes the application's TCP or UDP connection request and program execution continues. For many years now, the Internet community has been trying to migrate the current version of the Internet protocol (IPv4) to IPv6 to counter the diminishing number of available IP addresses. This has proven to be a difficult process mainly on the network side and is due, in part, to a lack of IPv6-capable applications. Most mobile operating systems like Linux, Windows Mobile and Symbian, however, already support IPv6.

5.10.7.6 Security

In the days when connected mobile devices were only used for phone calls, there was little danger from external attackers gaining access to the mobile phone and the data inside. The reason for this was that the network itself isolated the devices from each other via a switching center and all commands exchanged were originated, terminated or filtered on the switching node. In addition, hardware and software of such devices were simple (cf. Figures 5.1 and 5.2) and thus offered few if any opportunities for external attacks. Today's sophisticated mobile devices are connected the Internet, however, and are thus more and more exposed to the same kind of security threats as PCs and notebooks. It will therefore become increasingly important in the future for the operating system to defend the device against attacks or exploits. In practice there are many ways for malicious programs to gain access to a system.

- Malicious programs – a program should only be installed on a mobile device if its origin is known and trusted. It is therefore important that users realize that installing programs is a potential security risk and should only be done if the source is trusted. Once a program is installed, operating systems like Linux protect the integrity of the system by executing the program in user mode, which prevents programs from making changes to the system configuration. A malicious program, however, still has access to the user's data and thus could potentially destroy or modify data without the consent of the user. Spyware, sometimes, also referred to as a Trojan horse, goes one step further and send private data it has found to a remote server on the Internet. The Symbian operating system uses a slightly different approach to application security. As described in Section 5.10.2, application developers have to get a certification for their program from an independent body before they can be distributed. Programs which do not need direct access to drivers and other lower layer components of the operating system can be distributed without being certified. The user is then informed during the installation process that the program has not been certified, which actions it wants to perform (e.g. access to the network, access to the file system, etc.) and that this presents a certain security risk. The user can then choose to abort the installation or to proceed. As most programs do not require access to lower-layer operating system services, this is the most common distribution method. Noncertified Symbian applications have similar capabilities as described for Linux applications in user mode and can thus also potentially corrupt or steal user data.

- Peripheral software stack attacks – another angle of attack is trying to break the software stack of network peripherals. Several well-known attacks on the Bluetooth protocol stack used malformed Bluetooth packets. In this way it was possible to access the calendar and address book on some devices without the consent or knowledge of the user. While such an attack is still possible today, fewer reports about successful attacks have been published recently and it appears that most mobile device manufacturers have done their homework. If a new vulnerability is found it is important that the manufacturer can and does react quickly and provides a patch, as is done in the PC world today. While in the past, mobile operating systems could only be completely replaced if a problem was found, Symbian, Windows Mobile and others are now patchable. This is not only beneficial to strengthening the operating system and programs delivered by the manufacturer, but also to quickly and efficiently correcting software bugs.

- Attacks via MMS – occasionally, the press also speculates about potential attacks via the Multimedia Messaging Service (MMS). In most reported cases, an MMS contains an executable file which can be installed by the user when opening the message. All mobile operating systems, however, warn the user of the potential consequences of doing so. Another potential MMS attack, which has not been reported so far, is to exploit software bugs in the MMS implementation. This could be done in a similar way as described below for attacks on a Web browser.

- Web browser attacks – such attacks, which are quite common in the desktop PC world, try to exploit vulnerabilities of the browser software, for example of the JavaScript implementation, to break out of the browser environment to execute system commands. Other attacks aim at potential vulnerabilities of plugins such as PDF or Flash, which can be tricked into executing infiltrated code or launch system commands with malformed documents or video files.

- Attacks over the IP network – in the PC world, attacks aimed at server programs waiting for incoming connections are also widespread. Like in the Web browser example above, such attacks exploit an operating system and processor weakness known as stack overflow, sometimes also referred to as buffer overflow. When one function in a program calls another, the return address is stored in a part of the memory referred to as the stack. Once the function has performed all its tasks the program returns to the previous function via the memory address stored on the stack. In addition to storing the return address, the stack is also used to store temporary data used by the called function. If the function does not ensure that the amount of memory is sufficient for incoming data, for example from the network, then the return address and other variables can be overwritten by the incoming data. Under normal circumstances, this would result in a program fault as the program can no longer return to the previous function and the application would be terminated. This weakness, if not properly handled by the program, can be used by malicious exploits to send a specific stream of data that overwrites the return address with a value that points to the data that was sent over the network. Instead of returning to the original function, control is given to the code which was contained in the data sent over the network. This code can then exploit further weaknesses in the operating system to gain higher operating system privileges to load further program code and to install itself in the system. While such exploits have mainly hit Microsoft's Windows operating system due to its widespread use, other operating systems such as Linux and Mac OS are by no means immune to this issue. One of the few operating systems immune to this kind of attack is Symbian as it uses descriptors that prevent buffer overflow attacks [33].

For the moment, there have only been a few reports about widespread or planned attacks on connected mobile devices. One reason for this is that their number compared with PCs and notebooks on the Internet is still small. Therefore, programs tailored to attack a specific type of mobile device would not find many targets yet. This also limits the spread of a virus from one mobile device to another as the virus would not work if it attacked a device such as a PC that runs a different operating system. The wide variety of different devices and operating systems used in mobile devices is another reason why malicious programs and viruses have a difficult time spreading in the mobile domain as each combination of device and operating system might have different weaknesses. As the number of mobile devices grows, however, these reasons are no guarantee that mobile devices will not come under attack from viruses and other exploits in the future. Quick reaction from manufacturers to provide patches and systems automatically updating themselves might thus one day become equally important to such functions in the PC world today.

References

1. GSM Arena (2008) Panasonic GD 55 – full phone specifications, http://www.gsmarena.com/panasonic_gd55-372.php.
2. Asus (2008) eeePC, http://eeepc.asus.com/global/product.htm.
3. ARM (2008) http://www.arm.com/.
4. Krazit, T. (22 June 2007) ARM says it's ready for the iPhone, Cnet Newsmaker,http://www.news.com/ARM-says-its-ready-for-the-iPhone/2008-1006_3-6192601.html.
5. ARM (2008) ARM powered products, http://www.arm.com/markets/mobile_solutions/armpp/835.html.
6. Freescale Semiconductor (June 2006) i.2xx Platform Family Product Brief, document number I2XXPB, Revision 3, http://www.freescale.com/files/wireless_comm/doc/prod_brief/I2XXPB.pdf.
7. ARM (2008) ARM7TDMI, http://www.arm.com/products/CPUs/ARM7TDMI.html.
8. Sauter, M. (2006) Communication Systems for the Mobile Information Society, Chapter 1.7.3, John Wiley and Sons, Ltd, Chichester.
9. Texas Instruments (2006) OMAP 3 family of multimedia applications processor.
10. ARM (date unknown) Achieving stronger SIM-lock and IMEI implementations on open terminals using ARM Trustzone technology, www.arm.com/miscPDFs/15500.pdf.
11. Texas Instruments (2007) WiLink 6.0 single-chip WLAN, Bluetooth and FM solutions, http://focus.ti.com/pdfs/wtbu/ti_wilink_6.pdf.
12. Intel (January 2008) Intel Processor A100 and A110 on 90 nm process with 512-KB L2 Cache, http://www.intel.com/design/mobile/datashts/316908.htm.

13. Intel (July 2007) Mobile Intel 945 Express Chipset Family, http://download.intel.com/design/mobile/datashts/30921905.pdf.
14. Sauter, M. (2006) Communication Systems for the Mobile Information Society, Chapter 6.4.4, John Wiley and Sons, Ltd, Chichester.
15. Krishnadas, K.C. (January 2006) SiRF chip integrates GPS, Bluetooth, EE Times, http://www.eetimes.com/news/latest/showArticle.jhtml?articleID=177104014.
16. Krazit, T. (February 2008) Intel sheds a little bit more light on Silverthorne, CNet news, http://www.news.com/8301-13579_3-9865129-37.html?part=rss&tag=feed&subj=OneMoreThing.
17. Intel (2008) Microprocessor Quick Reference Guide, http://www.intel.com/pressroom/kits/quickref fam.htm.
18. Intel (April 2007) Intel news disclosures from day 2 of the Intel Developer Forum in Beijing, http://www.intel.com/pressroom/archive/releases/20070417supp.htm.
19. Polymer Vision (2008) http://www.polymervision.com/.
20. Anscombe, N. (August 2000) Methanol fuel cells seen as mobile phone power source, EETimes, http://www.eetimes.com/story/OEG20000208S0036.
21. Blass, E. (September 2007) Hands-on with Texas Instruments' cellphone projector, Engadget, http://www.engadget.com/2007/09/20/hands-on-with-texas-instruments-cellphone-projector/.
22. Sony (2008) FeliCa in Use, http://www.sony.net/Products/felica/csy/index.html.
23. 3GPP (2008) User Equipment (UE) radio transmission and reception (FDD), TS 25.101, Table 5.0, version 8.1.0.
24. Shandle, J. (December 2006) WiMAX Forum rolls frequency-availability database, EETimes.com, http://www.eetimes.com/showArticle.jhtml?articleID=196800096.
25. Airspan (2008) WiMAX – MicroMAX, http://www.airspan.com/products_wimax_micromax.aspx.
26. Sauter, M. (November 2007) Sony-Ericsson launches Tri-Band HSDPA phone, http://mobilesociety.typepad.com/mobile_life/2007/11/sony-ericsson-l.html.
27. Ziegler, C. (October 2006) T-Mobile details 3G plans, Engadget, http://www.engadgetmobile.com/2006/10/06/t-mobile-details-3g-plans/.
28. Sorensen, C. (January 2008) Telus considers dumping its 'Betamax' of wireless networks, The Toronto Star, http://www.thestar.com/Business/article/293353.
29. Varrall, G. (May 2007) RF cost economics for handsets, RTT, http://www.rttonline.com/Research/RF%20Cost%20economics-Handsets-white%20paper.pdf.
30. Sauter, M. (February 2008) Wireless now accounts for a third of Austria's Broadband connections, http://mobilesociety.typepad.com/mobile_life/2008/02/wireless-now-ac.html.
31. Nokia (2008) Open C, http://www.forum.nokia.com/main/resources/technologies/open_c/index.html.
32. Jaaksi, A. (December 2006) Building consumer products with open source, Nokia whitepaper, http://www.linuxdevices.com/articles/AT7621761066.html.
33. Wood, D. (date unknown) Insight 9: The keystone of security, http://www.symbian.com/symbianos/insight/insight9.htm.

6

Mobile Web 2.0, Applications and Owners

6.1 Overview

In addition to telephony services and mobile devices discussed in the previous chapters, Internet applications are another important driver for the evolution of wireless communication. After all, it is the use of applications and their demand for connectivity and bandwidth that drives network operators to roll out more capable fixed and wireless IP-based networks. This chapter looks at the application domain from a number of different angles.

In the first part of this chapter the evolution of the Web is discussed, to show the changes that the shift from 'few-to-many communication' to 'many-to-many' brought about for the user. This shift is often described as the transition from Web 1.0 to Web 2.0. However, as will be shown, Web 2.0 is much more than just many-to-many Web-based communication.

As this book is about wireless networks, this chapter then shows how the thoughts behind Web 2.0 apply to the mobile domain, that is, to mobile Web 2.0. Mobility and small-form factors can be as much an opportunity as a restriction. Therefore, the questions of how Web 2.0 has to be adapted for mobile devices and how Web 2.0 can benefit from mobility are addressed. During these considerations it is also important to keep an eye on how the constantly evolving Web 2.0 and mobile Web 2.0 impacts networks and mobile devices. Following on from this is an overview of different categories of mobile applications and a discussion of several existing applications per category from a technical point of view, identifying their potential to change the way we communicate and interact with each other.

In a world where users are no longer only consumers of information but also creators, privacy becomes a topic that requires special attention. It is important for users to realize what impact giving up private information has in the short and long term. Some Web 2.0 applications implicitly gather data about the actions of their users. How this can lead to privacy issues and how users can act to prevent this will also be discussed.

Beyond 3G – Bringing Networks, Terminals and the Web Together: LTE, WiMAX, IMS, 4G Devices and the Mobile Web 2.0   Martin Sauter   © 2009 John Wiley & Sons, Ltd

In practice, there are many different motivations for developing applications. Students, for example, create new applications because they have ideas they want to realize and can experiment without financial pressure or the need for a business model. Such an environment is quite different from the development environment in companies where deadlines, business models and backwards compatibility rule during the development process. With this in mind this chapter will also discuss how the different environments shape the development of Web 2.0 vs the development of mobile Web 2.0 and examine the impact.

6.2 (Mobile) Web 1.0 – How Everything Started

For most users the Internet age started with two applications: e-mail and the Web. While the first form of e-mail dates back to the beginnings of the Internet in the 1960s and 1970s, the World Wide Web, or Web for short, is much younger. The first Web server and browser date back to the early 1990s. Becoming widespread in the research community by the mid 1990s, it took until the end of the decade before the Web became popular with the general public. Popularity increased once computers became powerful enough and affordable for the mass market. Content proliferated and became more relevant to everyday life, as shops started to offer their products online, banks opened their virtual portals on the Web, companies started to inform people about their products and news started being distributed on the Web much faster than via newspapers and magazines. Furthermore, the availability of affordable broadband Internet connections via DSL and TV cable since the early 2000s helped to accelerate the trend. While the Web was initially intended for sharing information between researchers, it got a different spin once it left the university campus. For the general public, the Internet was at first a top-down information distribution system. Most people connected to the Internet purely used the Web to obtain information. Some people also refer to this as the 'read-only' Web, as users only consumed information and provided little or no content for others. Thus, from a distribution point of view, the Internet was very similar to the 'offline' world where media companies broadcast their information to a large consumer audience via newspapers, magazines, television, movies and so on. Nonmedia companies also started to use the Web to either advertise their services or sell them online. Amazon is a good example of a company that quickly started using the Web not only to broadcast information but also as a sales platform. However, what Amazon, and other online stores, had in common with media companies was that they were the suppliers of information or goods and the user was merely the consumer. Note that this has now changed, to some degree, as will be discussed in the next section.

In the mobile world, the Web had a much more difficult start. First attempts by mobile phone manufacturers to mobilize the Web were a big disappointment. In the fixed line world the Internet had an incubation time of at least a decade to grow, to be refined and fostered by researchers and students at universities before being used by the public, who already had sufficiently capable notebooks, PCs and a reasonably priced connection to the Internet. In the mobile world, things were distinctly different when the first Web browsers appeared on mobile phones around 2001:

- Mobile Internet access was targeted at the general public instead of first attracting researchers and students to develop, use and refine the services.
- Unlike at universities, where the Web was free for users, companies wanted to charge for the mobile service from day one.
- It was believed that the Web could be extended into the mobile domain solely by adapting successful services to the limitations of mobile devices, rather than looking at the benefits of mobility. That is like taking a radio play, assembling the actors and their microphones in front of a camera, and broadcasting them reading the radio play on TV [1].
- Little, if any, appealing content for the target audience was available in an adapted version for mobile phones.
- Mobile access to the Internet was very expensive so only a few were willing to use it.
- Circuit-switched bearers were used at the beginning, which were slow and not suitable for packet-switched traffic.
- The mobile phone hardware was not yet powerful enough for credible mobile Web browsers. Display sizes were small, screen resolutions not suited for graphics, there was no color, not enough processing power and not enough memory for rendering pages.
- The use of a dedicated protocol stack (the Wireless Application Protocol, WAP) instead of HTML required special tools for Web page creation and at the same time limited the possibilities to design mobile and user-friendly Web pages.

Any of the points mentioned above could have been enough to stop the mobile Web in its tracks. Consequently, there was a lot to overcome before the Internet on mobile devices started to gain the interest of a wider audience. This coincided with the emergence of the Web 2.0 and its evolution into the mobile domain, as described in the next section.

6.3 Web 2.0 – Empowering the User

While the Web 1.0 was basically a read-only Web, with content being pushed to consumers, advances in technology and thinking and market readjustment (with the bursting of the dot com bubble at the beginning of the century) have returned the Internet to its original idea: exchange of information between people. The ideas that have brought about this seismic shift from a read-only Web to a read/write Web are often combined into the term Web 2.0. Web 2.0, however, is not a technology that can be accurately defined; it is a collection of different ideas. With these ideas also being applicable to the experience of the Web, and the Internet in general, on mobile devices, it makes sense to first discuss Web 2.0 before looking at its implications for mobile devices and networks.

The following sections look at Web 2.0 from a number of different angles: from the user's point of view, from a principal point of view and from a technical point of view.

6.4 Web 2.0 from the User's Point of View

For the user, the Web today offers many possibilities for creating as well as consuming information, be it text-based or in the form of pictures, videos, audio files and so on. The following section describes some of the applications that have been brought about by Web 2.0 for this purpose.

6.4.1 Blogs

A key phenomenon that has risen with Web 2.0 is blogging. A Blog is a private Web page with the following properties:

- Dynamic information – Blogs are not used for displaying static information but are continuously updated by their owners with new information in the form of articles, also referred to as Blog entries. Thus, many people compare Blogs with online diaries. In practice, however, most Blogs are not personal diaries accessible to the public, but platforms on which people share their knowledge or passions with other people. Companies have also discovered Blogs as a means of telling their story to a wider audience in a semi-personal fashion. Blogs can also be valuable additions to books, giving the author the possibility to interact with her readers, go into details of specific topics and to share her thoughts. Figure 6.1, for example, shows the Blog that complements this book.

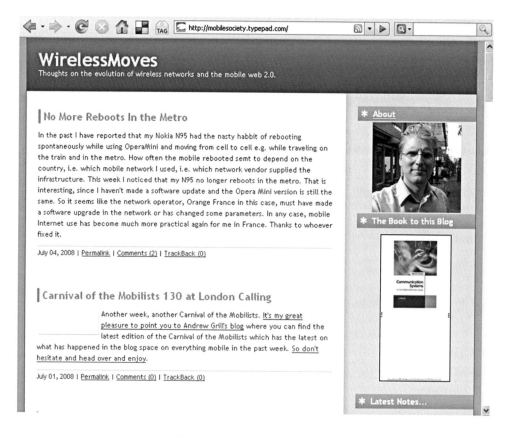

Figure 6.1   The author's Blog.

- Ease of use – Blogs are created, maintained and updated via a Web-based interface. No Web programming skills are required. Thus, Blogs can be created and used by everyone, not only technically skilled people.
- Blogs order content in a chronological fashion with the latest information usually presented at the top of the main page.
- Readers of a Blog can leave comments, which encourages discussion and interaction.
- Other people can subscribe to an automated news feed of a Blog. This way they can easily find the Blog again (bookmarking functionality) and be automatically informed when the author of the Blog publishes a new entry. This is referred to as aggregation and is discussed in more detail in Section 6.3.3.
- A Blog is often the central element for the online activities of a user. It may be used to link to other online activities, for example, links to accounts at picture sharing sites, the user's pages in social networks and so on. Readers of the Blog can thus easily discover additional information from or about the owner of the Blog.

6.4.2 Media Sharing

Blogs can also be used to share nontextual content such as pictures and videos. In many cases, however, it is preferable to share such content via dedicated sharing sites such as Flickr [2] for pictures, YouTube [3] for videos, del.icio.us [4] for bookmarks and so on. This has the advantage that users looking for a video or picture about a specific subject can go to such a sharing site and obtain a relevant list of videos that other people have made available to share. In private Blogs, links can then be used to point Blog readers to the content. It is also possible to embed pictures and videos from sharing platforms directly in Blog entries. Thus, no redirection is required for Blog readers, while people who are unaware of the Blog can still find the content.

6.4.3 Podcasting

Podcasting is another important form of media sharing. The word itself is a combination of the words iPod and broadcasting. Podcasting combines audio recording and making the recording available on Blogs, Web pages and via automated feeds. Automated feeds allow interested users to be informed about a new podcast in a feed and connected MP3 players can automatically download new podcasts from feeds selected by the user. Thus, distributing audio content is no longer an exclusive domain of radio stations. Radio stations, however, have also discovered the value of podcasting and today many stations offer their content as podcasts after the initial traditional broadcast. The advantage for listeners is that radio shows can now be downloaded and consumed at any time and any place.

While Web sites exist that offer podcast directories and podcast archives, many podcasters host the audio files themselves and only use podcast directories to make others aware of their podcasts.

6.4.4 Advanced Search

Being a publisher of information is only useful when a potential audience can find the content (Blog entries, pictures, videos, etc.). This is made possible by advanced search engines such as Google, Yahoo, Technorati and others, who are constantly updating their databases. The ranking of the search results is based on a combination of different parameters such as the number of other sites linking to a page and their own popularity, when the page was last updated and algorithms which are the well guarded secrets of search companies. While search engines can analyze text-based information, automated analysis of images and videos is still difficult. To help search engines find such nontextual information, users often add text-based tags to their multimedia content. Tags are also useful to group pieces of information together. It is thus possible to quickly find additional information on a specific topic on the same Blog or sharing site.

6.4.5 User Recommendation

In addition to ensuring a certain quality in reporting news, traditional media, such as newspapers and magazines, select the content they want to publish. Their selection is based on their understanding of user preferences and their own views. Consequently, a few people select the content that is then distributed to a large audience. Furthermore, mass media tailors content only for a mass audience and are thus not able to service niche markets. The Web 2.0 has opened the door for democratizing the selection process. User recommendation sites, such as Digg [5], let users recommend electronic articles. If enough people recommend an article it is automatically shown on the front page of Digg or in a section dedicated to a specific subject. This way the selection is not based on the preferences of a few but based on the recommendation of many.

6.4.6 Wikis – Collective Writing

Wikis are the opposite of Blogs. While a Blog is a Web site where a single user can publish their information and express their views, Wikis let many users contribute toward a common goal by making it easy to work on the same content in a Web-based environment. The most popular Wiki is undoubtedly the Wikipedia project. Within a short time the amount of articles and popularity has far surpassed other online and offline lexica of traditional media companies. Today, Wikipedia has hundreds of thousands of users helping to write and maintain the online encyclopedia. Participating is simple, since no account is needed to change or extend existing articles. The quality of individual articles is usually very good since people interested in a certain topic often ensure that the related articles on Wikipedia are accurate. As anyone can change any article on Wikipedia, entries on controversial topics sometimes go from one extreme to the other. In such cases, articles can be put under change control or set to immutable by users with administrator privileges. This shows that, in general, the intelligence and knowledge of the crowd is superior to the intelligence and knowledge of the few, but that the concept has its limits as well.

It is also possible to subscribe to Wiki Web pages in a similar fashion as subscribing to Blogs and podcasts. Thus, changes are immediately reported to interested people.

Apart from Wikipedia, a wide range of other Wikis exist on the Web today that are dedicated to specific topics. Starting a Wiki is just as easy as starting a Blog, since there are many Wiki hosting services on the net where new Wikis can be created by anyone with a few minutes to spare. Figure 6.2 shows a Wiki dedicated to the topic of how to access

Figure 6.2 A small Wiki running on a server of a Wiki hosting service. (Reproduced by Permission of Wetpaint.com, Inc., 307 3rd Ave. S., Suite 300, Seattle, WA 98104, USA. Photograph reproduced from Martin Sauter)

the Internet with prepaid SIM cards of 2G and 3G network operators. Started by the author of this book, many people have since contributed and added information about prepaid SIM cards and Internet access in their countries. As is the nature of a Wiki, articles are frequently updated when people notice that network operators have changed their offers.

Wikis are also finding their way into the corporate world, where they are used for collaboration, sharing of information or to help project teams to work together on a set of documents.

6.4.7 Social Networking Sites

While Blogs, Wikis and sharing sites make it easy to publish, share and discover any type of content, social networking sites are dedicated to connecting people and making it easy to find other people with similar interests. Famous social network sites are Facebook [6] and Myspace [7] in the private domain and LinkedIn [8] for business contacts. Being a member of a social networking site usually means sharing of private information, so one can be found by other people based, for example, on common interests. Many different types of social networks exist. Some focus on fostering professional contacts and offer few additional functionalities, while others focus on direct communication between people, for example, by offering Blogging functionality and automatically distributing new entries to all people who the user has declared as friends on the site. The Blogging behavior on social networking sites is usually different to dedicated Blogs, since entries are shorter, usually more personal and dedicated to the people in the friends list rather than a wider audience. Many social networking platforms also allow users to create personalized Web pages on which they present themselves to others.

6.4.8 Web Applications

In the days of Web 1.0 most programs had to be installed on a device and the Web was mostly used to retrieve information. Advanced browser capabilities, however, have brought about a wide range of Web applications which do not have to be installed locally. Instead, Web applications are loaded from a Web server as part of a Web page. They are then either exclusively executed locally or are split into a client and server part, with the server part running on a server in the Internet. Google has many Web applications, a very popular one being Google maps. While the maps application itself is executed in the browser, as a JavaScript application, the 'maps and search' databases are in the network. When users search for a specific location, or for hotels, restaurants and so on at a location, the application connects to Google's search database, retrieves answers and displays the results on a map that is also loaded from the network server. The user can then perform various actions on the map, like zooming and scrolling. These actions are performed locally in the browser until further mapping data is required. At this point the map's application running on the Web browser asynchronously requests the required data. During all these steps the initial Web page on which the map's application is executed is never left. The application processes all input information itself, updates the Web page and communicates with the backend server.

Today, even sophisticated programs such as spreadsheets and word processors are available as Web applications. Documents are usually not stored locally, but on a server in the network. This has the advantage that several people at different locations can work on a document simultaneously. Also, a user can work on documents via any device connected to the Web, without taking the document with him. Another benefit of Web applications is that they do not have to be deployed and installed on a device. This makes deployment very simple and changes to the software can be done seamlessly, when the application is sent to the Web browser as part of the Web page. The downside of Web applications and Web storage is that the user becomes dependent on a functioning network connection and relies on the service provider to keep their documents safe and private.

6.4.9 Mashups

Mashups are a special form of Web application. Instead of a single entity providing both the application and the database, mashups retrieve data from several databases in the network via an open API and combine the sources in a new way. An example of a mashup is a Web application that uses cartographic data from the Google maps database to display the locations of the members of a user's social network, where data about the members is retrieved from the social networking site of the user. This is something neither Google maps nor the social networking site can do on their own. The crucial point for mashups is that other Web services allow their data to be used without their own Web front end. This is the case for many Web services today, with Google maps just being a prominent example. Also important for mashups is that the interface provided by a Web application does not change, otherwise the mashup stops working. Mashups also depend on the availability of their data sources. As soon as one of the data sources is not available the mashup stops working as well.

6.4.10 Virtual Worlds

Another way the Internet has connected people over the last few years is with the concept of virtual worlds. The most prominent virtual world is Second Life by Linden Labs [9]. Virtual worlds create a world in which real people are represented by their avatars. Avatars can look like the real person owning them or, more commonly, how that person would like to look. Avatars can then walk through the virtual world, meet other avatars and communicate with them. Avatars can also own land, buy objects and create new objects themselves. While virtual worlds might have initially been conceived as pure games, they have, in the meantime, also become interesting for companies and many have opened virtual store fronts. Avatars of employees work as shop assistants and interact with customers. Also, some universities use virtual worlds for online learning by holding classes in the virtual world that are attended by real-life students, who visit the classroom with their avatars. Communication is possible via instant messaging but also via an audio channel. It should be noted at this point that most virtual worlds require a client application on the user's device. Therefore, they are not strictly a part of the Web 2.0, as they are not running in a Web browser. Nevertheless, in everyday life most people count virtual worlds as part of the Web 2.0.

6.4.11 Long-tail Economics

Web 2.0 services enable users to move on from purely being consumers to also become creators of content, which in turn considerably increases the variety of information, viewpoints and goods available via the Web. By using search engines or services such as eBay, Amazon, iTunes and so on, this information, or these goods, can also be found and consumed by others without having to be promoted by media companies and advertisements. The ability to find things 'off the beaten path' also facilitates the production and sale of goods for which, traditionally, there has been no market, because people were not aware of them. Chris Anderson has described this phenomenon as long-tail economics in [10]. The term long-tail is explained in Figure 6.3. The vertical axis represents the number of copies sold of a product, e.g. a book, and the horizontal axis shows its popularity. Very popular items start out on the left of the graph, with the long tail beginning when it is economically no longer feasible to keep the items in stock, that is, when only limited space and local customers are available.

While still making a fair percentage of their revenue with mainstream products, companies like Amazon today are successful because they can offer goods which only

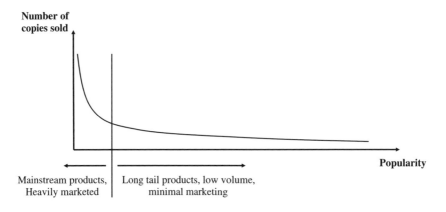

Figure 6.3   The long tail.

sell in quantities too small to be profitable when they have to be physically distributed and stored in many places. This in turn again increases the popularity of the site since goods are available which cannot be bought at a local store where floor space is limited and interest in stocking products which sell in small quantities is not high. As there are many more products sold in small quantities compared with the few products sold in very large quantities, a substantial amount of revenue can be generated for the company running the portal. eBay is another good example of long-tail economics. While not stocking any goods itself, eBay generates its revenue from auctions of goods from the long tail and not from those sold at every street corner. Whether it is possible to be profitable by producing goods or content on the long tail, however, is another matter [11]. For many, however, generating revenue is not the goal of providing content on the long tail, as their main driver is to express their views and give something back for the information, produced by others, that they have consumed for free.

6.5 The Ideas Behind Web 2.0

Most of the Web 2.0 applications discussed in the previous section have a number of basic ideas behind them. Tim O'Reilly, who originally coined the term Web 2.0, has written an extensive essay [12] about the ideas behind Web 2.0. Basically, he sees seven principles that make up Web 2.0 and points out that, for applications to be classified as belonging to Web 2.0, they should fulfill as many of the criteria as possible. This section gives a brief overview of these principles as they form the basis for the subsequent analysis, that is how these principles are enhanced or limited by mobile Internet access and if the mobile Web 2.0 is just an extension of Web 2.0 or requires its own definition.

6.5.1 The Web as a Platform

A central element of Web 2.0 is the fact that applications are no longer installed locally but downloaded as part of a Web page before being executed locally. Also, the data used by these applications is no longer present on the local device but is stored on a server in the network. Thus, both the application and the data are in the network. This means that software and data can change and evolve independently and the classic software release cycle which consists of regularly upgrading locally installed software is no longer necessary. As software and data change, Web 2.0 applications are not packaged software but rather a service.

6.5.2 Harnessing Collective Intelligence

User participation on Web 2.0 services is the next important element. Services that only exist because of user participation are, for example, Wikipedia and Flickr. While the organizations behind these services work on the software itself, the data (Wiki entries, pictures, etc.) is entirely supplied by the users. Users submit their information for free, working toward the greater goal of creating a database that everybody benefits from.

While in the traditional top-down knowledge distribution model the classification of information (taxonomy) was done by a few experts, having a countless number of people working on a common database and classifying information is often referred to as folksonomy. Classifying information is often done by tagging, that is by adding text-based information (catch words) to anything from articles to pictures and videos. This way it is possible to find nontextual information about a certain topic and to quickly correlate information from different sources.

Collective intelligence also means that software should be published as open source and distributed freely so everybody can build on the work of others. This idea is similar to contributing information to a database (e.g. Wikipedia) that can then be used by others.

Blogs are also a central element of the Web 2.0 idea, as they allow everyone with a computer connected to the Internet to easily share their views in articles, also referred to as Blog entries. Blog entries are usually sorted by date so visitors to a Blog will always see the latest entries first. In contrast to the above services, however, Blogs are not collecting information from several users but are a platform for individuals to express themselves. Therefore, powerful search algorithms are required to open up this 'wisdom', created by the crowds, to a larger audience as users first of all need to discover a Blog before they can benefit from the information. Some Blogs have become very successful because users have found the information so interesting that they have linked to the source from their own Blogs. When this is repeated by others a snowball effect occurs. As one input parameter for modern search algorithms is the relevance of a page based on the number of links pointing to it, this snowball effect gives such Web pages a high rating with search engines and thus moves them higher in the search result lists. This in turn again increases their popularity and creates more incoming links.

Less frequented Blogs, however, are just as important to Web 2.0 as the few famous ones. Many topics, such as mobile network technology, for example, are only of interest to a few people. Before Blogs became popular, little to no information could be found about these topics on the Web, since large media companies focus on content that is of interest to large audiences and not niche ones. With the rise of Web 2.0, however, it has become much simpler to find people discussing such topics on the net. Blogrolls, which are placed on Blogs and contain links to Blogs discussing similar topics, help newcomers to quickly find other resources.

As many Blogs are updated infrequently and thus interesting information is spread over many different sources, a method is required to automatically notify users when a Blog is updated. This is necessary since it is not practical to visit all previously found interesting Blogs every day to see if they have been updated. Automatic notification is done with feeds, to which a user can subscribe to with a feed reader. A feed reader combines all feeds and shows the user which Blogs have updated information. The Blog entries are then either read directly in the feed reader or the feed reader offers a link to the Blog.

6.5.3 Data is the Next Intel Inside

While users buy standalone applications like, for example, word processors because of their functionality, Web 2.0 services are above all successful because of their database in the background. If services offer both information and the possibility for users to enhance the database or be the actual creator of most of the information the service is likely to become even more popular, due to the rising amount of useful information that even the most powerful company could not put together. An example of this is a database of restaurants, hotels, theaters and so on. Directories assembled by companies will never be as complete or accurate as directories maintained by the users themselves. Control over such user-maintained databases is an important criterion for them to become successful, as the more information is in the database the harder it gets for similar services to compete. To stimulate users to add content it is also important to make the database accessible beyond the actual service, via an open interface. This allows mashups to combine the information of different databases and offer new services based on the result. This can in turn help to promote the original service. An example is Google maps. It allows other applications to request maps via an open interface. When mashups use maps for displaying location information (e.g. about houses for sale, hotels, etc.), the design of the map and the copyright notice always point back to Google.

6.5.4 End of the Software Release Cycle

As software is no longer locally installed, there are no longer different versions of the software that have to be maintained so users no longer need to upgrade applications. Errors can thus be corrected very quickly and it enables services to evolve gradually instead of in distinctive steps over a longer period of time. This concept is also known as an application being in perpetual beta state. This term, however, is a bit misleading as beta often suggests that an application is not yet ready for general use.

Running applications in a Web browser and having the database and possibly some processing logic in the network also allows the provider of the service to monitor which features are used and which are not. New features can thus be tested to see if they are acceptable or useful to a wider audience. If not, they can be removed again quickly, which prevents rising entropy that makes the program difficult to use over time.

Web 2.0 services often regard their users as co-developers, as their opinions of what works and what does not can quickly be put into the software. Also, new ideas coming from users of a service can be implemented quickly if there is demand and deployed much faster than in a traditional development model, in which software has to be distributed and local installations have to be upgraded. This shortens the software development cycle and helps services to evolve more quickly.

6.5.5 Lightweight Programming Models

Some Web 2.0 services retrieve information from several databases in the network and thus combine the information of several information silos. Information is usually accessed either via RSS feeds or a simple interface based on HTTP and XML. Both methods allow loose coupling between the service and the database in the network. Loose coupling means that the interface has no complex protocol stack for information exchange, no service description and no security requirements to protect the exchange of data. This enables developers to quickly realize ideas, but of course also limits what kind of data can be exchanged over such a connection.

6.5.6 Software above the Level of a Single Device

While in a traditional model, software is deployed, installed and executed on a single device, Web 2.0 applications and services are typically distributed. Software is downloaded from the network each time the user visits the service's Web page. Some services make extensive use of software in the backend and only have the presentation layer implemented in the software downloaded to the Web browser. Other Web 2.0 software runs mostly in the browser on the local machine and only queries a database in the network.

Some services are especially useful because they are device-independent and can be used everywhere with any device that can run a Web browser. Web-based bookmark services for example allow users to get to their bookmarks from any computer, as both the service and the bookmarks are Web-based.

Yet another angle to look at software above the level of a single device is that some services become especially useful because they can be used from different kinds of devices and not only computers. Instant messaging and social networks, for example, can be enhanced when the user does not only have access to the service and data when at home or at the office, but also when he roams outside and only has a small mobile device with him. As both the service and the data reside in the network and are used with a browser, no software needs to be installed and use of the service on both stationary and mobile devices is easy. This topic will be elaborated in more detail in the next section on mobile Web 2.0.

Some companies have also combined Web 2.0 services, traditional installable software and mobile devices to offer a compelling overall service to users. Apple for example offers iTunes, which is a traditional program that has to be installed. The media database it uses, however, is not only created by Apple and media companies but also includes a podcast catalog entirely managed by users. To make the service useful, an iPod is sold as part of the package, to which content can be downloaded via the software installed on the computer.

6.5.7 Rich User Experience

Web 2.0 services usually offer a simple but rich user experience. This requires methods beyond static Web pages and links. Modern browsers support JavaScript to create interactive Web pages in addition to XHTML and CSS for describing Web page content. The Extensible Markup Language (XML) is used to encapsulate information for the transfer between the service and the database in the network. This way, standard XML libraries can be used to encapsulate and retrieve information from a data stream without the programmers having to reinvent data encapsulation formats for every new service. All methods together are sometimes referred to as AJAX (Asynchronous JavaScript and XML). Asynchronous in this context means that the JavaScript code embedded in a page can retrieve information from a database in the network and show the result on the Web page, without requiring a full page reload. This way it is possible for services running in a browser to behave in a similar way to locally installed applications and not like a Web page in the traditional sense.

6.6 Discovering the Fabrics of Web 2.0

The previous sections have taken a look at Web 2.0 from the user's perspective and which basic ideas are shared by Web 2.0 services. This section now introduces the technical concepts of the most important Web 2.0 methods and processes.

6.6.1 Aggregation

The glue that holds Web 2.0 together is aggregation, or the ability to automatically retrieve information from many sources for presentation in a common place or for further processing. Blog or feed reading programs, for example, are based on aggregation. The idea of Blog or feed readers is to be a central place from which a user can check if new articles have been published on Blogs or Web pages supporting aggregation. For the user, using a feed reader saves time that has otherwise to be spent on visiting each Blog in a Web browser to check for news. Figure 6.4 shows Mozilla Thunderbird, an e-mail and feed reader program. On the left side, the program shows all subscribed feeds and marks those in bold which have new articles. On the upper right the latest feed entries of the selected Blog are shown. New entries are marked in bold so they can be found easily. On the lower right the selected Blog entry is then shown. The link to the article on the Blog is also shown, as it is sometimes preferable to read the article on the Blog itself rather than in the feed reader, as sometimes no pictures or only scaled down versions are embedded in the feed.

From a technical point of view, feed readers make use of Blog feeds, which contain the articles of the Blog in a standardized and machine readable form. When a user publishes a new article on a Blog the feed is automatically updated as well. Each time a user starts a feed reader, the feeds of all sources the user is interested in are automatically retrieved with a HTTP request, just like a normal Web page, and analyzed for new content which is then presented in the feed reader. In practice, there are two different feed formats, and feed readers usually support both:

- RSS (Real Simple Syndication), specified in [13];
- ATOM syndication format, specified in [14].

Both feed formats are based on XML, which is a descriptive language and a generalization of the Hypertext Markup Language (HTML), used for describing Web pages.

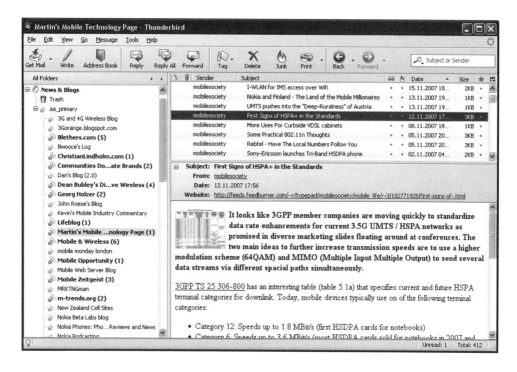

Figure 6.4 Mozilla Thunderbird used as a feed reader.

Figure 6.5 shows an extract of an Atom Blog feed. Information is put between standardized tags (e.g. <title> and </title>) so feed readers or other programs can search XML feeds for specific information. Besides the text of Blog entries, a lot of additional information is contained in feeds, such as the date an entry was created, information about the Blog itself, name of the author and so on. The text of the Blog entries can be formatted as HTML text and can thus also contain references to pictures embedded in the article or links to external pages. The feed reader can then request the pictures from the Blog for presentation in the Blog entry and open a Web browser if the user wants to follow a link in the article to another Web page.

In practice, users do not have to deal with the XML description delivered by an XML feed directly. The usual method to import a feed is by clicking on the orange feed icon that is shown next to the URL of the Web page, as shown in Figure 6.6. The Web browser then shows the URL of the feed which the user can then copy and paste into the feed reader.

Feeds are not only used for aggregating Blog feeds in a Blog reader. Today, other types of Web pages also offer RSS or Atom feeds so content from those pages can also be viewed in a feed reader program. Picture-sharing sites such as Flickr, for example, offer feeds for individual users or tags. Each time the feed reader requests updated information from a Flickr feed, Flickr includes the latest pictures of a user in the feed or the pictures for the specified tags.

```
File  Edit  View  Help
<content type="html" xml:lang="en-US" xml:base="http://mobilesociety.typepad.com/mobile_lif
&lt;div xmlns="http://www.w3.org/1999/xhtml"&gt;&lt;p&gt;&lt;a onclick="window.open(this.hr
After having had the pleasure of hosting the Carnival of the Mobilists last week, it has mc
</content>


        <feedburner:origLink>http://mobilesociety.typepad.com/mobile_life/2007/12/carnival-of-t
        <entry>
            <title>Mowser and a Good Mobile Search Experience</title>
            <link rel="alternate" type="text/html" href="http://feeds.feedburner.com/~r/typepad
            <link rel="replies" type="text/html" href="http://mobilesociety.typepad.com/mobile_
            <id>tag:typepad.com,2003:post-42301128</id>
            <published>2007-12-02T01:06:47+01:00</published>
            <updated>2007-12-02T01:06:53+01:00</updated>
            <summary>I don't search a lot on the net from my N93 as I don't really feel the nee
            <author>
                <name>mobilesociety</name>
            </author>


<content type="xhtml" xml:lang="en-US" xml:base="http://mobilesociety.typepad.com/mobile_l:
<div xmlns="http://www.w3.org/1999/xhtml"><p>I don't search a lot on the net from my N93 as
</content>


        <feedburner:origLink>http://mobilesociety.typepad.com/mobile_life/2007/12/mowser-and-a-
        <entry>
            <title>A Web 2.0 Story: esyURL Creates Short URLs And Detects Mobile Web Browsers</
            <link rel="alternate" type="text/html" href="http://feeds.feedburner.com/~r/typepad
            <link rel="replies" type="text/html" href="http://mobilesociety.typepad.com/mobile_
            <id>tag:typepad.com.2003:post-42265330</id>
Line 29, Col 37
```

Figure 6.5   An Atom feed of a Blog.

Figure 6.6   Feed icon of a Blog on the right of the URL of the Web page.

Feeds are also used by applications to automatically aggregate a user's information from different places. An example is social networking sites. Pictures from picture sharing sites or new Blog entries are thus automatically imported into the user's page on a social networking platform.

Feeds are also used in combination with podcasts. Apple's iTunes is a good example, which among other functionalities also works as a podcast directory. A podcast directory is in essence a list of podcast feeds. The feeds themselves contain a description of the podcasts available from a source, information about the audio file (e.g. size) and a link from which the podcast can be retrieved. It is also possible to use a podcast feed in a feed reader program, which will then present the textual information for the podcast and present a link from which the audio file can be retrieved. Most people, however, prefer programs like iTunes for podcast feeds.

6.6.2 AJAX

In Web 2.0 the Web browser is the user interface for services. The more capabilities Web browsers have, the better the user experience. In Web 1.0 most Web pages were static. Whenever the user, for example, put text into an input field or set a radio button and pressed the 'ok' or 'continue' button, the information was sent to a Web server for processing and a new Web page with the result was returned. The user experience of such an approach is relatively poor compared with local applications where the reaction to user input is displayed on the same screen without the typical reload effect of one Web page being replaced by another.

The solution to this problem comes in three parts. The first part is the support of JavaScript code on Web pages by the Web browser. The JavaScript code can interact with the user via the Web page by reading user input such as text input or when the user clicks on buttons on the Web page and so on. Unlike in the previous approach, where such actions resulted in immediate communication with the Web server in the network and the transmission of a new Web page as a result of the action, the JavaScript code can process the input locally and change the appearance of the Web page without the page reload effect.

The second part is allowing a JavaScript application on a Web page to send data to a Web server and receive a response without impacting what is shown on the Web page. The JavaScript can thus take the user input and send it to the Web server in the background. The Web server then sends a response and the JavaScript application embedded in the Web page will alter the appearance of the page without the need for loading a new Web page. Since this exchange of data is done in the background, it is also referred to as being asynchronous, since the exchange does not prevent the user interacting with the Web page (scrolling, pressing a button, etc.) while the JavaScript application is waiting for a response. The JavaScript application embedded in a Web page can modify the page in a similar way as a program running locally is able to modify the content of its window. Thus, for the user the behavior is similar to that of a local program.

The third part is a standardized way of exchanging information between the JavaScript application running in the Web browser and the program running on a Web server on the web. A format often used for this exchange is XML, which is also used as a descriptive language for Atom and RSS feeds, as shown in Figure 6.5. XML is a tag-based language that encapsulates information between tags in a structured way. The class used in JavaScript to exchange information with the backend includes sophisticated functions for extracting information from an XML formatted stream. This makes manipulation of the received data very simple for a JavaScript application embedded in a Web page. All three parts taken together are commonly referred to as Asynchronous JavaScript and XML, or AJAX for short.

An example of a very simple JavaScript application embedded in a Web page communicating asynchronously with an application hosted on the Web server is shown in Figure 6.6. The actual content of the Web page is very small and is contained in lines 25–27 between the <body> tags. The JavaScript code itself is embedded in the Web page before the visible content from line 0 to 23. On line 5 the JavaScript code instantiates an object from class XMLHttpRequest. This class has all the required functions to send data back to the Web server from which the Web page was loaded via the Hypertext Transfer Protocol (HTTP), asynchronously receive an answer and extract information from an XML formatted data stream. The XMLHTTP object is first used in line 18 where it is given the URL to be sent to the Web server. In this example, the JavaScript application sends the URL of a Blog feed. The application on the Web server then interprets the information and returns a result, for example it retrieves the Blog's feed and returns what it has received back to the JavaScript application running in the Web browser. This is done asynchronously as the send function on line 20 does not block until it receives an answer. Instead a pointer to a function is given to the XMLHTTP object, which is called when the Web server returns the requested information. In the example, this is done in line 19 and the function which is called when the Web server returns data is defined starting from line 7.

The JavaScript application therefore does not block and is able to react to other user input while waiting for the server response. Functions handling user input, however, are not part of the example in order to keep it short.

When the Web server returns the requested information, in the example the XML-encoded feed of a Blog, the 'OutputContent' function in line 7 is called. In line 12 the text from line 26 of the Web page is imported into a variable of the JavaScript application. In line 13, the 'getAttribute' function of the XMLHTTP object is used to retrieve the text between the first <title> tags of the feed. This text is then appended to the text already present in line 26 and put on the Web page without requiring a reload.

While the JavaScript application shown in Figure 6.7 is not really useful, due to its limited functionality, it nevertheless shows how AJAX can be used in practice. More sophisticated JavaScript applications can make use of the asynchronous communication to download much more useful information and draw graphics and other style elements on the Web page based on the data received.

6.6.3 Tagging and Folksonomy

While analyzing textual information on Blog entries and Web pages is a relatively easy task for search engines, classifying other available media such as pictures, videos and audio files (e.g. podcasts) is still not possible without additional information supplied by the person making the content available. A lot of research is ongoing to automatically analyze the content of nontextual sources on the Web. However, for the time being, search engines and other mechanisms linking content still rely on additional textual information. The most common way of adding additional information is by adding tags, that is, search words. As this form of classification is done by the users and not by a central instance, it is sometimes also referred to as folksonomy, that is, taxonomy of the masses.

```
00 <script language="JavaScript"
01    type="text/javascript">
02 // [!CDATA[
03 var XMLHTTP = null;
04
05 XMLHTTP = new XMLHttpRequest();
06
07 function OutputContent() {
08
09    var xml = XMLHTTP.responseXML;
10
11    if (XMLHTTP.readyState == 4) {
12      var d = document.getElementById("data");
13      d.innerHTML += xml.documentElement
         getAttribute("title");
14    }
15 }
16
17 window.onload = function() {
18    XMLHTTP.open("GET",
"getfeed?feed=mobilesociety.typepad.com/feed");
19    XMLHTTP.onreadystatechange = OutputContent;
20    XMLHTTP.send(null);
21 }
22 // ]]>
23 </script>
24
25 <body>
26    <p id="data">Data received from server: </p>
27 </body>
```

Figure 6.7   A Web page with a simplified embedded JavaScript application.

Flickr, an image hosting and sharing Web service, is a good example of a service that uses tagging and folksonomy. Tags can be added to pictures by the creator, describing the content and location as shown in Figure 6.8. Tags can also contain other information like, for example, emotion, event information and so on. Tags can also contain geographical location tags (latitude and longitude), which were generated automatically by the mobile device with which the picture was taken because it was able to retrieve the GPS position from a GPS device (internal or external) at the time the picture was taken. The tags are then used by the image-sharing service and other services for various purposes. The picture sharing service itself converts the tags into user clickable links. When the user clicks on a tag the service searches for other pictures with the same tag and presents the search result to the user. Thus, it is easy to find pictures taken by other users at the same location or about the same topic.

The picture sharing service treats the geographic location tags in a special way. Instead of showing the GPS coordinates, which would not be very informative for the user, it creates a special 'map' link. When the user clicks on the 'map' link a window opens up in

Figure 6.8 Tags alongside a picture on Flickr, an image-sharing service.

which a map of the location is shown. The user can then zoom in and out and move the map in any direction to find out more about the location where the picture was taken. The picture sharing site also inserts the location of other pictures the user has taken in the area which is currently shown and on request presents pictures other users have taken in this area, which are also stored together with geographical location tags. This functionality is a typical combination of the use of tags to find and correlate information, of AJAX for creating an interactive and user-friendly Web page and of open interfaces which allow information stored in different databases to be combined (pictures and text in the image database and the maps in a map database on the network).

The tags and geographical location information alongside images are also used by other services. Search engines such as Yahoo or Google periodically scan Web pages created by Flickr from its image/tag/user database. It is then possible to find pictures not only directly in Flickr but also via a standard Internet search. This is important since Flickr is not the only picture-sharing service on the net and searching for pictures with a general Web search service results in a wider choice, as the search includes the pictures of many sharing sites. It is important to note at this point that without tags the value of putting a picture online for sharing with others is very limited, since it cannot be found and correlated with other pictures.

6.6.4 Open Application Programming Interfaces

Many services are popular today because they offer an open API, which allows third-party applications to access the functionality of the service and the database behind it. Atom and RSS feeds are one form of open API to retrieve information from Blogs or Web pages. Requesting the feed is simple, as it only requires knowledge of the URL (Universal Resource Locator, e.g. http://mobilesociety.typepad.com/feed). Analyzing a feed is also possible since the feed is returned as an Atom or RSS formatted XML stream. How the XML file can be analyzed is part of the open RSS and Atom specifications. In Figure 6.4, Thunderbird, a locally installed feed reader was shown. There are also Web 2.0 feed readers which run as JavaScript applications in Web pages and which get feed updates and store information in a database in the network (e.g. which feeds the user has subscribed to, which Blog entries have already been read, etc.).

While feeds only deliver information and leave the processing to the Web service running on a user's computer, remote services can also share a library of functions with a JavaScript application running in the local Web browser. Examples of this approach are the APIs of Yahoo [15] or Google maps [16]. These APIs allow other Web 2.0 services to show location data on a map generated by Google or Yahoo. A practical example is a Web statistics service that logs the IP addresses from which a Web site was visited. When the owner later on calls the statistics Web page, the service in the background queries an Internet database for the part of the world in which the IP addresses are registered. This information is then combined with that of the mapping service and a map with markers at the locations where the IP addresses are registered is shown on the Web page. As the map is loaded directly from the server of the mapping service it is interactive and the user can zoom and scroll in the same way as if he had visited the map service directly via the mapping portal.

Figure 6.9 shows how this is done in principle. The statistics service comprises both a server component and a front-end component, that is a program or script running on the Web server and a JavaScript application executed in a Web page. The backend component on the Web server is called when people visit a Web site which contains an image that has to be loaded from the statistics server. Requesting the image then invokes a counting procedure. It is also possible to trigger an HTTP request to the statistics server for counting purposes with a tiny JavaScript application that is embedded in the Web page. The counting service on the statistics Web server processes the incoming request to retrieve the origin of the request and stores it in its database. When the owner of the Web site later on visits the statistics service Web page, the following actions are performed:

- After the user has identified himself to the service running on the Web server, the IP addresses from which the user's Web site was visited in the past are retrieved from the statistics database. The service running on the Web server then queries an external database to get the locations at which those IP addresses are registered.

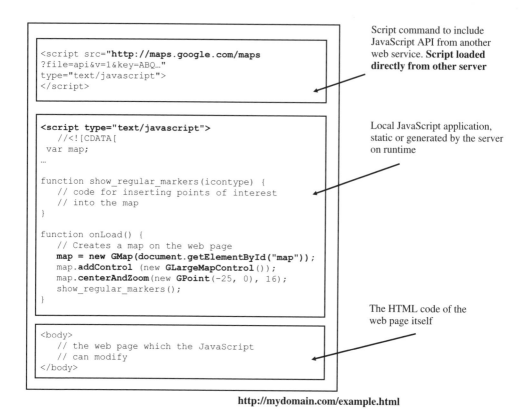

```
<script src="http://maps.google.com/maps
?file=api&v=1&key=ABQ…"
type="text/javascript">
</script>
```

Script command to include JavaScript API from another web service. **Script loaded directly from other server**

```
<script type="text/javascript">
   //<![CDATA[
 var map;
…

function show_regular_markers(icontype) {
   // code for inserting points of interest
   // into the map
}

function onLoad() {
   // Creates a map on the web page
   map = new GMap(document.getElementById("map"));
   map.addControl (new GLargeMapControl());
   map.centerAndZoom(new GPoint(-25, 0), 16);
   show_regular_markers();
}
```

Local JavaScript application, static or generated by the server on runtime

```
<body>
   // the web page which the JavaScript
   // can modify
</body>
```

The HTML code of the web page itself

http://mydomain.com/example.html

Figure 6.9   Remote JavaScript code embedded in a Web page.

- Once the locations are known the statistics service generates a Web page. At the beginning of the Web page, a reference to Goggle's mapping API is included. It is important to note that this is just a reference to where the Web browser can retrieve the API, that is the Web browser loads the API directly from Google's server and not from the Web statistics server.
- Next, the JavaScript code of the statistics service is put into the Web page by the server application. As the source code is assembled at run time, it can contain the information about where to put the location markers on the map either in variables or as parameters of function calls. In the 'onLoad' function shown in Figure 6.9, the JavaScript application embedded in the page then calls the JavaScript API functions of the mapping service that have been loaded by the script command above.
- As the API functions were loaded from the mapping service Web server, they have permission to establish a network connection back to the map server. They can thus retrieve all information required for the map.
- The map API functions also have permissions to access the local Web page. Thus, they can then draw the map at the desired place and react to input from the user to zoom and move the map.

- In the example above the local 'show_regular_markers' function is called afterwards to draw the markers on to the map with further calls to API functions. Note that the implementation of the function is not shown to keep the example short.

6.6.5 Open Source

In his Web 2.0 essay [12], Tim O'Reily also mentions that a good Web 2.0 practice is to make software available as open source. This way the Web community has access to the source code and is allowed to use it free of charge for their own projects. There are many popular open source license schemes and this section takes a closer look at three of the most important ones.

6.6.5.1 GNU Public License (GPL)

Software distributed under the GNU Public License (GPL) [17], originally conceived by Richard Stallman, must be distributed together with the source code. The company distributing the software can do this for free or request a fee for the distribution. The GPL allows anyone to use the source code free of charge. The condition imposed by the GPL is that in case the resulting software is redistributed this also has to be done under the GPL license. This ensures that software based on freely received open source software must also remain open source.

The GPL open source principle – to make the source code of derivate work available – only applies when the derivate work is also distributed. If open source software is used as the basis for a service offered to others, the GPL does not require the derivate source code to be distributed. The following example puts this into perspective: a company uses open source database software licensed under the GPL (e.g. a database system) and modifies and integrates it into a new Web-based e-mail service to store e-mails of users. The Web-based e-mail service is then made available to the general public via the company's Web server. Users are charged a monthly fee for access to the system. As it is the service and not the software that is made available to users (the software remains solely on the company's server), the modified code does not have to be published. If, however, the company sells or gives away the software for free to other companies, so they can set up their own Web-based e-mail systems, the distributed software falls under the terms of the GPL. This means that the source code has to be open and given away free of charge. Other companies are free to change the software and to sell or distribute it for free again.

The idea of the GPL is that freely available source code makes it easy for anybody to build upon existing software of others, thus accelerating innovation and new developments. The most successful project under the GPL license is the Linux operating system. The business model of companies using GPL software to develop and distribute their own software is not usually based on the sale of the software itself. This is why all Linux distributions are free. Instead, such companies are typically selling support services around the product such as technical support or maintenance.

Many electronic devices such as set-top boxes, Wi-Fi access points and printers with built-in embedded computers are based on the Linux operating system. Thus, the software of such systems is governed by the GPL and the source code has to be made available to the public. This has inspired projects such as OpenWrt [18], which is an alternative operating system for Wi-Fi routers based on a certain chipset. The alternative operating system, developed by the Web community, has more features than the original software and can be extended by anyone.

6.6.5.2 The BSD and Apache license agreements

Software distributed under the Berkeley Software Distribution (BSD) license agreement [19] is also provided as source code and the license gives permission to modify and extend the source code for derivate work. The big difference to source code distributed under the GPL license is that the derivate work does not have to be redistributed under the same licensing conditions. This means that a company is free to use the software developed by a third party under the BSD license within its own software and is allowed to sell the software and keep the copyright, that is to restrict others from redistributing the software. Also, it is not required to release the source code.

A license agreement similar to BSD is the Apache license [20], which got its name from the very popular Apache Web server – the first product to be released under this license. In addition to the BSD license, the Apache license requires software developers to include a notice when distributing the product that the product includes Apache licensed code.

Google's Android operating system, discussed in the previous chapter, makes use of the Apache license for applications created in the user space and the GPL license for the Linux kernel. This means that companies adopting the Android OS for their own developments do not need to publish the code for the software running on the application layer of Android if they do not wish to do so. It is likely that this decision was made in order to attract more terminal manufacturers to Android than would be the case if the whole system was put under the GPL, which would force companies to release their source code.

6.7 Mobile Web 2.0 – Evolution and Revolution of Web 2.0

The previous sections have focused on the evolution of the Web as it happens today on PCs and notebooks. With the rising capabilities of mobile devices, as discussed in Chapter 5, the Web also extends more and more into the mobile world. The following sections now discuss how Web 2.0 services can find their way to mobile devices and also how mobility and other properties of mobile devices can revolutionize the community-based services aspects of Web 2.0 and the possibilities for self expression.

As the extension of Web 2.0 into the mobile domain is both an evolution and revolution, many people use the terms Mobile Web 2.0 or Mobile 2.0 when discussing topics around the Internet and Web-based services on mobile devices.

6.7.1 The Seven Principles of Web 2.0 in the Mobile World

In Section 6.5 the seven principles of Web 2.0 as seen by Tim O'Reily [12] were discussed. Most of these principles also apply for Web 2.0 on mobile devices:

6.7.1.1 The Web as a platform

As on PCs and notebooks, services or applications can be used on mobile devices either via the built-in (mobile) Web browser or via local applications. Local applications can run entirely locally and store their data on the device. In this case they are 'nonconnected' applications and do not come into the Web 2.0 category. If local applications communicate with services or databases in the network and in addition incorporate several of the other principles, they can be counted as Web 2.0 applications. Local applications are deployed either as Java applications or as native applications. Java applications are portable to a certain extent over a wide variety of devices but are unable to use specific functionalities of a mobile device. Applications using specific features of a device or operating system are therefore implemented as native applications for mobile operating systems such as S60, UIQ, Windows Mobile, Android and so on.

As in the Web 2.0 world, many mobile Web 2.0 services use the Web browser as their execution and user interaction environment. At the moment, however, most mobile Web browsers do not yet support JavaScript. This makes it very difficult to develop interactive and user friendly user interfaces, as services depend on HTML forms and page reloads to communicate with a Web server.

The number of mobile devices with more sophisticated mobile Web browsers that include a JavaScript engine, however, is on the rise. As JavaScript is the basis for interactive Web applications, browsers such as the Nokia Web browser, which is based on Apple's WebKit [21], are an ideal platform for browser-based services. As mobile device hardware and operating systems are getting powerful enough to support more sophisticated Web browsers, it is therefore likely that in the mid-term most Internet capable mobile devices will include JavaScript in their mobile browsers. In the meantime, mobile Web 2.0 services should have both a JavaScript and a plain HTML version of their service front end, to reach as many users as possible in the most convenient way.

Another reason why it is more difficult to develop services for the mobile world is the wide range of different devices, operating systems and screen sizes. In the PC and notebook world, it is sufficient to support a small number of different browsers such as Internet Explorer, Firefox, Opera and Apple's Safari, which behave very similarly. In the mobile world, however, there are at least a dozen different mobile Web browsers available, running on a wide variety of mobile device hardware, especially concerning screen resolution and processing power. Each browser renders pages in a different way, which makes predictions of how the page will look very difficult. Furthermore, different screen resolutions make designing JavaScript applications more difficult than for PCs and notebooks, where the user interface layout is usually based on a single minimum screen resolution. If the browser window has a higher resolution, the user interface is scaled but does not usually use the additional space. In the mobile world, however, services should adapt to different screen sizes in order to make the best out of any display resolution.

When considering the Web as a platform for a service it has to be kept in mind that mobile devices are not always connected to the network when the user wants to use a service. When possible and desirable from a user's point of view, a service should have an online component but also be usable when no network is available. A distributed calendar application is a good example of a service that requires an online and an offline component. It is desirable to integrate a calendar application with a central database on a

Web server so people can share a common calendar – even for a single user a distributed calendar with a central database in the network is interesting as many users today use several devices – but the calendar must also be usable on a device even when no network is available. In the future, there will certainly be fewer places where no network is available and therefore, an offline component will become dispensable for some applications, while for others, such as calendars, it will remain an important aspect due to the required instant availability of the information, at any time and in any place. A number of different approaches are currently under development to make Web applications available in offline mode. This topic is discussed further in Section 6.7.3.

Another scenario which has to be kept in mind when developing Web-based mobile services is that a network might be available but cannot be used for a certain service due to the limited bandwidth (e.g. GPRS only) or high costs for data transfers. While checking the weather forecast is likely to cause only minimal cost no matter what kind of connection is used, streaming a video from YouTube should be avoided without a flat rate cellular data subscription if outside the coverage area of a home or office Wi-Fi network. Services which are aware of the connections they can use and which they cannot in terms of available bandwidth and cost are referred to as 'bearer aware' applications. This term is unfortunately inaccurate as it is usually not the bearer technology (UMTS, HSDPA, EvDO, etc.) that sets the limits but rather the cost for the use of the bearer set by the network operator. Therefore, the neutral term 'connection' is used in this chapter instead of 'bearer'.

Connection-aware applications are usually native applications, as Web-based and Java applications have limited or no control over the connection settings. Most connection aware applications only use a single connection defined by the user. More sophisticated applications allow the user to define a list of connections the service is allowed to use. An example of such an application is Shozu on S60 [22], a picture upload application which only uses connections for image transfers configured by the user. Figure 6.10 shows how this is implemented in practice. In the configuration menu the user can define which of the connections profiles (access points) that the user has previously configured, in the operating system's network settings, the program is allowed to use. When the user instructs Shozu to upload a picture it will try the connection profiles one by one until one of the connections can be successfully established. To the program itself the underlying bearer for each connection profile (access point) is transparent.

6.7.1.2 Harnessing Collective Intelligence

Many Web 2.0 applications and services enable users to share information with each other and break up the traditional model of top-down content and information distribution. This applies for mobile devices as well and is moreover significantly enhanced since mobile devices offer access to information in far more situations than desktop computers or notebooks, which rely on Wi-Fi networks and sufficient physical space around the user. Furthermore, users carry their mobile devices with them almost everywhere and access to information is therefore not limited to times when a notebook is available. Thus, it is possible to use the Internet in a context-sensitive way, for example to search for an address or to get background information about a topic in almost any situation.

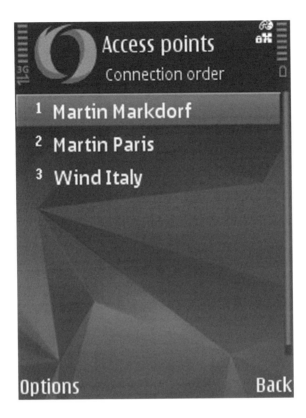

Figure 6.10 Bearer/connection awareness of Shozu [22], a mobile picture sharing application. (Reproduced by permission of Shozu.)

Mobility also simplifies the sharing of content, as mobile devices are used to capture images, videos and other multimedia content. Downloading content from a mobile device to the desktop computer or notebook before publishing is complicated and content is not shared at the time of inspiration. Connected mobile devices simplify this process as no intermediate step via a computer is necessary. Furthermore, users can share their content and thoughts at the point of inspiration, that is, right when the picture or the video was made or when a thought occurred. This will be discussed in more detail in the following sections.

On the software side, harnessing collective intelligence describes using open source software and making new developments available as open source again for others to base their own ideas on. The mobile world has little open source software to date, with Nokia being one of the few exceptions. The most notable of Nokia's open source projects are the use of open source software, for the operating system and some applications for the Nokia Internet tablets, and the S60 Web browser, which is based on open source software and again released as open source under a permissive BSD license [23]. Lately, the most comprehensive open source software approach in the mobile space has been made by Google with the Android operating system. The operating system kernel is based on

Linux and the GPL open source license and the application environment is distributed under the Apache open source license. In the same way as on the desktop, developers can now modify all layers of the software stack of mobile devices and are no longer confined to the application programming interface of an operating system.

6.7.1.3 Data is the Next Intel Inside

Another attribute of Web 2.0 is a network-based database. The database becomes more valuable as more people use it and contribute information. In the mobile world, network databases are even more important since local storage capacity is limited. Furthermore, databases supplying location-dependent information (e.g. restaurant information or local events) and up-to-date information from other people are very valuable in the mobile space as mobile search is often related to the user's location.

6.7.1.4 End of the Software Release Cycle

The idea behind the end of the software release cycle is that applications are executed in the Web browser and have a Web server-based backend and database. This way, software modifications can be made very quickly and new versions of an application are automatically distributed to a device when it loads the Web page of the service. This applies to mobile devices as well but, as discussed above for 'the Web as a platform' principle, many mobile applications have to include local extensions as network access might not always be available.

6.7.1.5 Lightweight Programming Models

Easy to use application programming interfaces are very important to foster quick development of applications using the services (APIs) of other Web-based services. As in the desktop world, Javascript applications running in mobile Web browsers and XML-based communication with network databases and services provide a standardized way across the many different devices of different manufactures to create new services. Whenever confidential user data is transmitted over the network, a secure connection (e.g. secure HTTP, HTTPS) should be used to protect the transmission. This is especially important when Wi-Fi hotspots are used, as data is transferred unencrypted over the air which makes it easy for attackers to intercept the communication of other hotspot users.

6.7.1.6 Software above the Level of a Single Device

Most services are not exclusively used on a mobile device but always involve other devices as well. A good example is a Web-based Blog reader application such as Bloglines [24], which can be used both on the desktop and also via a Web browser of a mobile device. Such Web 2.0-based applications have a huge advantage over software that is installed on a device and keeps information in a local database, as the same data is automatically synchronized over all devices. An example, while at home a user might read their Blog feeds with a Web-based feed reader and mark Blog entries as being read or mark them for later on. When out of the home, the user can use a version of the application adapted to mobile devices and continue from the point where he stopped reading on the desktop. Articles already read on the desktop will also appear as read on the mobile device since both Web-based front ends query the same database on the Web server. All actions performed on the mobile device are also stored in the network so the process also works vice versa.

Another example of software (and data) above the level of a single device is a music library and applications which enable the use of the music library via the network from many devices. If the music library is stored on a mobile device which allows other devices to access the library, the music files can be streamed to other devices over the network. This could be done over Bluetooth, for example, and a mobile device could output the music stream via a Bluetooth connection to a Bluetooth enabled hi-fi sound system. Music can also be streamed over the local Wi-Fi network to a network enabled hi-fi sound system which is either Universal Plug and Play (UPnP) capable or can access the music library of a mobile device via a network share.

Yet another example is streaming audio and video media files via the network to a mobile device. The Sling box [25] is such a device and adapts TV channels and recorded video files to the display resolution of mobile devices and sends the media stream over the network to the player software on a mobile device. Such services also have to take the underlying network into account and have to adapt the stream to the available network speed.

6.7.1.7 Rich User Experience

Early Web-capable mobile phones suffered from relatively low processing power and screen resolutions which made it difficult to develop an appealing user front end. Since then, however, display sizes and screen resolutions have significantly improved and enough processing power is available to run sophisticated operating systems and appealing graphical user interfaces. The user interface of Apple's iPhone is a good example of a mobile device with a rich user experience and fast reaction to user input. At the same time, the device is small enough to be carried around almost anywhere and battery capacity is sufficient for at least a full day of use.

6.7.2 Advantages of Connected Mobile Devices

The Internet cannot and should not be replicated piece by piece from the desktop onto mobile devices. This is partly because of the limitations of small devices, such as the need to scroll to see more than a few lines of text, a small keypad which makes it difficult to input text, no mouse for easy navigation, and a small screen size and lower display resolution then on the desktop. However, it is also because mobile devices are game changing, that is they are much more then just the 'small' Internet. Tomi Ahonen sees the Internet on mobile devices as the seventh mass media and explains in an essay that instead of looking at the disadvantages one should rather explore the unique elements of connected mobile devices and how they can be used to create new kinds of applications [1].

The current mass media channels are:

- print media (e.g. books, newspapers, magazines);
- various forms of discs or tapes (recording and music industry);
- movies and documentaries for entertainment;
- radio broadcasting;
- television broadcasting;
- personal computers and the Internet.

Each channel has unique elements, with the Internet being a bit of an exception since it universally embraces all other mass media types and adds interactivity and search.

New forms of mass media have been able to establish themselves alongside already existing media because they offered something the previous channels did not have. One of the advantages of radio broadcasting over print media is, for example, that news can be spread much faster than would ever be possible with newspapers.

The emergence of a new mass media usually does not lead to a complete demise of previous types of mass media, as some of their properties are not shared by the new media. Instead, it can be observed that media types usually adapt to the arrival of new media. In the case of the print media, newspapers adapted to the fact that they were no longer the source of breaking news once radio and television broadcasting became popular. They have still retained a roll in the media landscape, however, due to their ability to cover news in much more depth and because they are much more suitable for delivering background information. Even with the emergence of the Internet, the print media is still alive and well, as in some circumstances it is still more convenient to read an article in the newspaper than on a computer screen.

Ahonen describes the following advantages of connected mobile devices over previous mass media channels.

6.7.2.1 Mobile is Personal

Previous mass media channels were not personal. A single copy of a newspaper, a book, a movie, a CD or a television set is potentially used by more than one person. Therefore it is difficult to establish a direct relationship with a customer through a single copy or a single device. Even connected desktop PCs or notebooks at home are often not personal, as the device is usually used by several family members. The connected mobile device on the other hand is highly personal, as it is not shared with friends or even family members. For content creators one device therefore equals one user, which is ideal for assembling statistics about the use of a service, for marketing purposes, and also as a sales channel. As the number of personal devices per user increases, the downside is that the marketer can no longer assume that one device represents one person.

6.7.2.2 Always On

Connected mobile devices are the first type of mass media that is always online. Thus, users can be informed or alerted about events even more quickly than via radio or television – which users do not watch or listen to all of the time. Mobile devices are rarely switched off. A service that has capitalized on this is mobile e-mail, for example, which can be delivered instantly to mobile devices. RIM was probably the first company to fulfill this need with its Blackberry mobile e-mail devices. Another application that uses the instant reachability of subscribers is text alerts by companies such as CNN, who deliver breaking news via SMS [26].

6.7.2.3 Always Carried

Always on is so valuable because users tend to carry their mobile devices with them most of the time. Users can be informed instantly about breaking news and they can get access to information at any time. Studies show that many people even keep their phones switched on at night and have them on their bedside table, mostly to serve as alarm clocks. Equally, users have the ability to get or search for information at any time and at any place. Specialized search engines such as Taptu [27] and Mowser [28] are now appearing, which take into account that searching for information on a topic while on the move with a mobile device is usually different than searching for information via a desktop or notebook.

The idle screen or the screen saver of connected mobile devices is also an ideal place to display content. Opera's mobile widgets were one of the first programs to explore this area by displaying content retrieved from the Web, sending updates to users in real time from their favorite news feeds [29].

6.7.2.4 Built-in Payment Channel

Most of the traditional mass media channels use advertisements as a source of revenue. Companies advertise via mass media channels in the hope that people like a product and will buy it later on. The issue for advertisers is that there is a gap between users reacting to the advertisement and the opportunity to actually buy the product. This gap has been shortened considerably by advertising on the Internet but users usually still have to type in credit card details before the product is sold. As connected mobile devices are personal and as connecting to a cellular network requires identification and authentication, even this final step of typing in credit card details can be removed. The process from advertising something and giving the user the possibility to buy the product thus becomes a single click, or 'single click to buy'. This approach is used on Web portals of network operators for products such as ring tones or games. The user can browse a catalog and when deciding on a game, for example, can pay instantly by clicking on a purchase button. No identification is necessary since the mobile is personal and the transaction is performed in the background and either deducted immediately from the user's prepaid account or put on the telephone bill.

6.7.2.5 At the Point of the Creative Impulse

As connected mobile devices are personal and continuously carried, they are a unique tool to capture thoughts and impressions at the point of inspiration. An advantage of this is the capture of pictures and videos that would otherwise never have been recorded, as other nonconnected and single-purpose devices such as digital photo cameras are not always at hand. In addition, by being connected, mobile devices enable users to instantly send that picture, video or thought to a picture- or video-sharing platform, to a social network, to an instant messaging service, to a microblogging service like Twitter or to a personal Blog. This helps to inform others of big and small events as they break. Thoughts are also quickly recorded for personal use as, unlike PCs, mobile devices do not need a minute to boot before an application is available for note taking. Also, connected mobile devices simplify the sharing process as it is no longer necessary to first transfer pictures and videos from a digital camera to a PC and then upload the files to Web-based services. By simplifying this process it is much more likely that people will be willing to share content, as it can be made available with much less effort.

6.7.2.6 Summary

When looking at the unique properties of connected mobile devices it becomes clear that they are likely to become an interesting new mass media channel and that other channels will have to adapt, as print media had to, following the emergence of radio broadcasting at the beginning of the last century.

6.7.3 Offline Web Applications

On the desktop, many Web browser-based JavaScript applications exist today with well-designed user interfaces and connectivity to services and databases on the Web. Web-based applications such as GMail (e-mail reader), Bloglines (Blog reader), Writley (text processor) and so on are now even taking over some of the functionality of locally installed programs. On mobile devices, this trend is not yet widespread for a number of reasons:

• Web browsers on mobile devices need to become AJAX capable. This is well underway and in the mid-term it is likely that most mobile devices will include such a browser. As JavaScript and the functions for asynchronous communication with the Web server are standardized, developers can design applications which will work over a wide variety of mobile devices. This is an important key to success as services which only work on a limited number of mobile devices are unlikely to be successful, as they will be unable to attract a sufficiently large user base.
• Applications need to be available even without network coverage. This is crucial for applications such as calendars, address books and so on, which have to be accessible at any time and any place.
• Many applications require access to the file system, the camera and other services of the device. This is important for applications that upload information, such as pictures, to the Web or for applications that use the device's built-in GPS unit, in order to include information relevant to the user's current location with uploaded content.

With the traditional server–browser approach these things are not possible. If the connection to the network is lost, Web applications cannot be loaded onto the device from the Web server and user input cannot be sent back to the Web. Also, Web browser-based applications do not have access to local resources. A number of different initiatives have thus started to address these shortcomings.

6.7.3.1 Google Gears

Google's approach is based on a Web browser plugin which offers JavaScript applications on Web pages an API to store pages locally [30]. Furthermore, the plugin allows applications to store data locally. When the page is accessed while the device is not connected to the network, the local copy of the Web page including the JavaScript program is used instead. From a technical point of view this occurs as follows:

- A local server feeds the Web browser with Web pages which have been stored in the local cache in case the device is used in offline mode.
- A relational database is used, in which JavaScript applications on Web pages can store information. To stay with the example above, a calendar application could store a copy of the user's calendar entries locally which are then synchronized with the calendar entries in the Web, when the device is online and the application is used.
- As synchronizing the local and remote data depository after going online could take considerable time, it is important that such activities do not block the execution of the JavaScript application on the Web page. Thus, the Gears API contains a 'WorkerPool' that allows JavaScript code to be executed in the background.

At the beginning of 2008, Google Gears is only available for desktop-based browsers, as it depends on a Web browser plugin interface which is not yet available in any mobile browser. As the software is available as open source under a BSD license the way is clear for mobile browser developers to include it in future versions of their products.

While Google Gears enables Web applications to run offline, it is still not possible to use local resources (e.g. access to the file system, getting network information, GPS information, etc.).

6.7.3.2 Nokia S60 Web Widgets

Nokia's vision for local Web applications on mobile devices is widgets [31]. Unlike the Google Gear's approach, a widget is not executed inside the Web browser. Instead, it just uses the Web browser's core as a runtime environment and otherwise looks like a native application. This is shown in Figure 6.11. Widgets can even have their own icon appearing alongside local applications in the device's application menus. Having the full screen available, rather than running in the user interface of the Web browser, has a number of advantages. From the user's point of view the program can be started and controlled just like local applications. When the widget is started, by clicking on the icon in the program list, the Web browser core is loaded and the JavaScript and HTML page, which the JavaScript application can modify, is shown on the screen just like a local application.

The first version of the Web widgets packet does not offer local storage to applications and therefore its use is limited to the times when the device is connected to the network, in

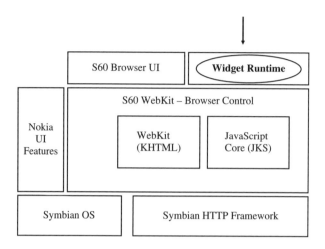

Figure 6.11   Nokia Web widgets execution environment.

case data needs to be stored between invocations of the widget. Also, there are no interfaces at the moment to access other data from the device such as calendar entries, address book entries and so on. Widgets do have access, however, to system information such as the current battery power level, network properties such as signal strength and network name, display light control, vibration control, speaker control, memory properties and so on.

6.7.3.3 HTML-5

For version 5 of the HTML the World Wide Web Consortium (W3C) is specifying tools to make Web pages and embedded JavaScript applications available offline [32]. The approach taken by the W3C is very similar to Google Gears and also features offline caching of Web pages and a data repository for JavaScript applications to store data locally. There are two ways of storing data locally. For simple data structures, data can be stored in name/value pairs. For more complex data, JavaScript applications can use an SQL database with standardized functions to query, insert, change and delete data. The advantage over Google Gears is that once the standardization of HTML-5 is finalized, browsers will be able to natively support offline Web applications and a plugin like Google Gears will no longer be required.

6.7.3.4 Data Synchronization

When storing data in both a local database and in the network, the databases have to be synchronized once the mobile device is connected to the network again. Even for single user applications local and remote database synchronization is not straight forward. The user could have first used the application on the mobile device while it was offline to change some data and then later on modified some data in the network via a PC with an

Internet connection. This means that, when the mobile device goes online, it cannot assume the database in the network has not been modified. Thus, the application needs an algorithm to compare the changes made in the database on the mobile device with the changes made in the database in the network and then modify the local and the network databases accordingly. The same applies to multiuser applications such as shared calendars.

6.7.3.5 Access to Local Information Outside the Browser Cache

A big advantage of mobile devices is the ability to support the user's creativity at the point of inspiration. When pictures, videos, podcasts, Blog entries and so on are created on the mobile device they can automatically be enriched with other information such as GPS location before being instantly sent to a picture- and video-sharing site, Blogs and so on. Current Web applications, however, do not have access to files and other local data, beyond what they can write into the local database themselves, while the device is not connected to the Web. In the future it is thus necessary to create an interface for Web applications to get access to locally stored user created content and additional information such as location information. Access to local information needs to be dealt with carefully due to security and privacy issues. This will be discussed in more detail below. From the technical point of view the main question is how Web applications could gain local access in a generic way to make them interoperable over a wide variety of devices. Some devices might have more sources of information than others. As using proprietary APIs to access such information would make Web applications device-specific, a standard set of functions must be designed which are then translated on each device into proprietary operating system commands.

6.7.3.6 Security and Privacy Considerations

Allowing a Web application to store data locally or to even access local information beyond the browser cache has a number of security and privacy implications which have to be considered when implementing such functionality. The main issue with having a local application and data cache in the Web browser is that JavaScript applications can use the local cache for user tracking. Advertisement services could use the local interface to write a JavaScript program that is included in all Web pages of their advertising partners and that records the pages from which it has been invoked in the local cache. Each time the same JavaScript application is executed from a different page it reports the last page back to the advertisement server, which can then generate a user profile. As most users are unlikely to agree with such methods the user should be informed that an application wants to store information locally and what kind of privacy issues this could bring with it. The user should then have the choice to allow or deny the use of local storage. If allowed, then this should be either temporarily or, where the JavaScript application is trusted, permanently.

Access to local device information outside the browser cache is even more sensitive, as malicious JavaScript applications could read the user's private data and send it to a server in the network. Therefore, the user has to be informed if a JavaScript application tries to access local data, to give the user the opportunity to allow or deny the request. The user then has to decide whether he trusts the application and thus allows access to local resources. Again, the user should have the choice to allow access temporarily or permanently as trusted applications should be able to access local data without further queries to the user once consent has been given. Otherwise, usability of the application would suffer.

6.7.4 The Mobile Web, 2D Barcodes and Image Recognition

In the desktop world, URLs of new Web pages found in print magazines, books and so on can usually be quickly entered into the Web browser via the keyboard. In the mobile world, however, this is much more difficult as there is usually just a numeric keypad available. The use of T9 (text on nine keys), a predictive text input technology to speed up the writing process of text on numeric keypads, is difficult to use with URLs as names of Web sites are often composed of words that do not exist in the local language. Even if short Web site names are chosen, entering the Web site name into a mobile Web browser is still time-consuming. This tends to keep users from viewing new Web pages on their mobile device unless there is a very strong motivation to do so.

Most mobile phones, however, are equipped with a camera which can be used to simplify the process. In the future, mobile phones are likely to become powerful enough to include text recognition algorithms which are able to extract URLs from a picture the user has taken from an advertisement or a Web site URL in a print magazine. The extracted URL can then be forwarded automatically to the mobile Web browser and the page opened without any further interaction with the user.

A current alternative is two-dimensional (2D) barcodes. 2D barcodes are similar to standard one-dimensional Universal Product Code (UPC) barcodes which have been in use for a long time on everyday products. These are used in combination with cash registers in supermarkets, for example, to speed up the checkout process. Two-dimensional barcodes have an advantage over one-dimensional UPC barcodes in that much more information can be encoded in them. Figure 6.12 shows a 2D barcode embedded in the Blog of this book that encodes the link to the mobile version of the Blog (http://winksite.com/msauter/wireless). Instead of typing this URL into a mobile browser, mid- to high-end mobile devices now contain a barcode reader as shown in Figure 6.11 on the right. These can be used to scan the 2D barcode and to automatically decode the content. After decoding, the barcode scanner forwards the URL to the mobile browser, where it is opened automatically.

Today, several competing types of 2D barcodes exist. In Figure 6.12, a 2D QR (Quick Response) barcode is shown, which has been developed by Denso-Wave, a Japanese company [33]. Today, these 2D barcodes are common in Japanese print magazines, street side advertising and in tourism, for pointing users to Web site addresses. A competing format with similar success is the data matrix barcode, which was standardized by the International Organization for Standardization (ISO) in document 16 022. Furthermore, a number of other 2D barcode specifications have appeared on the market. While in Japan QR codes seem to be the de facto standard for mobile devices, there is still much uncertainty in other parts of the world. The barcode scanner shown in Figure 6.12 thus supports several 2D barcode types to be as universally usable as possible.

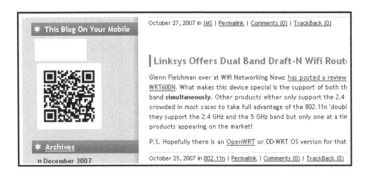

Figure 6.12 Two-dimensional barcode on a Web page and while being captured on a mobile device. (Reproduced by Permission of Nokia, Keilalahdentie 2-4, FI-02150 Espoo, Finland.)

Another advantage of 2D barcodes in magazines, on posters, on advertisements and so on is the strong message to the user that the Web site behind the barcode can be viewed with a mobile device and that no PC or notebook is required.

6.7.4.1 Image Recognition to Access Content

Instead of scanning a 2D barcode with a specialized application to get a URL of a Web page, applications are appearing which use image recognition technology in the network to find additional information for the user. For billboard or magazine advertising this works as follows: the user takes a picture and sends it via a standard MMS to an image recognition server. The server analyzes the picture, finds the company behind the advertisement and returns an SMS with a URL which can be accessed from the phone's browser. It is also possible to return an MMS with pictures and text-based information for users without an Internet subscription. Daem Interactive [34] is one of the companies developing such technology.

The main advantage of this approach compared with 2D barcodes is that no special software needs to be installed on the phone. In addition, it avoids the 2D barcode issue of many different types of 2D barcodes that are not compatible with each other and the resulting market fracturization and patent issues. Therefore, image recognition has a good chance to become established in the market next to 2D barcodes.

There are, however, also some disadvantages to image recognition compared with 2D barcodes. First, the server in the network might find it difficult to recognize an image because of poor image quality. Especially in low light conditions, pictures tend to be grainy and blurry. When taking pictures from magazine ads, users also tend to position the phone too close to the image and the picture will not be sharp.

Another disadvantage is that image recognition is performed in the network. This requires the user to assemble an MMS message, send it to the network and wait for the response. This is more time-consuming than starting a barcode reader and doing the analysis of the barcode on the device. In practice this might not be acceptable in some situations.

Another image recognition approach which overcomes these disadvantages is to use 3G video calls connected to an image recognition server in the network. Instead of analyzing single pictures the algorithm in the network analyzes the video stream. Once a positive match has been made, the server either presents further information as part of the video call or returns additional information in a picture or text message.

As technology progresses the use of network-based image recognition is likely to extend beyond advertising. Together with location information such as cell identity or GPS coordinates, picture databases with advanced image recognition software could be used in tourism, for example, to identify what the user is looking at and return relevant information back to the user (e.g. a link to a corresponding Wikipedia entry).

6.7.5 Walled Gardens, Mobile Web 2.0 and the Long Tail

In the early days of the mobile Web, mobile operators started building Web portal sites and offered content and services from within that portal. The term 'walled garden' was used by some for these portals because operators often restricted network access, only allowing users to connect to the portals because they could only control users and consequently revenue streams while the user stayed within the portal. Access to content and services outside the walled garden was either completely blocked or external data traffic was charged at rates that made it very unattractive to leave the portal site.

Walled gardens, however, are completely the opposite of mobile Web 2.0, Web services and long-tail economics, as discussed in this chapter. Popular Web services such as Flickr, Gmail, Myspace, Facebook, RSS feeds of Blogs, mobile Web presence of international publications, to just name a few, are difficult to integrate in operator portals. It has worked in a few cases where Internet companies and network operators have reached a financial and technical agreement of how to include the services as part of an operator portal. For Internet companies such alliances are difficult to set up and maintain, however, as there are hundreds of network operators in the world, each requiring its own contract and revenue sharing agreement. Also, mobile network operators often require an exclusive contract for their country, thus limiting the number of people who can use the service. As there are usually more than two mobile network operators per country, such exclusive contracts are even counter-productive for Internet companies since the majority of the potential user base cannot use the service from their mobile devices. This significantly reduces the attractiveness of services. Users who already use the service freely on the desktop are suddenly locked out in the mobile world. These users are likely to turn their back on the service they already use on the desktop and to look for alternatives that they can access from their mobile devices as well. Eventually, they will take the minority with them, as the service on the operator portal loses attractiveness.

A good historical example is the SMS service in the USA. For many years it was not possible to send text messages between users of different networks. Consequently only a few people used the service as it was not clear who they could reach. Since putting gateways between networks into place, use of the service has increased significantly as the user no longer has to wonder whether a message will reach the recipient.

More recently, many mobile operators have started to abandon their walled garden strategy as the number of users willing to use only services on a portal is very small and revenues are low. Some operators, such as '3' in the UK and other countries, have even made a 180 degree turn and are now promoting openness and the use of Internet services beyond their portal. Other big network operators such as T-Mobile have also changed tactics and their 'Web-and-Walk' offers also encourage users to make use of any service available on the Internet. Instead of requiring exclusive rights and revenue sharing, some operators are now partnering with Internet companies and promoting their portal by nonexclusive inclusion of external services into their mobile Web portal.

Despite this new openness, network operators are still trying to find ways to remain service providers in the Internet world. In their way of thinking, 'network' and 'service' should belong together in the same way as they see mobile networks and voice telephony belonging together. Their point of view is that the separation of network and service degrades them to mere 'bit pipes' for external companies and services. But maybe this is just a matter of perspective. In addition to the handful of killer applications born out of Web 2.0 (Google search, maps, Flickr, Amazon, eBay, Facebook, etc.) most people today use quite a number of services from the long tail of the Web, as described earlier and in [10]. It is these uncounted startup or niche segment services that are for many users the final incentive to go online with their mobile device. The problem mobile network operators have with this approach is that they cannot be the provider of the mainstream 'killer' applications, since they are already out there. In the best case they can only partner with already existing Internet companies. Also, they cannot be the provider of niche applications, since they are only used by comparatively few users, especially from a national perspective. Consequently there is not much revenue to be made with these long-tail applications as described in [11]. However, while network operators cannot make a lot of money IN the long tail, they can actually get considerable revenue *with* the long tail, as users pay to use their network to access these services. With this line of thinking the 'bit pipes' suddenly transform into the enablers of the mobile long tail.

With the introduction of the iPhone there has even been press speculation that mobile operators are willing to share some of their monthly revenue generated by iPhone users with Apple. In effect this means that the revenue generated by enabling the mobile long tail does not remain with the mobile network operators but has to be shared with the device manufacturer. Whether this sort of revenue-sharing model with device manufacturers will have a future is uncertain, but it does show, impressively, that there is flexibility on the mobile operator side if they see an opportunity for themselves.

6.7.6 Web Page Adaptation for Mobile Devices

Today, the overwhelming majority of Web pages are designed for desktop-based screen resolutions and often it is also assumed that the user is connected via a broadband connection, that is, that pictures and other content can be downloaded almost instantly. This is an issue for mobile Web browsing since the screen is usually much smaller. Advanced Web browsers on mobile devices such as the Nokia Web browser based on Webkit, Apple's iPhone Web browser and others are 'full screen' Web browsers as they can render the page exactly as intended for the desktop and only show part of the Web page on the screen. The part that is shown on the screen is either a particular area of the Web page or the text from a particular area, reformatted to fit into the mobile device's

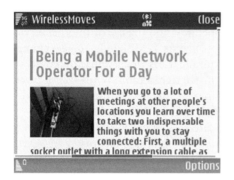
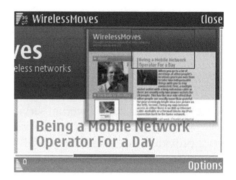

Figure 6.13 A standard Web page rendered on a mobile device. (Reproduced by Permission of Nokia, Keilalahdentie 2-4, FI-02150 Espoo, Finland.)

display. Figure 6.13 shows how a part of the full Web page of this book's Blog appears in Nokia's mobile browser. The browser has automatically reduced the size of the picture and the text is reformatted to fit on the screen. When scrolling through the Web page a window opens to show the full Web page to help with navigating to other parts. While this works quite well, there are a number of disadvantages to the approach depending on the situation. The first disadvantage is the processing power required to render a standard Web page. Long pages with a lot of content take quite some time to be rendered and require the user to wait until the end of the process. A second disadvantage is the time required to download all parts of a page if only a slow connection to the network is available, for example, downloading all pictures in full resolution. Also, depending on the network and type of subscription, downloading a single full page can be quite costly. In the future, however, the processing power and network coverage of fast B3G networks will surely increase and subscription costs will further decline. In most countries, data subscriptions are already available today such that the cost of downloading full Web pages via cellular networks is no longer an issue. The former two problems, however, remain, even when accessing the Internet from a mobile device via a Wi-Fi hotspot and DSL connection at home.

Less capable Web browsers on mobile devices require a different approach.

6.7.6.1 Mobile-friendly Pages

Some Web sites and Blogs are available in a full desktop browser format and also with a mobile layout. The mobile friendly layout usually does not use JavaScript and Flash, the page is formatted differently and advertisements are either missing or inserted in a different way, that is, not on a side bar since the screen of a mobile device is too small for Web pages to have sidebars. Mobile-friendly Web pages can be requested from the Web server in a number of different ways. One approach is to have different URLs for the desktop Web site and the corresponding mobile Web site. How this is done is not standardized. Google, for example, offers a mobile-friendly search engine page via http://google.com/m. Others put the indication that a Web site contains mobile content at the beginning of the URL such as http://mobile.domain.org or http://m.domain.org.

Since these methods are not standardized, the mobile industry has reacted and created a new top-level domain, '.mobi'. URLs like http://martin.mobi indicate that the Web pages on this Web site are specifically tailored for mobile devices. The creation of the .mobi domain was quite controversial at the time as the top-level domain now indicates the type of device for which it is intended. No other top-level domain has done this before and many had been against such a solution as they were in favor of the 'one Web' (for all devices) approach to prevent the World Wide Web from splitting into a desktop and a mobile sphere. One of the advocates of the 'one Web' movement was Sir Tim Berners-Lee, inventor of the world wide Web. In the article 'New Top Level Domains.mobi and.xxx Considered Harmful' [35] he gives some details of why he fears that a mobile-specific top-level domain might have a detrimental effect on the Web. In the end he did not prevail and the .mobi domain was created. Whether it will be successful and what kind of implications it has on the Web in the future remains to be seen.

6.7.6.2 Content Adaptation Services

As mobile optimized pages are the exception rather than the norm, there are some Web services that mobilize specific types of Web pages, such as Blogs. Winksite [36] is an example of such a service which is used for this book's Blog. Figure 6.14 shows how the Blog looks on a mobile device after being rendered by Winksite. Since Winksite takes the content of the Blog from the corresponding RSS feed, which only contains the pure text and pictures of the Blog entries, the style elements of the full Web page version of the Blog do not have to be removed. Instead, Winksite takes the text of the RSS feed, removes the pictures and creates mobile-friendly Web pages for an average to low-end mobile Web browser. Removing pictures also ensures that transmission costs over expensive networks are kept to a minimum. As Winksite URLs are quite long, it is not practical for users to type them in. Instead, there are a number of different ways to forward the URL to a mobile Web browser. Many Blog authors who use Winksite to 'mobilize' their Blogs include Winksite's logo and a 2D barcode on their Web page, as shown in Figure 6.12

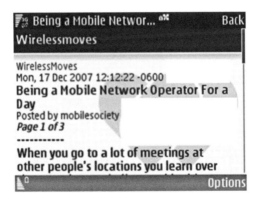

Figure 6.14   The Blog to this book mobilized by Winksite. (Reproduced by Permission of Nokia, Keilalahdentie 2-4, FI-02150 Espoo, Finland.)

above. Users visiting the Blog via their desktop PC or notebook can then use their mobile device and scan the 2D barcode, if the mobile device has a 2D barcode scanner application. Another option is to click on the Winksite logo to get to a menu where the user can then request an e-mail, containing the URL, to be sent to the mobile device. Winksite also sends SMS messages containing the URL, which is probably the quickest and easiest method to transfer the URL for most people. Once the SMS is received on the mobile device it is usually possible to start the Web browser from the SMS, by clicking on the link contained in the message.

6.7.6.3 Explicit Content Adaptation

A number of content adaptation solutions exist for Web sites not offering mobile-friendly versions of their pages that analyze the Web pages and deliver a mobile optimized page to the mobile device. To get these optimized pages a user does not go to the Web site directly but instead opens the front page of the content adaptation service. The service is usually coupled with a search function so the user can either type in the full domain name of the Web site to be converted into a mobile format or can specify keywords to get a list of search results. Popular content adaptation services are available, for example, from Google mobile [37], Mowser [29] (now acquired by DotMobi), Taptu [27] and Mippin [38]. Each of these products has its own approach to mobile content adaptation and additional services around it. Since these products also include mobile search services, they are also an ideal platform to start a Web search from a mobile device. Search results are then shown to the user in a similar way to results on a desktop from Yahoo, Google, Altavista and so on. When clicking on a search result, the corresponding Web page is optimized for smaller screens. Figure 6.15 shows how this book's Blog looks after being adapted to the small screen of a mobile device by Mowser.

As the screen resolution, processing power and Web browser capabilities of mobile devices are quite diverse, some mobile content adaptation services analyze the browser information that is part of HTTP request for a Web page. Based on this information, they

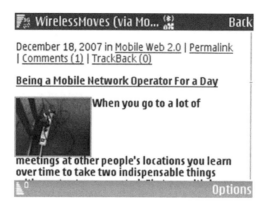

Figure 6.15 The Blog to this book adapted for mobile devices by Mowser. (Reproduced by Permission of Nokia, Keilalahdentie 2-4, FI-02150 Espoo, Finland.)

then decide how to format the page, if pictures can be included and so on. The mobile Web browser of Nokia N series devices, for example, sends the following Web browser identity information in a HTTP request which clearly identifies the source of the request as a mobile device:

```
"Mozilla/5.0 (SymbianOS/9.1; U; en) AppleWebKit/413
(KHTML, like Gecko) Safari/413"
```

This identity is clearly different to the identities of desktop browsers. The identity of Firefox, for example, is as follows:

```
"Mozilla/5.0 (Windows; U; Windows NT 5.1; de;
rv:1.8.1.11) Gecko/20071127 Firefox/2.0.0.11"
```

To be able to customize Web pages for specific mobile Web browsers, the content adaptation service requires a database of Web browser identification strings which needs to be kept up to date as new devices and new browser variants appear on the market.

Some content adaptation systems are also able to detect that a Web site is able to deliver a mobile-friendly version of their content from the same URL as the standard desktop page. This will be further described below. In such a case no content adaptation is performed and the page is delivered to the mobile phone as it was received from the Web site.

While content adaptation systems are a very beneficial service for the user, they remain a somewhat controversial topic of discussion in the Web community. Some argue that such services are a threat to Web sites financed by advertisements, as ads are usually removed from pages during the content adaptation process because they cannot be adapted well automatically. Others argue that Web sites that do not offer mobile device optimized versions of their pages are not interested in serving such users anyway and thus do not incur a financial loss. If Web site owners offer mobile optimized versions of their Web pages on their own, they can then also include advertisements in a mobile-friendly way. Some of the content adaptation systems are themselves financed by advertisements and these systems offer Web site owners a share of advertising revenue if owners are actively using their services to convert their Web pages into a mobile-friendly format.

6.7.6.4 Web Servers with Content Adaptation

It is also possible that Web sites themselves analyze the browser identification string in an HTTP request message and deliver a standard Web page for desktop browsers and an adapted version for mobile devices. From a theoretical point of view this approach is superior as:

- The Web is not fractured and the 'one Web' paradigm is fulfilled.
- No content adaptation services are needed and pages get rendered according to the capabilities of the requesting device.

Unfortunately the disadvantage of this approach in practice is that users have no way of determining from the URL that the content will be delivered in a mobile-friendly format and thus might not consider visiting the Web site from their mobile phone if they are not informed by other means like, for example, a 'mobile-friendly' logo on the Web site itself.

In addition, this approach does not take the type of network connection and the associated costs into account. A mobile Web browser might be capable of rendering sophisticated Web pages and thus the Web server can send the standard desktop page, but the user might prefer a scaled down version due to the high transmission costs.

6.7.6.5 Web Browser Content Adaptation Proxies

Yet another way to adapt Web pages for mobile devices is to combine a light-weight mobile Web browser with a content adaptation proxy in the network. Opera very successfully uses this strategy with its Opera minibrowser, a Java application available for most mobile phones and other mobile devices [39]. The Opera minibrowser on the mobile phone only communicates with the Web proxy, which requests the page on the user's behalf, modifies the page and leaves it to Opera mini to display it. Since the proxy and the Web browser work together, the resulting adaptation is usually excellent. This way it is also possible for the browser to show an overview of the full page without having to load the full content of the page. Figure 6.16 shows how this content adaptation approach looks in practice.

Figure 6.16 Opera mini with network-based proxy and light-weight browser. (Reproduced by Permission of Opera Software ASA, Waldemar Thranes gate 98, NO-0175 Oslo, Norway.)

6.7.6.6 Web-based Transparent Proxies

Some network operators use transparent proxies to reduce the amount of data transferred for a Web page and to decrease page loading times. This is done, for example, by compressing images and removing unnecessary HTML code from Web pages. As the proxy is transparent, the Web browser and Web server do not need to be modified. While the approach has some advantages, the downside is that the quality of the images embedded in a Web page is significantly reduced. As current 3G networks and future B3G networks are fast enough to serve full Web pages, most operators allow the user to deactivate this kind of content adaptation. Deactivation is done either via a Web page, or via proprietary software which was shipped together with the wireless network card or by sending a modified HTTP header [40].

6.7.6.7 Conclusion

In practice, a number of factors determine how the user can or prefers to surf the Web with their mobile device:

- the capabilities of the device itself (processing power, screen resolution, etc.);
- the mobile Web browser delivered with the device;
- the speed of the network connection;
- the cost of a particular network connection.

In theory the 'one Web' approach is a good idea, but unfortunately it does not work well in practice since only a few Web sites supply their Web pages in formats adapted to specific types of device. Thus, the user has to be aware of which sites support optimization techniques and consequently less ideal approaches than 'one Web' are preferable in practice. Content adaptation services are therefore useful tools. However, as devices get more powerful and costs for using cellular networks decrease, the need for content adaptation services in the future will probably diminish. However, it is also likely that even in the future there will be a significant number of devices that will still require content adaptation because some users will still prefer to have a device with a small display, to fit more easily into a pocket, or because mobile devices in developing markets will not have the same capabilities as devices sold in mature markets.

6.8 (Mobile) Web 2.0 and Privacy

The inherent goal of many (mobile) Web 2.0 social networking applications is to let users share private information with the rest of the world or with a group of other people. Privacy on this level can be controlled by the user, that is, by consciously deciding what kind of information they want to share with others. When making a decision of what to share and what not to share, one should not only think about the immediate implications of sharing private thoughts, ideas and so on, but also about what happens to this information over time. Everything that is shared publicly will remain on the Internet for an indefinite time and can be found via search engines such as Google and Yahoo by anyone and not only by those for whom the information might have initially been intended. Everything shared with a group of users is also shared with the company running the social networking site and so this information can be used for various things such as directed advertising or later on sold to other companies. Where the boundary lies between what to share and what to keep private is an individual decision and can be controlled at the time the information is shared.

There are, however, many mechanisms at work behind the scenes which track user behavior and collect sensitive private information without the direct consent of users at the time this information is gathered and stored. Thus, users cannot make individual and conscious decisions about such processes. Instead, users are required to know about these mechanisms and to actively take countermeasures in case they do not agree to this kind of private data collection and analysis. The following examples show which Web techniques are used by companies to collect information about user behavior and what users can do protect their privacy if they not agree with these methods. Unfortunately, these counter-measures require a fair amount of technical knowledge which the majority of Internet users do not have.

6.8.1 On-page Cookies

The oldest form of tracking users is a mechanism known as 'HTTP cookies'. When requesting a Web page, a Web server can, in addition to the requested Web page, return a cookie (a text string) to the Web browser. The Web browser stores the cookie in an internal database and returns it in future requests to the same Web server. This is one of the most popular mechanisms used today by Web sites to track a user through their Web pages. It is a useful tool during an order process, for example, which requires the user to go through several pages. Without cookies or similar technologies, the Web server could not correlate the information the user has supplied in the different pages. Cookies are also useful for online services such as Web-based e-mail, blogging tools or social net-working sites. Here, a Web-based environment is used which is spread over many different Web pages and it would not be practical to require the user to identify himself on each new page.

Cookies are also used by many companies to correlate user actions during site visits in order to specifically alter the content of a page based on previous behavior. Amazon is a good example. When browsing through the Amazon store the system tracks which items have been looked at and presents previous items as part of the new page. The system also tracks which articles the user has viewed during a session. Based on this knowledge, other users will later on get suggestions like 'other users who have viewed this item have also viewed the following item ...'. There are a number of privacy concerns with such behavior. First, the shop owner stores and uses information not only about which items a user has bought in the past, but also which items she has only looked at. Over time the shop owner can thus collect a huge amount of information about customers which can then be used to deduce preferences and to place specifically targeted ads on a Web page. Also, if this private information is sold to a third company or stolen, private data is revealed without the knowledge or consent of the user.

Second, cookies are stored in the Web browser's database for an indefinite amount of time by default and are stored even when the browser is closed. This is shown in Figure 6.17 for a cookie that has set its expiry time to the year 2035. By default, cookies are still present in a Web browser's database and are resent to the Web server when the user revisits the Web site at some later date. The advantage for the user is that he does not need to log in again as they are immediately recognized. This is useful in some instances, for example, for frequently used services such as Web-based e-mail or blogging, as the

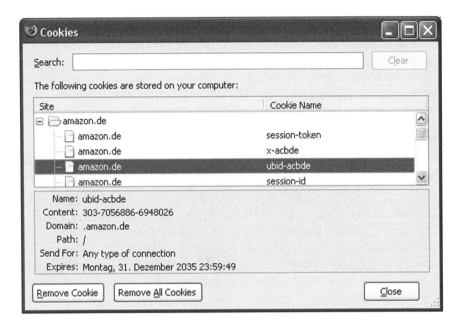

Figure 6.17   Cookie information stored in the browser.

user does not have to identify himself again. A potential downside side, however, is that user behavior can be tracked without a gap. Again, Amazon is a good example of a company that tracks user behavior across sessions. As soon as a user revisits the Amazon store, the user's browsing record is retrieved from a database and information and ads are presented based on the items the user has bought and viewed during their previous visits. The shop or service software logic in the background then also tracks the user's current visit and stores their browsing record for future analysis.

6.8.1.1 How can Cookie use be Controlled?

By default, Web browsers today accept cookies from all Web sites and keep them as long as requested by the Web servers. It is possible, however, to manually ban cookies from certain Web sites and to configure the browser to delete all cookies at the end of the session from the remaining sites. This ensures that long-term behavior tracking by Web sites is not possible unless the user identifies himself by logging in. For selected Web sites, where the user sees a benefit in remaining identifiable over a restart of the browser, the cookie can be exempted from deletion when the browser is closed. The author, for example, only allows about a dozen Web services to permanently store their cookie while the rest get deleted. To ensure that shopping Web sites do not use the information they have gathered about a user while browsing the shop, the user should manually delete the cookie before logging in or restart the browser before logging in to buy an item.

6.8.1.2 Applicability to the Mobile Web

Cookies are also used to preserve a session state for Web services adapted to mobile use. Unlike desktop Web browsers, however, mobile Web browsers do not yet give users the ability to define which cookies are acceptable, which are not, which should be deleted when the browser is closed and which are allowed to be stored between sessions. It remains to be seen when the configuration options of mobile browsers will be extended to allow this.

6.8.2 Inter-site Cookies

Cookies can also be used to track a user over several Web sites. All that is required for this is that Web sites include a picture or other element from the same third-party Web server. When this picture or page element is retrieved from the third-party server the cookie previously stored when visiting another site will be sent alongside the request. This can be used for marketing purposes, for example, to track how the user navigates through the Web. A precondition is, of course, that the Web sites the user visits use the same external marketing partner. As there are only a few large online marketing companies, inter-site user tracking and use of the gathered data for targeted marketing and other purposes is easily achieved.

Inter-site cookies can be controlled in the same way as described above for single-site cookies that are only used on one Web site, since from the Web browser point of view there is no difference. They are also applicable for mobile Web browsing where privacy issues today are greater than for standard Web browsing, as most mobile Web browsers do not allow the user to view and manage cookie information properly.

6.8.3 Flash Shared Objects

Interactive Web page content and animations based on Adobe's Flash environment [41] are found on many Web pages for various purposes today. These range from social Web site flash plug-ins to advertisement banners. The Flash environment allows applications to store information on the local hard drive. Information is not removed when a Web page is left or when the browser session ends. In effect, Flash animations/applications have a cookie-like mechanism which cannot be controlled via the browser. The same advantages and disadvantages discussed for HTTP cookies thus also apply for Web page Flash applications.

6.8.3.1 How to Control Shared Objects

Depending on the kind of Flash player installation, the environment can be configured to deny applications the right to store local information. Another possibility to prevent Flash applications from tracking user behavior is to use freely available Web browser plug-ins such as Adblock [42]. With such plug-ins the user can control which Flash applications and other page elements are loaded and executed.

6.8.3.2 Applicability to Mobile Web Surfing

Except for a few mobile Web browsers such as the Nokia NSeries browser, most mobile Web browsers do not support Flash today. As mobile devices get more powerful, however, it is likely that in the future Adobe's Flash environment will be used in more devices.

6.8.4 Site Information Sharing, Social Distribution

Automated information sharing between Web sites is a new way of automatically aggregating user supplied information from different Web pages, also referred to as social distribution. Facebook was the first social network to start such a program [43] for their users and advertising partners. The principle is as follows: Web site owners embed a JavaScript application, supplied by the social network platform, in their Web pages. This tiny application on the Web page then gathers information about the actions of the user on the Web page, in the background, and interacts with the social networking site to see if the user is a member of the social networking platform. If so, the user actions can be correlated with a user ID on the social network platform. At some point the Javascript application then asks the user if they would like to share this information (e.g. that they have bought a new electronic gadget in an online shop) with their friends on the social networking platform, e.g. via a dialog box. If they agree, the gathered information is then sent to the social networking site and included in the information update stream the user shares with their friends. Furthermore, the information can also be used for advertising purposes.

As long as the user has control over what kind of information is forwarded to the social network platform, such an application can be beneficial. Should the Javascript application, however, forward information to the social network site without explicit consent (e.g. because the dialog box is ignored), this approach becomes a privacy issue, even if the information is not shown to their friends but only stored in the service's database. If this occurs then the social network operator will become aware of a user's activities outside the social network without the user's consent. This allows the social network platform to further refine the user's social profile for their own purposes (e.g. for advertising).

6.8.4.1 How to Control Social Distribution

In cases where it is not clear what the social networking platform operator's policy is toward automatic transmission of private information from other Web sites it is advisable to block the communication between the social networking platform and external Web sites. This can be done with a Web browser plug-in such as BlockSite [44] but requires knowledge of the URL that is used by the JavaScript application. An alternative that requires less technical knowledge is using a different Web browser for social networking sites than for other online activities. In this case the social networking site is not able to identify the user when a third-party Web site sends it information, as it cannot include a cookie or other type of identification with the information that allows correlating the information with a certain user.

6.8.4.2 Applicability to Mobile Web Surfing

As JavaScript and cookies are supported by most modern mobile Web browsers today, automated sharing of private information is also a concern in the mobile Web.

6.8.5 Session Tracking

One of the big benefits of Blogs is that readers can discuss articles with the author and other readers by leaving a comment on the Web page. Some startups have now extended this principle to allow background discussions on any Web page. This works by installing a browser plug-in so that for every Web page the user visits an indication is provided of whether other people have left a comment for this page. While this functionality is quite interesting, the issue with this approach is that the plug-in queries a database in the background for every Web page the user visits, to find out if other people have left a 'virutal' comment. In effect the database in the network can thus record every step made by all users who have installed the plug-in. From a privacy point of view this is problematic as it potentially allows an external service to track a user's every move on the Web.

6.8.5.1 Countermeasures

The only effective countermeasure is not installing Web browser plug-ins which contain such a functionality, even if it is only a part of their overall functionality.

6.8.5.2 Applicability to Mobile Web Surfing

Most mobile Web browsers do not yet support plug-ins of this type and thus there is no threat to mobile Web surfing today.

6.9 Mobile Applications

Up to this point this chapter has looked at the technical side of the mobile Web and has given some examples of mobile services to help explain the different concepts. This section now takes a closer look at the different categories of mobile services and lists a number of applications per category. Note that the categories are not exhaustive and the applications are just taken as examples from a wide range of applications with similar functionalities. Also, it should be noted that, in the spirit of Web 2.0, these applications are evolving and the descriptions in this book only reflect their state at the time of publication.

With only a few exceptions, the applications discussed in this section are just emerging and are several years away from mass market adoption. Studies such as those described in [45] have shown that even the most successful services take at least four years to become mainstream. In the case of GSM, which is unarguably a successful service, mass market adoption took over six years from the date of its launch in 1992. An important early indication of whether a service will be successful in the long term is how popular the service is among early adopters. Popularity at an early stage, however, is not the only criterion that determines the fate of an application. During the evolution of an application or service toward mass market use, a viable business model has to be found and the application or service has to be adapted for a more general audience by either simplifying its use or by adding features not required by early adopters but which are indispensable for a less technically savvy user base. Also, new services sometimes break into the territory of already established services which will defend their position by, for example, lowering their prices and using other tactics to slow down the take-up of the new product. Circuit-switched voice telephony is a good example, where network operators lowered prices and used strategies like only selling DSL lines in combination with circuit-switched voice services to make it financially unattractive to use the VoIP telephony services of other companies. While this approach was not able to stop the eventual spread of VoIP telephony, it has nevertheless considerably slowed down the adoption of the service by a mass market audience for several years.

6.9.1 Web Browsing

Browsing the Web is already one of the most popular Internet applications on mobile devices today. Throughout this chapter a number of different mobile Web browsers have been introduced as well as their concepts. Some browsers, such as the Nokia Nseries Web browser or Apple's iPhone Web browser, load full Web pages, show the user an overview and allow users to zoom in to any place on the Web page. When zooming in the page is reformatted if required so the text and graphics fit within the boundaries of the smaller display. Most mobile Web browsers do not support JavaScript and Flash yet, with the Nokia Nseries Web browser being a notable exception. As mobile devices are becoming more powerful, such functionality is likely to be included in the majority of devices in the future. Consequently, AJAX-based Web applications and widgets will become more common on mobile devices, thus opening up the mobile device for rich Web-based applications.

The rise of more sophisticated browsers is likely to occur in combination with a continuing decline in prices for cellular Internet access. Until the right combination is in place and until fast data rates are ubiquitous, it will sometimes be preferable to only download mobile adapted versions of Web pages to mobile devices. Services such as Mowser, Taptu, Mippin, Google mobile and so on can be used to reformat Web pages for these purposes today and they are likely to remain important services for Internet access from mobile devices for the foreseeable future.

Other browsers such as Opera Mini are based on the concept of only downloading reduced versions of Web pages to mobile devices with small screens. Such browsers do not communicate with Web servers directly. Instead, all pages are loaded via a proxy server in the network that takes care of the downsizing process. This is done by reducing image resolution and by removing content from Web pages such as JavaScript and other elements which the browser on the mobile device cannot interpret.

While mobile Web browsers certainly are applications themselves, they are only the means to an end, that is they help to access Web-based information and services. Mobile devices are often used to access the Internet in situations where desktop-based browsing is not an option, for example, at the bus stop, in the metro or while queuing up at the supermarket checkout. In such cases the Web is more often used for accessing news Web sites and Blogs which the user has discovered and bookmarked previously, rather than for researching a specific topic. Search on mobile devices is often location- and context-dependent. Web sites such as Wikipedia are an ideal source of background information for ongoing discussions or to get background information while sightseeing. Web services such as Wapedia [46] search on the main Wikipedia site and reformat pages and images into a mobile friendly format. As links in the pages are kept it is also possible to easily find related information from other Wikipedia articles in the same way as with a desktop browser.

To search for other information, such as the address or phone number of a business while on the move, mobile device adapted versions of services such as telephone books and yellow pages exist. While these are quite useful for many people, they are of only limited use to international travelers as these services are country-specific. It is therefore often better to perform a general Web search with mobile search services from, for example, Mowser, Yahoo or Google, especially for business addresses and telephone numbers. As many businesses today have a Web site or are at least listed in local portals, it is usually possible to find this information quickly with a general search.

Automated location-based search is still difficult for Web-based applications since they do not have access to network or GPS information, as at this point in time there are no standardized local interfaces available on mobile devices for this purpose. Thus, Java-based or native programs are usually used for this purpose today as described later on.

6.9.1.1 Network Requirements

For accessing full Web pages a fast B3G network is required and the cost of the network service must be reasonably low as the amount of data transferred per page ranges from several tens of kilobytes to several hundreds of kilobytes. With content adaptation most pages can be compressed to a few kilobytes or a few tens of kilobytes. This way, accessing Web sites is possible over slower types of cellular networks (e.g. GPRS) and the cost of the network service is less of an issue due to the smaller amounts of data being downloaded.

6.9.1.2 Required Mobile Device Capabilities

Mobile Web browsers are embedded in almost all types of mobile phones and other mobile devices today, even at the low end of the spectrum. Freely available Java-based light-weight Web browsers such as Opera Mini, with a file size of around 100 kb, are compatible across a wide range of devices and often offer a better Web browsing experience than the built-in browser. High-end mobile devices with fast processors and high-screen resolutions often include browsers that are capable of processing and rendering Web pages designed for the desktop. The user can then zoom into particular areas of interest. With rising processor speeds, more memory and higher network speeds it is likely that the use of such mobile Web browsers will increase.

6.9.2 Audio

6.9.2.1 The Mobile Music Player

With the success of Apple's iPod and the iTunes software on the desktop, downloading music from the Internet has become mainstream over the last few years. The music, however, is not directly downloaded to the mobile device. Instead, iTunes or other software on the PC is used to download and pay for the content and the files are only copied afterwards to the mobile device. This approach is known as sideloading.

Mobile devices with Wi-Fi and fast cellular B3G network interfaces, on the other hand, can download content directly and do not require a PC. As there are already a wide range of devices with both kinds of network interfaces available today, it is likely that in the future the trend will be toward downloading audio content directly to the mobile device. By having both Wi-Fi and cellular network capabilities, music can be downloaded both at home and on the move while always using the best, that is the cheapest and the fastest, network available. Network operators offering both high-speed cellular Internet access and fixed-line DSL or cable Internet access coupled with Wi-Fi in the home or office can benefit from this by partnering with device, content and software companies such as Apple and Nokia. As people usually use their devices at home, most of the downloading can take place over Wi-Fi. By having both interfaces in a single device, usability can be improved as the user can interact with the service at home and on the go. For the network provider the extra network traffic on the cellular network is likely to be small from an overall perspective since most users are likely to use the service primarily at home and only sporadically while on the go.

6.9.2.2 Network Requirements

The best way to transfer music to a mobile device from the Internet is over Wi-Fi since it is cheap and fast. Cellular B3G networks are also capable of supporting music downloads at high speeds and in an acceptable time. Compared with Web surfing, however, downloading music tracks requires a significant amount of bandwidth. A flat rate data plan is required for this application.

6.9.2.3 Required Mobile Device Capabilities

Using a mobile device as a music player requires a large amount of storage capacity. Current high-end mobile devices have at least 8 Gb of flash memory. Some mobile devices also have small hard drives for even more storage space. In the future the amount of flash memory that can be put into a device will continue to rise, thus making hard disk-based solutions obsolete. Devices can then become smaller and power consumption can be reduced. For downloading music tracks directly to the mobile device, a combination of Wi-Fi and cellular network interfaces is required for a superior user experience.

6.9.2.4 Mobile Podcast and Audiobook Player

Another type of mobile audio application is mobile podcast reception. Podcasts are audio streams in MP3 or other formats produced by private and professional podcasters and radio stations. While already very popular with technology enthusiasts on PCs, podcasts are much better suited to being consumed on mobile devices and will be a strong competitior to broadcast radio in the future. Many radio stations have already discovered podcasts as an alternative distribution medium and now offer their shows not only over the air but also as podcasts. The big advantage over traditional broadcasting is that the user can listen to a particular show at any convenient time.

 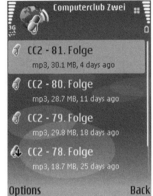 

Figure 6.18 Mobile podcast player. (Reproduced by Permission of Nokia, Keilalahdentie 2-4, FI-02150 Espoo, Finland.)

With a mobile podcast receiver program, podcasts are directly downloaded to the mobile device, instead of first being downloaded to a PC via a Web page or a feed reader and then sideloaded onto the mobile device. Thus, mobile podcast receiver programs act as a podcast feed reader, file downloader and MP3 player. Figure 6.18 shows such a podcast program on a Nokia NSeries phone. After installation or at any time thereafter, the user selects the podcast feeds he wants to subscribe to from a list. It is also possible to type in the feed's URL or to import feed descriptions from the PC. The user can then instruct the podcast receiver to update the feeds at any time and can select the podcast files of interest. The podcast receiver program then downloads the selected podcast files and once the download is finished the user can consume the content by clicking on the corresponding podcast show icon.

A similar application provides audiobooks via a mobile device. Audible, for example, has developed a player for Symbian S60 phones that is capable of downloading audio-books directly from their store on the Web. There are also freeware players available like the one from Nokia [47], for listening to freely available audio books. LibriVox [48], for example, is a good source for a wide variety of free audio books. While free audiobooks can be consumed with an ordinary MP3 player, dedicated audiobook applications offer specific functionality such as:

- audio compression codecs which achieve much higher compression gains than codecs optimized for music compression;
- browsing books by chapters;
- users can set bookmarks so they can return to specific parts of the book later.

6.9.2.5 Network Requirements

Network requirements are mostly the same as for the mobile music player application above. The consumption of podcasts, however, is done actively; that is, the user concentrates on the content of the podcast or audiobook. This is quite different from music, which is often consumed in the background while performing other activities. As a consequence, most users will only download a limited number of podcasts or audiobooks compared with music, since the goal is not to build a large audio database but to have specific podcasts and audiobooks available on the mobile device for active consumption. Mobile podcast and audiobook downloads thus do not require as much bandwidth per user as mobile music downloads.

6.9.2.6 Required Mobile Device Capabilities

Usually users only listen to a podcast once. Therefore, the amount of storage required on the mobile device is less than for storing music since podcasts will usually be deleted once they have been listened to.

6.9.2.7 Audio Streaming

Another audio application is audio streaming or Internet radio, which is similar to radio broadcasting. An audio streaming client on the mobile device lets the user select from a wide range of online radio stations. When the user selects a station, a connection to a streaming server in the network is established and the channel is then streamed in real time to the device. Data rates of audio streams range from 32 to 192 kbit/s for high quality audio. Unlike the previous audio applications, the content of the channel is consumed immediately and no information is stored on the mobile device. The advantage over the traditional radio broadcast approach is that the user can listen to their favorite radio station independently of their current location. Radio stations or other broadcasters are thus no longer geographically limited but can reach a worldwide audience. Figure 6.19 shows an audio streaming application on a Nokia Internet tablet.

Figure 6.19   Internet radio application. (Reproduced by Permission of Nokia, Keilalahdentie 2-4, FI-02150 Espoo, Finland.)

6.9.2.8 Network Requirements

From a user point of view, listening to audio broadcasts over cellular networks requires at least a 3G network to achieve the required data rates. Initially with UMTS, audio streaming was a particular issue since dedicated resources on the air interface for a user could not be tailored effectively around the required data rates. This resulted in a high waste of transmission resources. With 3.5G (e.g. HSDPA), the approach on the air interface shifted to a common shared channel, which resulted in much less overhead for such applications. Even though data rates for audio streaming are low compared with the capacity of a B3G base station, networks can get saturated quickly if many people use such services, since users often tend to listen for long periods of time. Compared with Web browsing, audio streaming requires a significantly higher overall bandwidth since the audio stream is continuous. Mobile devices with both a cellular and a Wi-Fi interface are ideal devices for this application since the overall load on the cellular network is significantly reduced when people use the application over Wi-Fi while at home. To support mobility for audio streaming the network coverage of cellular B3G networks must be excellent, since gaps in the coverage quickly lead to an interruption of the audio stream.

6.9.2.9 Required Mobile Device Capabilities

Mobile devices should have at least a Wi-Fi or a 3.5G network interface or better yet a combination of both. As audio streams are consumed in real time, the amount of storage capacity required for the service is low.

6.9.3 Media Sharing

Media sharing is one of the most interesting future applications for connected mobile devices due to the fact that mobile devices are always carried and can be used to capture thoughts, ideas, pictures, videos and so on at the point of inspiration. With mobile devices, the user participation and contribution aspect of Web 2.0 is significantly enhanced as it becomes much easier to share information. Figure 6.20 shows Shozu,

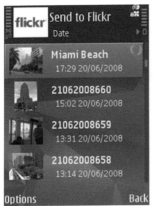

Figure 6.20   Media sharing with Shozu. (Reproduced by Permission of Shozu.)

a picture and video sharing program on a Nokia N93 [22]. While not actively used it runs in the background and detects when the user takes a pictures or a video. It then automatically attempts to get the current location by communicating with a GPS receiver or, where it is already built into the mobile device, from the on-board GPS chip. The user can then send the picture or video to a sharing Web site such as Flickr, YouTube, Facebook, Blip.tv, Picasa and so on right away. It is also possible to add additional information such as text, a more detailed description and tags, so other people using the sharing Web site can more easily discover the picture or video. If GPS coordinates can be retrieved from the GPS unit, latitude and longitude are added as additional tags, as shown in the middle and right screen shot in Figure 6.20. Some Web-based sharing services such as Flickr detect these special tags and automatically place the picture on a map. The map shows the location of where the picture was taken and additionally includes markers for other pictures the user or other people have taken in the surrounding area.

Text-based mobile blogging is another media-sharing application. One way to create a new Blog entry from a mobile device is by writing the Blog entry with a mobile e-mail program and sending the mail to a special address of the Blog hosting service. The Blog hosting service then correlates the e-mail address to the Blog of the user and creates a new Blog entry. Mobile devices with larger screens can even use the standard Web-based Blog user interface to create a new Blog entry. While blogging by e-mail requires little bandwidth, using the Web-based interface requires a fast network connection. There are several ways to input the text for a Blog entry. With small devices such as mobile phones the numeric keypad is used, often in combination with typing assistance such as T9. Some mobile phones have a miniaturized version of a standard keyboard that some people find quite helpful for faster typing. There are also lightweight foldable keyboards that connect to mobile devices via Bluetooth. These offer the fastest way of inputting text, not only for Blog entries but also for e-mails, SMS and other applications that require text input.

6.9.3.1 Network Requirements

Picture uploading requires a fast network connection since picture sizes range from 300 kb to several megabytes with increasing camera resolutions. Video uploading demands even more capacity as even relatively low-resolution 320×200 pixel videos encoded with an MP4 algorithm at 15 frames per second generate about 3 Mb of data per minute. Fast B3G networks with higher uplink speeds (e.g. HSUPA, WiMAX, LTE) are very beneficial for this application. Mobile blogging on the other hand is less bandwidth-intensive, especially when done via e-mail and if no pictures are included in the Blog entry. When pictures are included or if devices are used with larger screens, bandwidth requirements increase.

6.9.3.2 Required Mobile Device Capabilities

Except for text-based mobile blogging, media sharing requires mobile devices with a reasonably good camera function and enough storage space to be a replacement for a standalone digital camera in many situations. As prices for Wi-Fi and B3G network chips are declining, it is quite possible that such chips will be included in mobile devices such as cameras and video cameras in the future.

6.9.4 Video and TV

While the media-sharing category above contained applications to share content created with a mobile device with other people via a Web service, this category includes applications for consuming content. This includes user-generated content as well as content produced by broadcasting companies.

For several years, mobile network operators have offered mobile TV to their 3G subscribers. First offers were based on IP unicast, which means that each subscriber got an individual copy of a live broadcast stream or video. This is quite disadvantageous, especially for distribution of a live television program to a mobile device as the network load is much higher compared with broadcasting where all subscribers receive the same signal. As a result, a number of different mobile TV broadcasting systems have emerged in recent years. In Europe, DVB-H (Digital Video Broadcasting – Handheld) has emerged as the most popular system. It is directly derived from the terrestrial digital television standard DVB-T, which has succeeded analog television in many countries. DVB-T receivers with special software are even able to decode DVB-H signals. For receiving a DVB-H signal, mobile devices require DVB-H receiver hardware as the signal is not carried over B3G networks but over a separate broadcast network. This requires extra hardware in the mobile device but allows more users to receive a stream simultaneously compared with a solution in which each subscriber receives an individual data stream via a B3G network.

For video services such as mobile YouTube, for example, broadcasting is not an option due to the personalized nature of the service. With increasing screen resolutions, mobile video streaming requires a significant amount of bandwidth that far exceeds the bandwidth required for Web browsing. While one hour of mobile Web browsing, for example, results in a data volume in the order of a few megabytes, a single YouTube video with a length of 5 min requires the same amount of data to be transferred. Figure 6.21 shows video streaming on a Nokia Internet tablet.

Another interesting personal video application is streaming videos and music from a home storage system to a mobile device. Sling Media for example has developed the Sling box for this purpose [25]. The Sling box is connected to the TV aerial and to the home network and enables the user to remotely watch TV channels either with a notebook or from a mobile device via the Internet. The bandwidth of the signal is mostly limited by the uplink capacity of the Internet connection at home and the requirements of the receiving device. The resolution and thus the bandwidth of the signal required for receiving the stream on a PC is of course significantly higher than that for receiving the stream on a mobile device. The approach of a home gateway streaming content to a remote device is also useful to access recorded video content stored on a media server at home. As the service is private, broadcasting the stream is not feasible and bandwidth requirements for video streaming are much higher than for Web surfing, as already described above.

6.9.4.1 Network Requirements

Streaming video and other multimedia content is the main reason for rising capacity requirements in fixed-line and wireless networks, as more and more people use such services. For wireless networks, streaming applications are a particular challenge. As

Figure 6.21   YouTube Video streaming to a Nokia Internet tablet. (Reproduced by Permission of Nokia, Keilalahdentie 2-4, FI-02150 Espoo, Finland.)

described in Chapter 3, the main difference between fixed and wireless networks is the location of the bottleneck. In fixed-line networks, the last mile connection with a high capacity is dedicated to a single subscriber with DSL or to a small group of subscribers in the case of cable. In these networks the bottleneck is mainly the connection between the access concentrator and the core network of the operator. More capacity can be added easily by using additional fiber links, which are usually already in place. In wireless networks, on the other hand, many subscribers have to share the capacity of a single base station, that is the last mile is shared and not dedicated as in fixed-line networks. In addition, the backhaul connection between the base station and the core network is similarly limited. Adding new capacity is thus much more difficult than in fixed line networks. When cellular traffic increases by an order of a magnitude, additional base stations are required since an operator only has a limited frequency band available that can be used by a base station. By adding base stations, the coverage area per base station and thus the number of subscribers per base station decreases and the overall capacity increases. Network operators with both fixed and wireless networks have a major advantage over wireless only network operators as it is easy for them to deploy femtocells (cf. Chapter 3) to reduce the load on the cellular network or encourage users to use Wi-Fi networks at home by selling attractive bundles of cellular and DSL/cable access with dual mode devices.

6.9.4.2 Required Mobile Device Capabilities

Video streaming is quite a demanding application for mobile devices in a number of ways. First, displays must be large enough for video consumption. Second, mobile devices need a cellular B3G network interface, integrated Wi-Fi or preferably both interfaces. Another important aspect is power consumption as receiving and decoding video streams is a computationally intensive task and thus drains batteries quickly. Graphics chips with built-in decoding functionality help to reduce power requirements and to extend the operation time.

6.9.5 Voice and Video Telephony

The main application mobile networks were built for was voice calls. It might thus be a bit surprising to discuss voice telephony in this chapter. With the evolution of mobile networks to 3G and B3G, voice telephony is undergoing a significant change as well. While in 2G and 3G systems circuit-switched voice and video telephony were firmly integrated into the network, B3G systems have enough capacity to migrate these circuit-switched services to the IP domain. In practice, the migration to IP will bring about a number of important changes. As voice and video telephony are no longer integrated into the network by design but just one of many applications using IP packets, the door is open for third-party companies to offer voice telephony services as well. In Chapter 4, this topic was discussed in more detail. From the user perspective, porting voice telephony to the IP domain must first of all be transparent. If visible at all, it should only be because of additional functionality that cannot be offered by traditional circuit-switched telephony. Integrated presence and instant messaging are such differentiators.

Today, there are already a number of mobile phones available with both Wi-Fi and B3G interfaces and a well-integrated SIP protocol stack for Voice over IP (cf. Chapter 4). With integrated SIP telephony, which works both over Wi-Fi and B3G networks such as HSDPA, WiMAX and LTE, the user can call a contact from the address book either with the circuit-switched telephony functionality of the device or via VoIP. In the future such dual mode phones will become more common both in the sense of circuit and VoIP as well as Wi-Fi and B3G. Today, however, most people use the VoIP functionality of their mobile devices mainly in combination with their Wi-Fi network at home to replace a fixed line or cordless phone and use circuit-switched cellular voice telephony outside.

Proprietary VoIP solutions such as Skype are also in the process of being ported to mobile devices. Additionally, there are newcomers in the mobile VoIP domain such as Fring [49] who combine the voice and instant messaging capabilities of many proprietary solutions into their mobile device client. For many people this is quite appealing since it has become common to use more than a single Instant Messaging and VoIP solution on the desktop today. This is much more difficult on mobile devices due to processing power constraints and limited memory capacity.

The IP media subsystem and applications using the IMS framework are by design not Web 2.0 services (cf. Chapter 4). However, they are likely to play an important role in the migration from circuit-switched voice to VoIP as well. With the IMS, network operators have the possibility to continue offering voice and supplementary voice centric services over their wireless broadband networks. As the IMS framework is well integrated into the network environment, it enables network operators to offer voice-centric services over network technology boundaries. The same IMS account can, for example, be used in the B3G network and in Wi-Fi networks at home. In addition, the IMS-based Voice Call Continuity service can hand over an ongoing voice call between an IP network and a circuit-switched wireless network if the user roams out of the B3G or Wi-Fi coverage area.

Such cross-network transitions are difficult to achieve with nonoperator voice solutions, as Internet companies have little experience and control over wireless networks.

6.9.5.1 Network Requirements

On the one hand, network vendors are estimating that VoIP capacity of B3G networks is equal to or even surpasses the capabilities of 2G circuit-switched networks. Consequently, as optimized B3G networks are rolled out, the reluctance of network operators to offer VoIP services will certainly decrease. On the other hand, it has to be taken into account that video calling from notebooks and other devices with higher screen resolutions requires a bandwidth which is at least an order of magnitude higher than a standard voice call. A Skype video call, for example, uses a bandwidth of 400–700 kbit/s for video calls if the network connection offers this bandwidth in the uplink and/or the downlink. Video calls from smaller mobile devices will certainly be less demanding but will still exceed pure voice telephony bandwidth requirements. While current wireless networks are optimized for voice telephony, ensuring the same quality of experience for VoIP calls, especially during handovers from one cell to another, is much more difficult. This is due to the fact that IP telephony applications are only interfacing with the application layer and can no longer directly influence the network layer below. It will thus take considerable time before both VoIP applications and networks offer the same quality of service for mobile IP voice telephony while the user is moving as they do for circuit-switched mobile telephony today. Another issue for voice telephony is weak indoor coverage of B3G networks as they are usually deployed on higher frequency bands than 2G networks. In the future it is thus important to also offer B3G network coverage in the 900 MHz band (850 MHz band in the USA) to improve the situation.

6.9.5.2 Required Mobile Device Capabilities

Circuit-switched wireless voice telephony is very well understood today and deeply embedded in the hardware. Wireless VoIP on the other hand is still only available in high-end devices and requires significant processing resources and a Wi-Fi or B3G network interface. It is thus likely that it will take considerable time before wireless VoIP telephony will be as seamless and integrated as circuit-switched voice technology today. Compared with cordless phones, using digital standards such as DECT (Digital Enhanced Cordless Telecommunications), Wi-Fi VoIP phones have a shorter range. This is due to the fact that, while the output power of a Wi-Fi access point is similar to that of a DECT base station, it has to be spread over a much wider band. This issue might be reduced in the future with improvements in antenna technology or multiple antennas in devices and enhancements in the Wi-Fi standard such as beamforming (cf. Chapter 2). In addition, Wi-Fi repeaters can increase the range of a wireless network. However, it is unlikely that consumers will be willing to install repeaters.

6.9.6 Widgets

Widgets are small Web and AJAX-based programs which are not executed in a Web browser but in a widget environment on the desktop of a PC. Examples of desktop widgets are weather forecast widgets, feed reader widgets that show the user the latest Blog entries of feeds that have been subscribed to, up-to-date stock quotes, the latest news from news agency Web sites, Internet radio players, system information, sticky notes, translation and so on. The advantage over full applications is that they become part of the background screen and that they run all the time. Since they are programmed using JavaScript, in a similar way to Web browser-based JavaScript applications, creating a new widget is much simpler than programming a normal application. Furthermore, the developer community is much larger.

In the mobile world widget-like services can be found at a number of places. Nokia has pioneered widgets as a means of rapid development of small applications for their S60 platform. They are executed in a widget environment and are programmed in JavaScript in the same way as Web browser-based JavaScript applications and desktop widgets. Since creating a mobile widget is so similar to developing a desktop widget or a JavaScript Web application, they open a door into the mobile world for programmers who have so far only developed applications for the desktop. Widgets are also interesting for activating the idle screen of a mobile device to show up-to-date information retrieved from a Web site or Blog. Another interesting place for mobile widgets is the screen saver, to display the latest news, RSS feed items, pictures retrieved from a picture sharing site based on tags the user is interested in and so on. Opera is one of the companies active in this space [50] and was the first company to implement active screen savers in mobile phones in cooperation with KDDI, a mobile network operator in Japan.

6.9.6.1 Network Requirements

As widgets are mainly used to show small snippets of information, the amount of bandwidth required by mobile devices running widgets is small.

6.9.6.2 Required mobile device capabilities

At the application layer, mobile widgets require a Web browser-like environment and a JavaScript interpreter to run the widget. Today's medium- to high-end phones are thus capable of supporting this type of application. As widgets, especially on the idle screen and in the screen saver, will run most of the time, battery consumption of executing script code requires special attention. Furthermore, most widgets are only useful when they can retrieve information from Web sites. This requires a permanent logical connection to the network. Most of the time the connection is only logical and data is only exchanged when the widget requests an update. Nevertheless, it can be observed in practice that even just maintaining a logical connection to the network drains batteries in mobile devices faster than being in idle mode waiting for an incoming voice call or short message page. Mobile devices and network protocols in the future thus have to be further enhanced to support always-on functionality that has minimal impact on power consumption while no data is being exchanged. For details on this topic see the sections on power saving modes of the different B3G technologies.

6.9.7 Social Media

Social networking Web sites such as Myspace and Facebook have expanded rapidly in the desktop world and are hugely popular among the younger generation for communicating and exchanging information with friends online. Some social networking Web sites also have a mobile adapted version of their service. A trend seen on some of these platforms is that information exchange between people is becoming more and more real-time. Thus, accessing the social networking site from a mobile device will become more important or even integral to the service, especially when mobile Internet access becomes more commonplace. It can also be imagined that, once social networking sites discover the possibilities of location information, supplied by mobile devices or networks, services will be created around this information, for example, informing people about friends who are close to their current location or finding new contacts based on location information.

Another example of social media is MyStrands [51], which combines social recommendation with discovery of new music titles and social networking. Available as a music player, downloadable application and recommender service, on both the desktop and mobile devices, the application can recommend to users other titles of the same genre selected by other people who have previously listened to the same track. The recommendation can then be used to find other music tracks or to find new friends with similar music tastes. The service also reports to users which music their friends listen to. Web page and Blog plugins are available to share information about the music one listens to with a wider audience. On a mobile device the service can also be used to influence the choice of music played at parties and to exchange messages with other people at the event. The service thus integrates both the desktop and the mobile world and is an interesting example of how the fixed-line and mobile Internet can complement each other.

6.9.7.1 Network Requirements

As long as social media applications do not stream media, the impact on network resources is low.

6.9.7.2 Required Mobile Device Capabilities

This is very application-dependent. For mobile extensions of social networking platforms, only a mobile Web browser is required. In the future, however, such applications will benefit from location information made available to Web browser-based applications which in turn requires a more feature-rich mobile Web browser with built-in AJAX capabilities.

6.9.8 Microblogging

Microblogging mixes a number of existing services such as blogging and instant messaging into a new product while at the same time radically simplifying and reducing the feature set. Microblogging applications such as Jaiku [52], Twitter [53] and Pownce [54] let users publish new thoughts, ideas, interesting links and so on to their friends via a simple text-based interface. Microblogging applications usually limit messages to 140 characters, to keep them short and easily readable at a glance. By limiting messages to this size, it is possible to easily create and receive new messages on mobile devices. Microblogging is a type of application that integrates the mobile and the desktop PC world. A user can create a new message via a Web-based interface on the PC or via SMS or a mobilized version of the Web-based interface on a mobile device. The message is then immediately forwarded to friends, also referred to as followers, who have registered to receive updates from certain users. Messages are forwarded via the service's Web-based interface to desktop PCs or by SMS or the mobilized version of the Web interface to mobile devices. Figure 6.22 shows how the Jaiku Web interface looks on the desktop and on a mobile device.

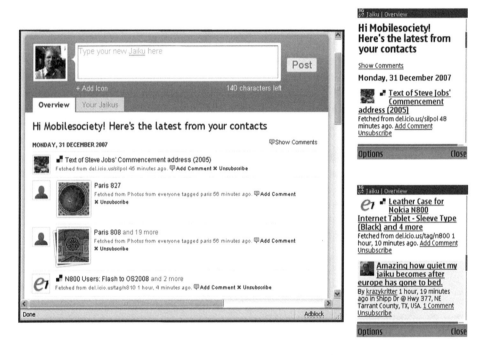

Figure 6.22 Jaiku on a desktop and a mobile device browser. (Reproduced by Permission of Nokia, Keilalahdentie 2-4, FI-02150 Espoo, Finland.)

Others can respond to messages either directly to the originator or openly to all others also subscribing to the originator's messages. Jaiku adds a second dimension to this process by grouping responses to an original message into a thread. It thus includes some elements of instant messaging as conversations can develop around a topic in a thread. Unlike instant messaging, however, the developing conversations are not between two people but between many. The disadvantage of allowing threads is that it gets difficult to follow these conversations via SMS. This limits the usefulness of the service to people who mostly use it on the desktop and to people with a mobile Internet connection.

Microblogging can also be used to quickly publish thoughts, ideas, links and so on to a wider audience, that is to people not part of the microblogging service. This can be done, for example, via plugins for Web pages and Blogs. Figure 6.23 shows such a plugin on this book's Blog. As messages can be composed quickly, at any time and at any place via a mobile device, they are an interesting complement to the much longer Blog posts, for interacting or sharing information with readers.

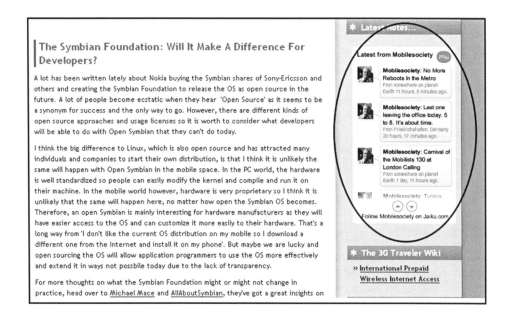

Figure 6.23   A Jaiku microblog plugin on a Web page.

Also, some microblogging services such as Jaiku can be used as an aggregator for RSS feeds which document the Web activities of a user with other services. Thus, when the user, for example, adds a new picture on Flickr, a new bookmark on Del.icio.us or a new Blog entry on their Blog, the service detects these activities and automatically creates a message in the name of the user. New pictures are shown to others as thumbnails that lead to the user's Flickr page when clicked on. New bookmarks on Del.icio.us are shown as links with a description and new Blog entries are shown with their titles in a new message.

Some microblogging services also allow users to form groups, also referred to as channels, in which certain topics can be discussed. A message posted to a group thus reaches members of the service who have not directly subscribed to the message flow of an individual user. Such groups can also be configured to use RSS feeds as inputs to serve as an aggregator for new information posted about a topic on different Web services. The Maemo (Internet tablet) group on Jaiku, for example, aggregates content it finds on Flickr, YouTube, Del.ico.us, several Blogs of team members and a number of news Web sites. News items are found because those services offer tag-based RSS feeds, that is they automatically generate feeds based on keywords which are attached to Blog entries, videos, pictures and so on.

6.9.8.1 Network Requirements

As microblogging services use both a Web-based interface and SMS for message delivery, the overall data traffic of such services is low. Today, SMS is an ideal interface for microblogging since users can reach their audience very quickly. Also, users can be informed quickly about updates from others. From a technical point of view this is much simpler than writing a dedicated client application for a variety of mobile devices that receive messages via an IP connection and alert the user. Another reason for using SMS today is that a much greater audience can be reached this way than by an approach that requires the mobile to be connected to the Internet all the time for a Web-based service or a dedicated application installed on the device. As mobile Internet connectivity becomes more popular, however, dedicated client applications and browser-based services can show off their strengths. In this way, the service can be made interactive and more sophisticated user interface functionalities can be integrated, such as opening threads and the possibility to open embedded links in a browser. In the short and medium term, network operators will benefit from increased SMS revenue. In addition, microblogging increases awareness of the possibilities offered by connected mobile devices and thus might be a catalyst for many people to consider going online with their mobile devices.

6.9.8.2 Required Mobile Device Capabilities

Even entry level mobile phones are capable of SMS-based microblogging. For Web-based microblogging, only a simple mobile Web browser is required today as current Web-based interfaces do not require Ajax or other sophisticated functionality.

6.9.9 Location

Today, there is already a wide range of mobile applications using location information provided by mobile devices. Navigation applications are one of the most widely known in this category. Some years ago, navigation applications became popular with affordable dedicated mobile devices, which were usually not connected to the network. Maps and other information such as points of interest had to be stored on the device. More recently, high-end mobile phones with built-in GPS functionality can now also be used for navigation. In addition, such applications can be used on mobile phones for street navigation, for example while on a sightseeing trip through a city. No additional equipment is required since the mobile phone is usually carried anyway. Maps are either pre-installed or can be downloaded in real time over the cellular network. The advantage of being connected is that it is possible to access additional information such as:

- up-to-date information on traffic conditions for car navigation, to dynamically modify the route;
- detailed information about nearby points of interest, including pictures and videos that require too much memory to be stored on the device;

- location information entered during trip planning via the PC;
- additional information other users have provided for locations, for example, restaurant reviews or pictures taken at nearby places by other users.

Examples of navigation services on connected mobile devices implementing various features described above are Nokia Smart2Go, as shown in Figure 6.24 [55], Google maps for mobiles [56] and products from traditional navigation companies such as TomTom [57].

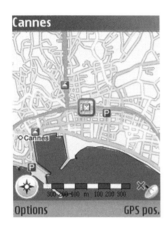 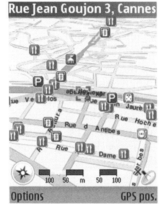 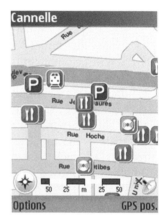

Figure 6.24   Nokia Smart2Go navigation with points of interest on an S60 mobile phone. (Reproduced by Permission of Nokia, Keilalahdentie 2-4, FI-02150 Espoo, Finland.)

Other popular mobile applications exploiting location information are fleet tracking and steering of transportation companies. Trucks with an embedded onboard computer equipped with a GPS device and a cellular modem keep dispatchers up to date on their location so last-minute orders can be better coordinated. Taxi companies have also started using GPS and cellular data connectivity instead of voice-based methods to contact the nearest taxi for a request and to send information about where to pick up the next customer.

Most of the programs described above use location information provided by satellites of the GPS. However, there are two other possibilities for devices which do not have built-in GPS functionality to report their location:

- Cell identities – each transceiver of a cellular base station has a unique identity which is broadcast to all mobile devices nearby. A database in the network then maps the cell identity to a geographical location. While network operators know the locations of their base stations, Internet-based services need to build cell identity databases from scratch. This is difficult to do in practice as it requires users to report the location of base stations. As embedded GPS chips become more commonplace, a possible solution could be that applications used mainly for another purpose, such as navigation, collect cell identity information and send it back to a service in the network together with the GPS location. The information is then stored in a network database and can be used for non-GPS-capable devices. An advantage of using cell identities for positioning is that this positioning method also works in buildings as long as the device can receive the broadcast signal of the network.

- Wi-Fi Access Point hardware addresses – the cell identity approach described above also works in combination with MAC hardware addresses of the Wi-Fi chip of wireless LAN access points. The advantages and disadvantages are the same. In most cities in developed countries there are only a few places left where Wi-Fi access point transmissions cannot be received. The advantage over using cell identities is that locating the user is more precise because Wi-Fi access points have a much shorter range. Skyhook wireless [58] is a company specialized in linking WLAN access point MAC addresses to locations. For initial location information the company drives through cities with GPS/Wi-Fi equipment to find and record the position of Wireless LAN access points. Devices and applications using the Skyhook database (e.g. Google Maps on Apple's iPhone) report the found WLAN Access Point IDs to the service in the network which then calculates the position of the devices and returns the information to the application. The query from a mobile device is in addition used by the service to update its database as the application might report IDs not yet known to the database or fail to report IDs which the database knows should be in the neighborhood. IDs can thus be added and removed from the database.

Some applications combine GPS with Cell-ID or Wi-Fi-Access Point ID positioning methods. In this way, a first position can be available within only a few seconds after the request to a Cell-ID or Wi-Fi-Access Point ID database in the network. This helps to bridge the time until the GPS chip provides a more precise location.

The Internet connectivity required to access the databases mentioned before can also be used to download satellite location information from a network-based GPS database to speed up the time to the first GPS fix. This method is referred to as Assisted-GPS and enables GPS chips to calculate the location of the user within 10–15 s. Nokia, for example, provides such a network operator independent database for the use with its devices with built-in GPS chips.

6.9.10 Shopping

As discussed above, a connected mobile device is ideal for online shopping as it is usually owned and used by a single person and, in addition, it can be identified by the network. Network operators have experimented with shopping applications for some time on their portals, where users can buy ring tones, games, news subscription and other things by simply clicking on a confirmation link or button. Buying goods this way is much simpler than, for example, typing in a credit card number, as it is implicit and quick. The approach, however, has not yet developed much further as operators are not yet offering a convenient and affordable way for external stores to use this payment method. Owing to its simplicity, however, such a payment method holds a lot of potential and might become more widespread once better and cheaper interfaces for external companies are available. Companies most likely to benefit from such a payment method are national rather than international companies, since the number of mobile network operators with whom an agreement would have to be reached on an international level is significant and thus this does not hold a lot of appeal.

Apart from goods and services offered on network operator portals, there are some Internet companies with interesting mobile extensions of their services. A good example is eBay, with a mobile version of their auctioning platform. Here, mobile access makes particular sense since auctions often end at times when the user has no access to a desktop PC. It is not uncommon towards the end of an auction for bids to rise significantly. Being informed about higher bids via a mobile device and being able to react immediately, independently of the current location, means the mobile Web portal is very appealing for eBay users. In practice it can thus be observed that many mobile network operators include eBay as one of the services in their advertisements for mobile Internet access offers.

6.9.11 Mobile Web Servers

All applications discussed above enable the user of a mobile device to either consume information or to upload user-generated content to a Web service. The concept of mobile Web servers reverses this concept and enables the device to directly deliver content to other users and devices without a Web server or a service in the network. Nokia was the first company to explore this concept on mobile phones [59] and released a port of the Apache Web server for their S60 mobile phone operating system. Applications delivered with the mobile Web server include the following:

- Camera application – permitted users can invoke an application that takes a picture with the phone's camera which is then returned to the Web browser.
- Share photo albums – pictures stored on the phone can be assigned to photo albums that users can access once the owner of the phone gives permission. Access permissions can be used to create individual photo albums for friends, family members, business partners and so on.
- Contact list browser – instead of searching for a contact on the mobile phone, permitted users can search addresses and phone numbers stored on the phone via a Web browser. This is probably an application that the owner wants to restrict to himself. By default all applications are restricted to the owner and thus access to any sensitive information is not given to others without express permission.
- Calendar spplication – the phone's calendar can be viewed and new entries can be created via the Web browser on a desktop PC. Figure 6.25 shows how the application presents the monthly view.
- Send SMS messages – gone are the days of fiddling around with the keys on the phone when writing an SMS. With the mobile Web server, short messages can be written in the Web browser whether the phone is next to the notebook or 5000 km away.
- WebDAV – with the http-based Web-based Distributed Authoring and Versioning standard the phone can be used as a network storage device to share files and folders. Files can be copied to and from the WebDAV drive, renamed and deleted. As with all

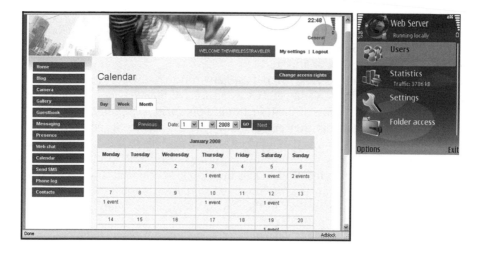

Figure 6.25 Calendar information stored on a mobile device served via a desktop browser. (Reproduced by Permission of Nokia, Keilalahdentie 2-4, FI-02150 Espoo, Finland.)

other server-side applications the owner of the phone can assign access rights to individual users. For Windows XP no additional software is required.

• Platform for experiments – the mobile Web server is open for additional server-side modules and a Python interpreter is also included. This makes it very easy for external programmers to use the embedded Web server for their own projects.

Mobile Web server applications can be used over the Internet and also locally over Wi-Fi. Use via the Internet makes most sense for offering services to other people or for access to the device from a PC connected to the Internet, while no Wi-Fi network is available to which both devices are connected. When the service is used over the Internet a gateway server is required since it cannot be predicted which IP address is assigned to the mobile device when it connects to the network. Furthermore, many mobile network operators perform network address translation at their Internet gateways and only assign private IP addresses to mobile devices that cannot be directly reached from the Internet. This is similar to DSL or cable routers at home that also assign local IP addresses to devices because the Internet provider only assigns a single IP address per line.

If both the mobile device and the PC, or desktop computer, are in the same Wi-Fi network, the mobile Web server can be accessed directly without the need for a gateway. By communicating directly, no sensitive information leaves the local network and response times and transmission speeds are quicker. This is especially important when using the mobile device as a WebDAV network storage device.

6.9.11.1 Network Requirements

In the future, mobile Web server applications could be very interesting for network operators with integrated cellular B3G and DSL/cable Wi-Fi offerings. Such operators can offer converged services which can be used both at home and while on the move, without burdening the macro cell layer with local data traffic. With an intelligent network selection scheme on the mobile device all traffic to and from the device from people outside the local network will also be handled over Wi-Fi, and not over the cellular network, while the device is within reach of a configured Wi-Fi home or office network.

6.9.11.2 Required Mobile Device Capabilities

In an ideal scenario the mobile Web server is automatically started when the device is powered on, runs continuously in the background and is capable of detecting which networks are currently available and dynamically attaches to the most suitable one. Running a full Web server in the background with the ability to execute applications written in a script language requires significant processing power, memory space and storage capacity on the mobile device. Furthermore, a multitasking operating system is required as the Web server needs to run in the background. Also, being continuously attached to a network requires that the Web server, the operating system and the device hardware are optimized to keep power consumption as low as possible while connected but idle.

References

1. Ahonen, T. (February 2007) Mobile: the 7th mass media is to internet like TV is to radio, Communities Dominate Brands Blog, http://communities-dominate.blogs.com/brands/2007/02/mobile_the_7th_.html.
2. Flickr (2008) www.flickr.com.
3. YouTube (2008) www.youtube.com.
4. Del.Icio.Us (2008) http://del.icio.us/.
5. Digg (2008) Digg/all news, videos & images, http://www.digg.com.
6. Facebook (2008) http://www.facebook.com/.
7. Myspace (2008) http://www.myspace.com/.
8. LinkedIn (2008) Relationships matter, http://www.linkedin.com/.
9. Linden Lab (2008) Makers of Second Life & virtual world platform Second Life Grid, http://lindenlab.com/.
10. Anderson, C. (October 2004) The long tail, Wired Magazine, http://www.wired.com/wired/archive/12.10/tail.html.
11. Iskold, A. (2007) There's no money in the long tail of the Blogosphere, ReadWriteWeb, http://www.read-writeweb.com/archives/blogosphere_long_tail.php.
12. O'Reily, T. (September 2005) What is Web 2.0 – design patterns and business models for the next generation of software, http://www.oreillynet.com/pub/a/oreilly/tim/news/2005/09/30/what-is-web-20.html?page = 1.
13. The RSS Advisory Board (October 2007) RSS 2.0 specification, http://www.rssboard.org/rss-specification.
14. Nottingham, M. and Sayre, R. (2005) The Atom Syndication Format, RFC 4287, The Internet Society, http://tools.ietf.org/html/rfc4287.
15. Yahoo (2008) Yahoo! Maps Web Service – AJAX API getting started guide, http://developer.yahoo.com/maps/ajax/index.html.
16. Google (2008) Google Maps API reference, http://code.google.com/apis/maps/documentation/reference.html.
17. The Free Software Foundation (2007) GNU general public license, version 3, http://www.gnu.org/licenses/gpl-3.0.html.
18. OpenWrt, http://www.openwrt.org/.
19. The Open Source Initiative (OSI) (31 October 2006) The BSD license, http://www.opensource.org/licenses/bsd-license.php.

20. The Apache Software Foundation (January 2004) Apache license, Version 2, http://www.apache.org/licenses/LICENSE-2.0.txt.

21. Nokia and S60 (2008) Web Browser for S60, http://www.s60.com/business/productinfo/builtinapplications/webrowser/.

22. Shozu. http://www.shozu.com.

23. Nokia S60 (2008) WebKit FAQ. http://opensource.nokia.com/projects/s60browser/s60-oss-browser-faq.html.

24. Bloglines. http://www.bloglines.com/.

25. Slingbox. http://www.slingmedia.com/.

26. Cable News Network (2008) CNN.com mobile services, http://edition.cnn.com/mobile.

27. Taptu (2008) A search engine for your mobile phone, http://taptu.com.

28. Mowser (2008) Mobilizing the web, http://mowser.com.

29. Investors.Com (October 2007) OPERA SOFTWARE: first widgets for mobile: Opera widgets on new 3G handsets from KDDI, http://investors.com/breakingnews.asp?journalid = 62205494.

30. Google (2008) Google Gears API, http://code.google.com/apis/gears.

31. Soininen, P. (September 2007) S60 Web Widgets, Slides available at: http://www.slideshare.net/romek/s60-3rd-fp2-widgets.

32. W3C (12 December 2007) HTML 5 – W3C editor's draft, http://www.w3.org/html/wg/html5/#offline.

33. Denso-Wave (2008) QR codes – 2 dimensional barcodes, http://qrcode.com.

34. Deam-Interactive (2008) Deam Interactive delivers intelligence to camera phones, http://www.daeminteractive.com/eng/tecnologia.jsp.

35. Sir Berners-Lee, T. (April 2004) http://www.w3.org/DesignIssues/TLD.

36. Winksite. http://winksite.com.

37. Google mobile content adaptation (2008) http://google.com.

38. Mippin mobile content adaptation (2008) http://mippin.com.

39. Opera (2008) Opera Mini – Free mobile Web browser for your phone, http://www.operamini.com/.

40. Sauter, M. (July 2007) Deactivating the Vodafone Websession Compression Proxy, http://wirelessmoves.com/mobile_life/2007/07/deactivating-th.html.

41. Adobe Flash Player (2008) http://www.adobe.com/products/flashplayer.

42. Adblock (2008) https://addons.mozilla.org/en-US/firefox/addon/10.

43. Facebook (2008) Leading websites offer facebook beacon for social distribution, http://www.facebook.com/press/releases.php?p = 9166.

44. BlockSite (2008) https://addons.mozilla.org/en-US/firefox/addon/3145.

45. Web, W. (2007) Wireless Communications – The Future, Chapter 5.2, John Wiley and Sons, Ltd, Chichester.

46. Wapedia. http://wapedia.mobi/.

47. Nokia Audiobooks (2008) http://www.nokia.com/betalabs/audiobooks.

48. LibriVox (2008) Acoustical liberation of books in the public domain.

49. Fring (2008) http://www.fring.com/.

50. Opera (October 2007) First widgets for mobile: Opera widgets on new 3G handsets from KDDI, http://www.opera.com/pressreleases/en/2007/10/24_2/.

51. Mystrands (2008) Social recommendation and discovery, http://www.mystrands.com/.

52. Jaiku. http://www.jaiku.com.

53. Twitter (2008) http://www.twitter.com.

54. Pownce (2008) Send stuff to your friends, http://www.pownce.com.

55. Nokia (2008) Smart2Go – share the world with your friends, http://www.smart2go.com/en/.

56. Google (2008) Google maps on mobile phones, http://www.google.com/gmm.

57. TomTom (2008) TomTom, portable GPS car navigation systems, http://www.tomtom.com/.

58. Skyhook Wireless (2008) http://www.skyhookwireless.com/.

59. Vilkamo, T. (2008) Is that a Web Server in your pocket, or are you just happy to see me? http://blogs.s60.com/tommi/2007/06/is_that_a_web_server_in_your_p.html.

7

Conclusion

This book has looked at mobile networks, mobile devices and the mobile Web 2.0 from a number of different perspectives. The mobile Web 2.0 both extends and improves Web 2.0 applications from the desktop world. Like Web 2.0, mobile Web 2.0 is also about user participation and new services and information are no longer only distributed top-down but are created by the users themselves. This is possible because of a decentralized approach, because of the openness of services to be included in other new services, freely available information in network databases, open interfaces and open source software. Furthermore, applications and services are continuously evolving and the user is part of the development process. Changes to services and products can be made quickly and global reach ensures there are enough users contributing and sharing information to create a critical mass for the service. The main driver for users to share information on Blogs, Wikis, picture-sharing sites and so on is the wish to communicate, the need for self-expression and the desire to return something to the community for using other services and information for free. Most Web services generate revenue by including advertisements on Web pages or do not require revenue at all because they are driven by enthusiasts. New services are often introduced to the Web community without a firm business plan in mind at first. Most properties described above apply to both Web 2.0 and the mobile Web 2.0. In addition, mobile devices are transforming Web 2.0 due to their unique properties such as always on, always carried, being available at the point of inspiration and so on.

Services deployed by traditional mobile network operators, on the other hand, have little in common with mobile Web 2.0 services described in this chapter. A main requirement for operator deployed services is to generate revenue from day one, which requires a business model and a significant customer base right from the start. This is difficult to achieve in practice and there are only a few services that can develop successfully under such a business model. Services that might prosper under these conditions are voice-based services of the IP Multimedia Subsystem as described in Chapter 4. Creating

Beyond 3G – Bringing Networks, Terminals and the Web Together: LTE, WiMAX, IMS, 4G Devices and the Mobile Web 2.0   Martin Sauter   © 2009 John Wiley & Sons, Ltd

successful voice-centric applications is a difficult task, especially in the wireless domain, since the network layer has a profound impact on the quality of service of this type of application. Unlike Internet companies, wireless network operators have a lot of experience running and optimizing their networks, in particular for circuit-switched voice services. With B3G networks being deployed, circuit-switched voice services need to be migrated to the packet domain and network operators are in a good position to capitalize on this transfer and use it to develop applications around today's traditional person-to-person voice service. Furthermore, voice is a convenience service, that is it must be very easy to use, it must work under all circumstances and it must be instantly available at any time. Operators with both fixed and wireless access networks will have a big advantage in the future as most users will have some sort of fixed broadband Internet connection at home in combination with Wi-Fi access points and, in addition, will use wireless services with their mobile devices. Those network operators deploying voice services which integrate well and as seamlessly as possible in both domains will have a major competitive advantage in the future. Migrating cellular voice service to packet-switched next generation wireless networks is still in its infancy and IMS services are still not very well tested or widely deployed. It is therefore likely that some sort of intermediate technology such as Unlicensed Mobile Access, which was introduced in Chapter 4, will have an interesting role to play in fixed wireless convergence before IMS-based voice services can take over as a long term solution.

In the future it is most likely that the fixed and mobile operators who will be the most successful will be those that are able to combine fixed-line and wireless access networks in an intelligent way to deliver a smooth end to end user experience. Furthermore, successful operators will take advantage of the fact that, while they can offer some services themselves, there will be countless others which are developed and run by Internet companies and private individuals. In the end, both types of applications will generate revenue for them if they can find a balance between:

- being a network provider for the user and thus the enabler of popular Internet-based applications and applications in the long tail;
- providing operator-centric services such as voice-based applications themselves and tying them into the mobile Web 2.0 ecosphere by using open APIs available from these services;
- getting the right combination of cellular coverage complemented with very high-speed Wi-Fi 'bubbles' at homes and offices to offload traffic from the cellular network layer.

As there is no single wireless technology for all scenarios, it is likely that in the future the number of mobile devices with several wireless interfaces increases. On the cellular macro level, beyond 3G technologies such as UMTS/HSPA, LTE and WiMAX will be the dominant technologies. As UMTS/HSPA and LTE address incumbent operators and WiMAX alternative new operators, the mix of access technologies will stimulate competition and in turn better prices and network coverage for the customer. In homes and offices, Wi-Fi is likely to remain the dominant radio technology and increasing data rates will further reduce the need for network cables. The high bandwidths and simple design make it the ideal technology to interconnect household or office devices with each other and to keep the data exchanged between these devices in the local network. By having a limited coverage area, it also ensures that neighboring networks on the same channel can coexist and the overall bandwidth available via Wi-Fi networks connected to the Internet, via DSL or cable will be much higher than what can be delivered over the cellular macro layer. A combination of cellular and Wi-Fi also ensures that cellular networks will not be overloaded as despite increasing speeds the overall capacity of cellular networks will remain limited, as discussed in Chapter 3. Mobile devices supporting both a cellular network technology and Wi-Fi will become more commonplace in the future. Applications on such devices will thus be able to use the best network available for a task at any time, seamlessly using home and office Wi-Fi networks when available and cellular networks when out of Wi-Fi reach.

As this book has shown, there is a very dynamic relationship between networks, terminals and the Web 2.0. With new developments announced almost daily, readers of this book are invited to keep up to date by subscribing to articles published by the author on this book's Blog at http://www.wirelessmoves.com.

Index

2D barcodes, 308
3G network coverage, 136
3GPP frequency bands, 253
802.11 protocol stack, 88
802.11e, 103
802.16, 70
802.16d, 76
802.16e, 79
802.16j, 87

Access Service Network Gateway, 72
Adaptive Antenna Systems, 81
Adaptive Multi Rate codec, 137
ADSL, 16
ADSL2+, 4
Advanced receivers, 117
Aggregation, 286
AJAX, 205, 289
All over IP, 8
AMR, 137
Analog networks, 1
Android, 262
Antenna, 120
    indoor, 134
    outdoor, 134
Apache license, 296
Application Servers, 182
ARM, 237
ARPU, 111–12
ASN-GW, 71
Asynchronous Transfer Mode, 8
ATM, 8, 16, 46, 149, 150
ATOM, 286

Audio streaming, 327
Avatar, 281
Average Revenue
    Per User, 111

Backhaul, 114, 148
    bandwidth, 150
Back-to-back User Agent, 184
Base Station Controller, 159
Battery, 249
    capacity, 120
    evolution, 249
Beamforming, 113
Bearer Independent Call Control, 201
Beta version, 284
BICC, 201
Bill of Materials, 238–239
Billing, 19–20
Blogs, 276
Bluetooth, 153
BOM, 239
BREW, 259
Broadband Wireless Access, 117
BSC, 159
BSD license, 296
Busy hour, 128
BWA, 117

Call Server, 8
CAPEX, 108
Capital Expenditure, 108
Carrier, 126
CDMA principle, 23

Cell capacity, 112, 115, 128
Cell Global ID, 192
Cell Update, 34
Cell-DCH state, 26
Cell-FACH state, 26
Cell-PCH state, 27
CGI, 192
Channel capacity, 115
Charging, 109
Chipset, 247
    evolution, 247
    power consumption, 239
Circuit-switched telephony, 158
Circuit-switching, 158
CMOS, 248
Codec, 170
Coding, 113
Collective intelligence, 283
Command sequence number, 167
Content adaptation, 313
Continuous Packet Connectivity, 38
Cookies, 318
CPC, 38
Customer Premises Equipment, 72

Development cycle, 284
DHCP, 182
    Server, 90
Digital Signal Processor, 240
Digital Subscriber Line, 4
Display, 249
    evolution, 249
    resolution, 235
DMB, 252
Doppler effect, 54
Downlink capacity, 112
DPCCH, 26
DPDCH, 26
Drift-RNC, 28
DSL, 4
    Access Multiplexer, 10
DSLAM, 10, 145
DSP, 240
DTMF, 202
Dual Tone Multiple Frequency, 202
DVB-H, 252

E-1, 15,149
    timeslots, 149
E.164, 200
EAP-AKA, 210
EAP-SIM, 210
EDCA, 104
EDGE, 129
EFR, 137, 141

Emergency calls, 177
Emerging markets, 136
End of software release cycle, 284
Enhanced Cell-FACH, 42
Enhanced Full Rate codec, 137
Enhanced IMT-2000, 7
eNodeB, 46
ENUM, 200
EPS, 14
Equipment refresh, 135
Ethernet, 16, 46
    microwave backhaul, 152
Evolved Packet System, 14
Extensible Markup Language, 286

FDD, 77, 119
F-DPCH, 139
FeliCa, 250
Femtocell, 224
FFT, 54
File systems, 267
File transfer test, 122
Filter criteria, 188
Fixed-line Internet replacement, 6
Flash shared objects, 320
Flat rate, 109
FM radio chip, 244
Folksonomy, 290
Fractional Dedicated Physical Control
    Channel, 139
Frequency bands, 114, 118, 252
Frequency Division Duplex, 77, 119
Future capacity estimation, 132

G.711, 138, 141
GAN, 228, 257
Gateway GPRS Support Node, 22
Gateway MSC, 17
General Packet Radio Service, 3
Generic Access Network, 228
GGSN, 22, 161
GNU Public License, 295
Google Gears, 305
GPL, 295
GPRS, 3, 129
GPS, 244, 246, 250, 291, 339
Graphics accelerator, 242
GSM, 2
    network launch, 1

H.248, 202, 215
Handover, 34
Hardware impact on data use, 255
Harnessing collective intelligence, 283
HARQ, 34

Header compression, 139
High-speed downlink shared channel, 32
High-speed mobility, 124
High-speed shared control channel, 33
HLR, 8, 49, 158
Home Location Register, 8, 49, 158
HSDPA, 4, 31
  categories, 34
HS-DSCH, 32
HSPA, 14, 31
  scheduling, 34
HSPA+, 14
HSS, 49, 184
HSUPA, 31
HTML-5, 306

I2C, 268
IARI, 218
I-CSCF, 184
ICSI, 218
Idle state, 26
IFFT, 51
iLBC, 141
Image recognition, 308
IMEI, 243
IMS
  application interoperability, 221
  Application Reference ID, 218
  Application Servers, 182
  challenges, 219
  clients, 216
  Communication Service ID, 218
  conference bridge, 187
  confidentiality, 181
  frameworks, 221
  instant messaging, 203
  Java JSR-281, 218
  media gateway, 199
  presence, 203
  private user id, 188
  public user id, 188
  resource reservation, 195
  signaling compression, 181
  subscription profile, 188
  traffic flow template, 196
  Wi-Fi, 209
IMSI, 18
IMT Advanced, 7
IMT-2000, 7
IMT-2000 band, 117
Initial filter criteria, 188
Interference, 114
International Telecommunication
    Union, 7
Internet tablets, 236

Inverse multiplexing over ATM, 150
IP over Ethernet, 152
IP tunnel, 153
I-perf, 143
ISM band, 91, 142
ITU, 7
ITU-R M.1457-6, 7
ITU-R M.1645, 7
I-WLAN, 210

J2ME, 259
Java, 258
  2 Micro Edition, 259
  Script, 290
  Virtual Machine, 242
Jitter buffer, 172
JVM, 242

Keyboard, 250

License, 295
Linksys, 143
Location, 338
Long-tail economics, 281
Low capacity deployment, 130
LTE, 45
  advanced, 69
  bandwidths, 55
  channels, 56
  cyclic prefix
  frames, 55
  reference symbols, 56
  RRC states, 63
  scheduling, 60
  signals, 56
  slots, 55
  subcarrier spacing, 54
  symbol duration, 54

Maemo, 262
Mashups, 280
Mass storage, 267
MBMS, 252
Media
  attributes, 170
  description parameters, 170
  gateway, 8, 172
  sharing, 277, 328
  stream, 170
Media Gateway Control Protocol, 201
Media Resource Function, 187
MediaFlo, 252
MEGACO, 201, 215
Memory management, 266
Micro blogging, 335

Microwave, 152
MIMO, 38, 60, 80, 96, 113
MME, 47
MMS attacks, 270
MOAP, 260
Mobile
  broadband, 4
  device processing power, 127
  device security, 269
  Internet, 2
  operating system fracturization, 265
  social networks, 6
  Switching Center, 8, 17
  Virtual Network Operators, 112
  Web 2.0, 273
  web server, 341
Mobility impact on transfer speed, 123
Mobility Management Entity, 47
Modulation, 33
MPEG-4 compression, 127
MRF, 186
MSC, 8, 17
Multitasking, 265
MVNO, 112, 157

NASS, 214
NAT, 90, 175
Network Address Translation, 90, 175
Network and application separation, 160
Network Attached Storage, 153
Network Attachment Subsystem, 214
Network capacity per km^2, 129
Network refresh, 136
NodeB, 14

OFDM, 77
  carrier, 126
OFDMA, 51, 80
Offline web applications, 304
OLPC, 108
OMA device management, 206
OMC client provisioning, 206
One laptop per child, 108
One-tunnel, 43
Open application programming
    interfaces, 293
Open source, 262, 295
Operating system security, 269
Operating system tasks, 265
Operational Expenditure, 108
Operator monopoly, 163
OPEX, 108
Optical Ethernet, 152
OTA, 190
Over air download, 190

Packet Mobility Management, 30
Packet-switched telephony, 159
PAPR, 52
P-CCPCH, 25
PCF, 181
PCM codec, 172, 177, 201
P-CSCF, 181
PDA, 2
PDN Gateway, 49
PDP context activation, 161
Peak to Average Power Ratio, 52
Per minute charging, 109
Personal Digital Assistant, 2
PES, 215
PMM, 30
PoC, 204
Podcasting, 277
Policy Decision Function, 181
PRACH, 25
Prepaid Internet Access, 110–11
Privacy, 317
Protocol stack, 160
Proxy authentication, 168
Pseudo-wire, 151
PSTN emulation subsystem, 216
PSTN gateway, 174
Pulse Code Modulation, 170
Push-to-talk, 203

QoS, 161
Quality of Service, 161

RACS, 215
Radio Link Control Protocol, 124
Radio Network Controller, 16
Radio Resource Control, 63
RCEF, 215
Real Simple Syndication, 286
Real-time data traffic, 123
Real-time Transport Protocol, 139, 171
Reduced Instruction Set Computer, 239
Resource and Admission Control Subsystem, 215
Resource Control Enforcement Function, 215
Revenue, 128
  per base station, 128
RFID, 250
Rich user experience, 285
RISC, 239
RLC, 124
RNC, 16
Roaming, 136
Robust Header Compression, 65, 139
ROHC, 65, 139
RSS, 286
RTP, 139

S1 interface, 48
S60, 260
S-CCPCH, 25
SC-FDMA, 53
Scheduling, 140
SCP, 19
S-CSCF, 182
SDH, 152
SDP, 169
Search, 277
Secondary PDP-context, 124
Service Control Points, 19
Serving Gateway, 47
Serving GPRS Support Node, 21
Serving-RNC, 28
Session Description Protocol, 169
Session tracking, 322
SGSN, 21, 161
SGW, 200
Shannon-Hartley capacity equation, 115
Shared Channels, 32
Shared XML Document Management
    Server, 205
Short Message Service, 5
    Center, 18
SIGCOMP, 181
Signaling gateway, 172, 200
Signaling System Number 7, 172
Signal-to-noise ratio, 115
SIM card, 18, 158
Simple Traversal of UDP through NAT, 175
Single Carrier-Frequency Division
    Multiple Access, 53
SIP, 90, 164
    authentication, 168
    call, 167
    identity, 164
    location server, 164
    provider, 165
    proxy, 164
    registrar, 164
    URI, 165, 183
    web configuration, 174
Skype, 124, 141
SLF, 186
Small-screen devices, 154
Small-screen flat rates, 109
Smartphone, 235
SMS, 5, 17, 112, 160
SMSC, 18–19
SNR, 115
Social distribution, 321
Social media, 335
Social networking sites, 279
Software release cycle, 284

Spectral efficiency, 115
Split charging, 109
Spreading codes, 23, 32
SS-7, 172
STUN, 175
Subscription Locator Function, 186
Switching matrix, 158
Symbian, 260
Synchronous Digital Hierarchy, 152

T-1, 15, 148
Tagging, 290
Taggs, 291
Talk Burst Control Protocol, 205
TBCP, 205
TDD, 55, 77
TDM, 149
TEL URL, 183
Terminal capabilities, 114
Throttling, 110
Time Division Duplex, 55, 77, 120
Time Division Multiplexing, 149
Timeslot, 126
TISPAN, 213
Traffic estimation per user, 127
Traffic Flow Template, 196
Transcoding and Rate Adaptation
    Unit, 159
Transmission power, 120
TRAU, 159
Trojan horse, 269

UART, 268
UDP, 139
UICC, 189
Ultra-mobile PC, 236
UMA, 228
UMPC, 236
UMTS, 14
    mobility management, 27
Universal Integrated Circuit Card, 189
Universal Resource Identifier, 165
Unlicensed Mobile Access, 228
Uplink capacity, 118
URA-PCH state, 27
USB 2.0, 268
User(s)
    Agent, 165
    Datagram Protocol, 139
    recommendation, 278
    throughput, 120
    throughput in downlink, 120
    throughput in uplink, 125
    per km^2, 133
UTRAN, 14

VCC, 206
VDSL, 16, 145
Video clip download, 122
Virtual worlds, 281
VLR, 18
Voice call, 24
Voice Call Continuity, 206
Voice calls per access point, 140
Voice centric communication, 1
Voice minute, 112
Voice optimized protocol stack,
    157
Voice over IP, 6
Voice over wireless, 157
Voice-centric, 111
VoIP, 332
    capacity, 136, 140
    data rate, 139
    over Wi-Fi, 140
    over Wi-Fi capacity, 140
Volume charging, 109

Walled gardens, 310
WAP, 3
Web 1.0, 273
Web 2.0, 273
Web applications, 280
Web as platform, 283
Web browser attacks, 270
WebDAV, 341
Widgets, 305
Wi-Fi, 88
    40 MHz channel, 144
    access point, 89

bandwidth, 93
capacity, 143
frame size, 93
full overlapping, 143
interference, 141
medium access, 97
modulation, 94
partial overlapping, 143
Quality of Service, 103
scheduling, 140
security, 102
sleep mode, 100
transmission time, 93
Wiki, 278
WiMAX, 70
    authentication, 72
    bandwidths, 79
    cyclic prefix, 79
    frequency bands, 253
    handover, 81
    macro mobility management, 75
    micro mobility management, 74
    Mobile Multihop Relay, 87
    subcarrier spacing, 79
    symbol duration, 79
Windows Mobile, 262
WMM, 104
WPA, 102
WPA-2, 102

X.509 certificate, 72
x86, 238
XDMS, 205
XML, 286